Real Analysis with an Introduction to Wavelets and Applications

Real Analysis with an Introduction to Wavelets and Applications

Don Hong*
Jianzhong Wang†
Robert Gardner*

*Department of Mathematical and Statistical Sciences
East Tennessee State University
Johnson City, Tennessee

†Department of Mathematics and Statistics
Sam Houston State University
Huntsville, Texas

ELSEVIER
ACADEMIC
PRESS

AMSTERDAM • BOSTON • HEIDELBERG • LONDON
NEW YORK • OXFORD • PARIS • SAN DIEGO
SAN FRANCISCO • SINGAPORE • SYDNEY • TOKYO

Senior Acquisition Editor	Barbara Holland
Project Manager	Kyle Sarofeen
Associate Editor	Tom Singer
Marketing Manager	Phillip Pritchard
Cover Design	Eric DeCicco
Composition	Kolam Information Services
Cover Printer	Phoenix Color Corporation
Interior Printer	Maple-Vail Book Manufacturing Group

Elsevier Academic Press
30 Corporate Drive, Burlington, MA 01803, USA
525 B Street, Suite 1900, San Diego, California 92101–4495, USA
84 Theobald's Road, London WC1X 8RR, UK

This book is printed on acid-free paper. ∞

Library of Congress Cataloging-in-Publication Data
Application submitted.

British Library Cataloguing in Publication Data
A catalogue record for this book is available from the British Library

ISBN   978-0-12-354861-0

For all information on all Elsevier Academic Press Publications
visit our Web site at www.books.elsevier.com

To our wives: Ying, Hui-Lan, and Kathryn

Contents

Preface

Real Analysis is based on the real numbers, and is naturally involved in practical mathematics. On the other hand, it has taken on subject matter from set theory, harmonic analysis, integration theory, probability theory, the theory of partial differential equations, etc., and has provided these areas with important ideas and basic concepts. This relationship continues even today. This book contains the basic matter of "real analysis" and an introduction to "wavelet analysis," a popular topic in "applied real analysis."

Today, students studying real analysis come from different backgrounds and have diverse objectives. A course in this subject may serve both undergraduates and graduates, students not only in mathematics but also in statistics, engineering, etc., and students who are seeking master's degrees and those who plan to pursue doctoral degrees in the sciences. Written with the student in mind, this book is suitable for students with a minimal background. In particular, this book is written as a textbook for the course usually called Real Analysis as presently offered to first- or second-year graduate students in American universities with a master's degree program in mathematics.

The subject matter of the book focuses on measure theory, the Lebesgue integral, topics in probability, L^p spaces, Fourier analysis, and wavelet theory, as well as on applications. It contains many features that are unique for a real analysis text. This one-year course provides students the material in Fourier analysis, wavelet theory and applications and makes a very natural connection between classical pure analysis and the applied topic — wavelet theory and applications. The text can also be used for a one-semester course only on Real Analysis (I) by covering Chapters 1–3 and selected material in Chapters 4 and

5 or a one-semester course on *Applied Analysis* or *Introduction to Wavelets* using material in Chapters 5–9. Here are a few other features:

• The text is relatively elementary at the start, but the level of difficulty increases steadily.

• The book includes many examples and exercises. Some exercises form a supplementary part of the text.

• In contrast to the classical real analysis books, this text covers a number of applied topics related to measure theory and pure analysis. Some topics in basic probability theory, the Fourier transform, and wavelets provide interesting subjects in applied analysis. Many projects can be developed that could lead to quality undergraduate/graduate research theses.

• The text is intended mainly for graduates pursuing a master's degree, so only a basic background in linear algebra and analysis is assumed. Wavelet theory is also introduced at an elementary level, but it leads to some updated research topics. We provide relatively complete references on relevant topics in the bibliography at the end of the text.

This text is based on our class notes and, a primary version of this text has been tested in the classroom of our universities on several occasions in the courses of *Real Analysis, Topics in Applied Mathematics*, and *Special Topics on Wavelets*.

Though the text is basically self-contained, it will be very helpful for the reader to have some knowledge of elementary analysis — for example, the material in Walter Rudin's *Principles of Mathematical Analysis* [25] or Kirkwood's *An Introduction to Analysis* [16].

The Outline of the book is as follows.

Chapter 1 is intended as reference material. Many readers and instructors will be able to quickly review much of the material. This chapter is intended to make the text as self-contained as possible and also to provide a logically ordered context for the subject matter and to motivate later development.

Chapter 2 presents the elements of measure theory by first discussing measure on the rings of sets and then the Lebesgue theory on the line.

Chapter 3 discusses Lebesgue integration and its fundamental properties. This material is prerequisite to subsequent chapters.

Chapter 4 explores the relationship of differentiation and integration and presents some of the main theorems in probability which are closely related to measure theory and Lebesgue integration.

Chapters 5 and 6 provide the fundamentals of Hilbert spaces and Fourier analysis. These two chapters become a natural extention of the Lebesgue theory and also a preparation for the later wavelet analysis.

Chapters 7 and 8 include basic wavelet theory by starting with the Haar basis and multiresolution analysis and then discussing orthogonal wavelets and the construction of compactly supported wavelets. Smoothness, convergence, and approximation properties of wavelets are also discussed.

Chapter 9 provides applications of wavelets. We examine digital signals and filters, multichannel coding using wavelets, and filter banks.

A Web site is available which contains a list of known errors and updates for this text:

http://www.etsu.edu/math/real-analysis/updates.htm.

Please e-mail the authors if you have any input.

We are deeply grateful to Professor Charles K. Chui for his constant encouragement, kind support, and patience. Our thanks are due to, in particular, Jiansheng Cao, Bradley Dyer, Reneé Ferguson, Wenyin Huang, Joby Kauffman, Jun Li, Anna Mu, Chris Wallace, David Atkins, Panrong Xiao, and Qingbo Xue for giving us useful suggestions from students' point of view. We thank our colleagues who have reviewed the manuscript: Doug Hardin (Vanderbilt University), Pete Johnson (Auburn University), Ram Mohapatra (University of Central Florida), Dave Ruch (Metropolitan State College of Danver), and Wasin So (San Jose State University). It is a pleasure to acknowledge the great support given to us by Robert Ross, Barbara Holland, Mary Spencer, Kyle Sarofeen, and Tom Singer at Elsevier/Academic Press.

Don Hong and Robert Gardner
Johnson City, TN
Jianzhong Wang
Huntsville, TX

Chapter 1

Fundamentals

The concepts of real analysis rest on the system of real numbers and on such notions as sets, relations, and functions. In this chapter we introduce some elementary facts from set theory, topology of the real numbers, and the theory of real functions that are used frequently in the main theme of this text. The style here is deliberately terse, since this chapter is intended as a reference rather than a systematic exposition. For a more detailed study of these topics, see such texts as [16] and [25].

I Elementary Set Theory

We start with the assumption that the reader has an intuitive feel for the concept of a "set." In fact, we have no other choice! Since we can only define objects or concepts in terms of other objects and concepts, we must draw the line somewhere and take some ideas as fundamental, intuitive starting blocks (almost *atomic* in the classical Greek sense). In fact, one of the founders of set theory, Georg Cantor (1845–1918), in the late 1800s wrote: "A set is a collection into a whole of definite, distinct objects of our intuition or our thought [13]." With this said, we begin. ...

A *set* is a well-defined collection of objects. The objects in the collection will be called *elements* of the set. For example, $A = \{x, y, z\}$ is a set with three elements $x, y,$ and z. We use the notation $x \in A$ to denote that x belongs to A or, equivalently, x is in A. The set A, in turn, will be said to *contain* the element

x. By convention, a set contains each of its elements exactly once (as opposed to a *multiset*, which can contain multiple copies of an element).

Sets will usually be denoted by capital letters: $A, B, C, \ldots$ and elements by lowercase letters: $a, b, c, \ldots$. If a does not belong to A, we write $a \notin A$. We sometimes describe sets by displaying the elements in braces. For example, $A = \{2, 4, 8, 16\}$. If A is the collection of all x that satisfy a property P, we indicate this briefly by writing $A = \{x \mid x \text{ satisfies P}\}$. A set containing no elements, denoted by $\emptyset$, is called an *empty set*. For example, $\{x \mid x \text{ is real and } x^2 + 1 = 0\} = \emptyset$. We say that two sets are *equal* if they have the same elements.

Usually, we use $\mathbb{R}$ to denote the set of real numbers, i.e., $\mathbb{R} = \{x \mid -\infty < x < \infty\}$. Using the interval notation, we have the real line $\mathbb{R} = (-\infty, \infty)$, closed interval $[a, b] = \{x \mid a \leq x \leq b\}$, open interval $(a, b) = \{x \mid a < x < b\}$, and half-open and half-closed intervals $[a, b) = \{x \mid a \leq x < b\}$ and $(a, b] = \{x \mid a < x \leq b\}$.

As usual, the set of natural numbers is denoted by $\mathbb{N}$, the set of integers by $\mathbb{Z}$, the set of rational numbers by $\mathbb{Q}$, and the set of complex numbers by $\mathbb{C}$:

$$\mathbb{N} = \{1, 2, 3, \ldots\}$$

$$\mathbb{Z} = \{\ldots, -2, -1, 0, 1, 2, \ldots\}$$

$$\mathbb{Q} = \left\{ \frac{m}{n} \;\middle|\; m, n \in \mathbb{Z}, n \neq 0 \right\},$$

and

$$\mathbb{C} = \{z \mid z = x + yi \text{ for } x, y \in \mathbb{R} \text{ and } i^2 = -1\}.$$

Given two sets A and B, we say A is a *subset* of B, denoted by $A \subset B$, if every element of A is also an element of B. That is, $A \subset B$ means that if $a \in A$ then $a \in B$. We say A is a *proper subset* of B if $A \subset B$ and $A \neq B$. Notice that $A = B$ if and only if $A \subset B$ and $B \subset A$. The empty set $\emptyset$ is a subset of any set. The *union* of two sets A and B is defined as the set of all elements that are in A or in B, and it is denoted by $A \cup B$. Thus, $A \cup B = \{x \mid x \in A \text{ or } x \in B\}$. The *intersection* of two sets A and B is the set of all elements that are in both A and B, denoted by $A \cap B$. That is, $A \cap B = \{x \mid x \in A \text{ and } x \in B\}$. If $A \cap B = \emptyset$, then A and B are called *disjoint*. We can generalize the ideas of unions and intersections of sets from a pair of sets to an arbitrary collection of sets. For a family of infinitely many sets A_λ, $\lambda \in \Lambda$ (Λ is called the *indexing set*), union and intersection are defined as

$$\bigcup_{\lambda \in \Lambda} A_\lambda = \{x \mid x \in A_\lambda \text{ for some } \lambda \in \Lambda\} \text{ and}$$

$$\bigcap_{\lambda \in \Lambda} A_\lambda = \{x \mid x \in A_\lambda \text{ for all } \lambda \in \Lambda\}.$$

We define the *relative complement* of B in A, also called the *difference set* and denoted $A \setminus B$, to be $A \setminus B = \{x \mid x \in A \text{ and } x \notin B\}$. The *symmetric difference* of A and B, denoted by $A \Delta B$, is defined as $A \Delta B = (A \setminus B) \cup (B \setminus A)$. The following properties are easy to prove.

Theorem 1.1.1 Let A, B, and C be subsets of X. Then
- (i) $A \cup B = B \cup A$, $A \cap B = B \cap A$;
- (ii) $(A \cup B) \cup C = A \cup (B \cup C)$, $(A \cap B) \cap C = A \cap (B \cap C)$;
- (iii) $(A \cup B) \cap C = (A \cap C) \cup (B \cap C)$, $(A \cap B) \cup C = (A \cup C) \cap (B \cup C)$.

When it is clearly understood that all sets in question are subsets of a fixed set (called a *universal set*) X, we define the *complement* A^c of a set A (in X) as $A^c = X \setminus A$. In this situation we have *de Morgan's Laws*:

$$(A \cup B)^c = A^c \cap B^c \text{ and } (A \cap B)^c = A^c \cup B^c.$$

In general, we have the following.

Theorem 1.1.2 For any collection of sets A_λ, $\lambda \in \Lambda$, we have:

$$\left(\bigcup_{\lambda \in \Lambda} A_\lambda \right)^c = \bigcap_{\lambda \in \Lambda} A_\lambda^c \text{ and} \tag{1.1}$$

$$\left(\bigcap_{\lambda \in \Lambda} A_\lambda \right)^c = \bigcup_{\lambda \in \Lambda} A_\lambda^c. \tag{1.2}$$

Proof: We only show (1.1). The proof of (1.2) is similar. Let $P = (\bigcup_{\lambda \in \Lambda} A_\lambda)^c$ and $Q = \bigcap_{\lambda \in \Lambda} A_\lambda^c$.

For $x \in P$, we have that $x \notin \bigcup_{\lambda \in \Lambda} A_\lambda$. Thus, for every $\lambda \in \Lambda$, $x \notin A_\lambda$ and so $x \in A_\lambda^c$. Therefore, $x \in Q$. This implies that $P \subset Q$.

On the other hand, for $x \in Q$, we have that for any $\lambda \in \Lambda$, $x \notin A_\lambda$. Thus, $x \notin \bigcup_{\lambda \in \Lambda} A_\lambda$. Therefore, $x \in (\bigcup_{\lambda \in \Lambda} A_\lambda)^c = P$. This means that $Q \subset P$. Hence, $P = Q$. ∎

Let $(A_n) = \{A_n\}_{n \in \mathbb{N}}$ be a sequence of subsets of X. (A_n) is said to be *increasing* if

$$A_1 \subset A_2 \subset A_3 \subset \cdots .$$

(A_n) is called *decreasing* if

$$A_1 \supset A_2 \supset A_3 \supset \cdots .$$

For a given sequence (A_n) of subsets of X, we construct two new sequences (B_n) and (C_n) of sets as follows:

$$B_n = \bigcup_{k=n}^{\infty} A_k \text{ and } C_n = \bigcap_{k=n}^{\infty} A_k.$$

Clearly, (B_n) is decreasing and (C_n) is increasing. The set of intersection of B_n, $n \in \mathbb{N}$ is called the *limit superior* of (A_n) and is denoted by $\overline{\lim}_{n \to \infty} A_n$ or $\lim \sup_{n \to \infty} A_n$. The set of union of C_n, $n \in \mathbb{N}$, is called the *limit inferior* of (A_n) and is denoted by $\underline{\lim}_{n \to \infty} A_n$ or $\lim \inf_{n \to \infty} A_n$. Therefore,

$$\overline{\lim}_{n \to \infty} A_n = \bigcap_{n=1}^{\infty} \left(\bigcup_{k=n}^{\infty} A_k \right) \text{ and } \underline{\lim}_{n \to \infty} A_n = \bigcup_{n=1}^{\infty} \left(\bigcap_{k=n}^{\infty} A_k \right).$$

It can be seen that the limit superior is the set of those members that belong to A_n for infinitely many values of n, while the limit inferior is the set of those members that belong to A_n for all but a finite number of subscripts. For this description it can be seen that $\underline{\lim}_{n \to \infty} A_n \subset \overline{\lim}_{n \to \infty} A_n$. If $\underline{\lim}_{n \to \infty} A_n = \overline{\lim}_{n \to \infty} A_n$, then we say the *limit* of (A_n) exists and is denoted by $\lim_{n \to \infty} A_n$.

EXAMPLE 1.1.1:   Let E and F be any two sets. Define a sequence (A_k) of sets by

$$A_k = \begin{cases} E, & \text{if } k \text{ is odd,} \\ F, & \text{if } k \text{ is even.} \end{cases}$$

Then, $B_n = \bigcup_{k=n}^{\infty} A_k = E \cup F$, so $\overline{\lim}_{k \to \infty} A_k = \bigcap_{n=1}^{\infty} B_n = E \cup F$. From the fact that $C_n = \bigcap_{k=n}^{\infty} A_k = E \cap F$, $n = 1, 2, 3, \ldots$, we obtain $\underline{\lim}_{k \to \infty} A_k = \bigcup_{n=1}^{\infty} \left(\bigcap_{k=n}^{\infty} A_k \right) = E \cap F$.

For a given set A, the *power set* of A, denoted by $\mathcal{P}(A)$ or 2^A, is the collection of all subsets of A. For example, if $A = \{1, 2, 3\}$, then $\mathcal{P}(A) = \{\emptyset, \{1\}, \{2\}, \{3\}, \{1, 2\}, \{1, 3\}, \{2, 3\}, \{1, 2, 3\}\}$. There are 8 distinct elements in the power set of $A = \{1, 2, 3\}$. In general, for any finite set A containing n distinct elements, we find that A has 2^n subsets and therefore, $\mathcal{P}(A)$ has 2^n elements in it.

Exercises

1. For sets A, B, and C, the operations $\cap$ and $\cup$ satisfy the properties of commutation, association, and distribution:
   (a) $A \cup B = B \cup A, A \cap B = B \cap A$;
   (b) $(A \cup B) \cup C = A \cup (B \cup C), (A \cap B) \cap C = A \cap (B \cap C)$;
   (c) $(A \cup B) \cap C = (A \cap C) \cup (B \cap C), (A \cap B) \cup C = (A \cup C) \cap (B \cup C)$.

2. Show that S and T are disjoint if and only if $S \cup T = S \Delta T$.

3. Prove that $A \Delta (B \Delta C) = (A \Delta B) \Delta C$ and $A \cap (B \Delta C) = (A \cap B) \Delta (A \cap C)$.

4. For any sets A, B, and C, prove that
   (a) $A \Delta B \subset (A \Delta C) \cup (B \Delta C)$, and show by an example that the inclusion may be proper.
   (b) $(A \setminus B) \cup B = (A \cup B) \setminus B$ if and only if $B = \emptyset$.

5. For a given integer $n \geq 1$, let $A_n = \left\{ \dfrac{m}{n} \,\middle|\, m \in \mathbb{Z} \right\}$. Show $\overline{\lim}_{n \to \infty} A_n = \mathbb{Q}$, the set of rational numbers and $\underline{\lim}_{n \to \infty} A_n = \mathbb{Z}$, the set of integers.

6. Let (A_n) be a sequence of sets. Prove the following properties.
   (a) $x \in \overline{\lim}_{n \to \infty} A_n$ if and only if for any $N \in \mathbb{N}$, there exists $n \geq N$ such that $x \in A_n$.
   (b) $x \in \underline{\lim}_{n \to \infty} A_n$ if and only if there exists $N_x \in \mathbb{N}$, such that for all $n \geq N_x$, we have $x \in A_n$.
   (c) If (A_n) is increasing, then $\lim_{n \to \infty} A_n$ exists and

   $$\lim_{n \to \infty} A_n = \bigcup_{n=1}^{\infty} A_n.$$

   (d) If (A_n) is decreasing, then $\lim_{n \to \infty} A_n$ exists and

   $$\lim_{n \to \infty} A_n = \bigcap_{n=1}^{\infty} A_n.$$

7. If $A = \underline{\lim}_{n \to \infty} A_n$ and $B = \overline{\lim}_{n \to \infty} A_n$. Show that $B^c = \underline{\lim}_{n \to \infty} A_n^c$ and $A^c = \overline{\lim}_{n \to \infty} A_n^c$.

8. Let (A_n) be a sequence of sets defined as follows:

   $$A_{2k+1} = \left[0, 2 - \frac{1}{2k + 1} \right], k = 0, 1, 2, \cdots$$

   $$A_{2k} = \left[0, 1 + \frac{1}{2k} \right], k = 1, 2, \cdots.$$

   Show that $\underline{\lim}_{n \to \infty} A_n = [0, 1]$ and $\overline{\lim}_{n \to \infty} A_n = [0, 2)$.

9. Let (A_n) be a sequence of sets which are mutually disjoint, i.e., $A_k \cap A_\ell = \emptyset$ for $k \neq \ell$. Then $\varliminf\limits_{n \to \infty} A_n = \varlimsup\limits_{n \to \infty} A_n = \emptyset$.

10. Let A be a nonempty finite set with n elements. Show that there are 2^n elements in its power set $\mathcal{P}(A)$.

2 Relations and Orderings

Let X and Y be two sets. The *Cartesian product*, or *cross product*, of X and Y, denoted by $X \times Y$, is the set given by

$$X \times Y = \{(x, y) \mid x \in X, y \in Y\}.$$

The elements of $X \times Y$ are called *ordered pairs*. For $(x_1, y_1), (x_2, y_2) \in X \times Y$, we have $(x_1, y_1) = (x_2, y_2)$ if and only if $x_1 = x_2$ and $y_1 = y_2$.

Any subset R of $X \times Y$ is called a *relation* from X to Y. We also write $(x, y) \in R$ as xRy. For nonempty sets X and Y, a *function*, or *mapping* f from X to Y, denoted by $f : X \mapsto Y$, is a relation from X to Y in which every element of X appears exactly once as the first component of an ordered pair in the relation. We often write $y = f(x)$ when (x, y) is an ordered pair in the function. In this case, y is called the *image* (or the *value*) of x under f and x the *preimage* of y. We call X the *domain* of f and Y the *codomain* of f. The subset of Y consisting of those elements that appear as second components in the ordered pairs of f is called the *range* of f and is denoted by $f(X)$ since it is the set of images of elements of X under f. If $f : X \mapsto Y$ and $A \subset X$, the *image* of A under f is the set

$$f(A) = \{y \in Y \mid y = f(x) \text{ for some } x \in A\}.$$

If $B \subset Y$, the *inverse image* of B is the set

$$f^{-1}(B) = \{x \in X \mid y = f(x) \text{ for some } y \in B\}.$$

A *real-valued function on* A is a function $f : A \mapsto \mathbb{R}$, and a *complex-valued function on* A is a function $f : A \mapsto \mathbb{C}$.

The proofs of the following properties are left to the reader as exercises.

Theorem 1.2.1 Let $f : X \mapsto Y$ and $\mathcal{A}$ a collection of subsets of X, $\mathcal{B}$ a collection of subsets of Y. Then

(i) $f(\bigcup_{A \in \mathcal{A}} A) = \bigcup_{A \in \mathcal{A}} f(A)$;

(ii) $f(\bigcap_{A \in \mathcal{A}} A) \subset \bigcap_{A \in \mathcal{A}} f(A)$;

(iii) $f^{-1}(\cup_{B\in\mathcal{B}}B) = \cup_{B\in\mathcal{B}}f^{-1}(B)$;

(iv) $f^{-1}(\cap_{B\in\mathcal{B}}B) = \cap_{B\in\mathcal{B}}f^{-1}(B)$.

A function $f : X \mapsto Y$ is called *one-to-one*, or *injective*, if $f(x_1) = f(x_2)$ implies that $x_1 = x_2$. Therefore, by the contrapositive of the definition, $f : X \mapsto Y$ is one-to-one if and only if $x_1 \neq x_2$ implies that $f(x_1) \neq f(x_2)$. A function $f : X \mapsto Y$ is called *onto*, or *surjective*, if $f(X) = Y$. Thus, $f : X \mapsto Y$ is onto if and only if for any $y \in Y$, there is $x \in X$ such that $f(x) = y$. $f : X \mapsto Y$ is said to be *bijective* or a *one-to-one correspondence* if f is both one-to-one and onto.

Let f be a relation from a set X to a set Y and g a relation from Y to a set Z. The *composition* of f and g is the relation consisting of ordered pairs (x, z), where $x \in X, z \in Z$, and for which there exists an element $y \in Y$ such that $(x, y) \in f$ and $(y, z) \in g$. We denote the composition of f and g by $g \circ f$. If $f : X \mapsto Y$ and $g : Y \mapsto Z$ are functions, then $g \circ f : X \mapsto Z$ is called the *composite function* or *composition* of f and g. If $f : A \mapsto B$ is bijective, there is a unique inverse function $f^{-1} : B \mapsto A$ such that $f^{-1} \circ f(x) = x$, for all $x \in A$, and $f \circ f^{-1}(y) = y$ for all $y \in B$.

Definition 1.2.1 For $E \subset \mathbb{R}$, a function $f : E \mapsto \mathbb{R}$ is said to be *continuous at a point* $x_0 \in E$, if for every $\epsilon > 0$, there exists a $\delta > 0$ such that

$$|f(x) - f(x_0)| < \epsilon$$

for all $x \in E$ with $|x - x_0| < \delta$. The function f is said to be *continuous on E* if and only if f is continuous at every point in E. The set of continuous functions on E is denoted by $C(E)$. When $E = [a, b]$ is an interval, we simply denote $C(E)$ by $C[a, b]$.

EXAMPLE 1.2.1: Consider the functions $f : \mathbb{Z} \mapsto \mathbb{N}, g : \mathbb{Z} \mapsto \mathbb{Z}, h : \mathbb{Z} \mapsto \mathbb{Z}$, defined by $f(n) = n^2 + 1, g(n) = 3n, h(n) = 1 - n$. Then f is neither one-to-one nor onto, g is one-to-one but not onto, h is bijective, $h^{-1}(n) = 1 - n$, and $f \circ h(n) = n^2 - 2n + 2$. All of these functions are continuous on their domains.

EXAMPLE 1.2.2: A real-valued *polynomial function* of degree n is a function $p(x)$ of the form

$$p(x) = a_0 + a_1x + \cdots + a_nx^n$$

defined on $\mathbb{R}$, where n is a fixed nonnegative integer and the coefficients $a_k \in \mathbb{R}, k = 0, 1, \cdots, n$ with $a_n \neq 0$. We usually denote by $\mathcal{P}_n$

the set of polynomials of degree at most n. $p(x) = x^3 \in \mathcal{P}_3$ is bijective from $\mathbb{R}$ to $\mathbb{R}$. However, $q(x) = x^2 + x - 2$ is neither one-to-one nor onto. Polynomial functions are continuous.

Definition 1.2.2 Let $-\infty < a = x_0 < x_1 \cdots < x_n = b < \infty$. Then the set $\Delta := \{x_k\}_{k=0}^n$ of knots x_k, $k = 0, 1, \cdots, n$, is called a *partition* of the interval $[a, b]$. A function $s : [a, b] \mapsto \mathbb{R}$ is called a *spline* of degree d defined over the partition Δ if s restricted to each subinterval (x_k, x_{k+1}) is a polynomial in $\mathcal{P}_d$. We denote by $S_d^r(\Delta)$ the set of splines of degree d with smoothness order r defined over Δ. That is, $s \in S_d^r(\Delta)$ if and only if s is a spline of degree d defined over Δ and its rth derivative is continuous. As usual, the derivative of $f : [a, b] \mapsto \mathbb{R}$ at a point $x \in [a, b]$, denoted by $f'(x)$ or $\frac{d}{dx}f(x)$, is defined as $f'(x) = \lim_{h \to 0} \frac{f(x+h)-f(x)}{h}$ (under the restriction that $x, x + h \in [a, b]$) and the rth derivative is defined by $f^{(r)}(x) = \frac{d}{dx}f^{(r-1)}(x)$ for any integer $r > 1$. We denote by $C^r[a, b]$ the set of functions with continuous rth derivatives over $[a, b]$.

EXAMPLE 1.2.3:   Let

$$s(x) = \begin{cases} 0, & x \notin [0, 3], \\ \frac{x^2}{2}, & 0 \le x \le 1, \\ -x^2 + 3x - \frac{3}{2}, & 1 \le x \le 2, \\ \frac{x^2}{2} - 3x + \frac{9}{2}, & 2 \le x \le 3. \end{cases}$$

Then $s(x)$ is a $C^2 \to C^1$ quadratic spline function.

Any subset of $X \times X$ is called a *binary relation* on X. Thus, a binary relation on X is a relation from X to X. A relation R on X is called *reflexive* if for all $x \in X$, $(x, x) \in R$. R is said to be *symmetric* if $(x, y) \in R$ implies $(y, x) \in R$ for all $x, y \in X$. R is called *transitive* if for all $x, y, z \in X$, $(x, y), (y, z) \in R$, implies that $(x, z) \in R$. R is called *antisymmetric* if for all $x, y \in X$, $(x, y) \in R$ and $(y, x) \in R$ implies that $x = y$.

Definition 1.2.3 A relation R on X is called an *equivalence relation* if it is reflexive, symmetric, and transitive. A relation R on X is called a *partial ordering*, denoted by $\le$, if R is reflexive, antisymmetric, and transitive. R is called a *total ordering* if R is a partial ordering and for any $x, y \in X$, either $x \le y$ or $y \le x$. A *well ordering relation* on X is a total ordering on X such that for any nonempty subset A of X, there exists an element $a \in A$ such that $a \le x$ for all $x \in A$. Such an a is called the *smallest element* of A.

EXAMPLE 1.2.4:

(a) For a given $s \in \mathbb{N}$ define a relation on $\mathbb{Z}$ as $nR_s m$ if $n \equiv m$ (mod s). Then R_s is an equivalence relation on $\mathbb{Z}$.

(b) Define the relation on $\mathbb{N}$, denoted by nRm, or $(n, m) \in R$, if $n \leq m$. Then R is an equivalence relation. R is also a partial ordering, and a total ordering. In addition, R is a well ordering on $\mathbb{N}$. We write $R = (\mathbb{N}, \leq)$.

(c) $(\mathbb{Z}, \leq)$ is a total ordering relation but it is not a well ordering.

(d) For any set X, its power set $\mathcal{P}(X)$ is partially ordered by inclusion.

Corresponding to the notation of partial orderings $\leq$, we write $x < y$ to mean that $x \leq y$ and $x \neq y$. Notice that a partial ordering on X naturally induces a partial ordering on every nonempty subset of X. If X is partially ordered by $\leq$, a *maximal (minimal) element* of X is an element $x \in X$ such that $x \leq y$ $(y \leq x)$ implies $y = x$. Maximal and minimal elements may or may not exist, and they need not be unique unless the ordering is a total ordering. If $E \subset X$, an *upper (lower) bound* for E is an element $x \in X$ such that $y \leq x$ $(x \leq y)$ for all $y \in E$. An upper bound for E need not be an element of E. Unless E is totally ordered, a maximal element of E need not to be an upper bound for E.

Earlier we defined the Cartesian product of two sets. Similarly, we can define the Cartesian product of n sets in terms of ordered n-tuples. For infinite families of sets, $\{X_\lambda\}_{\lambda \in \Lambda}$, we define their *Cartesian product* $\Pi_{\lambda \in \Lambda} X_\lambda$ to be the set of all maps $f : \Lambda \mapsto \cup_{\lambda \in \Lambda} X_\lambda$ such that $f(\lambda) \in X_\lambda$ for every $\lambda \in \Lambda$. For $x \in \Pi_{\lambda \in \Lambda} X_\lambda$, we write $x_\lambda = f(\lambda)$ and call it the λth coordinate of x. For a proof of the following theorem, see [14].

Theorem 1.2.2 The following statements are equivalent:

(i) *The Hausdorff Maximal Principle.* Every partially ordered set has a maximal totally ordered subset.

(ii) *Zorn's Lemma.* If X is a partially ordered set and every totally ordered subset of X has an upper bound, then X has a maximal element.

(iii) *The Well Ordering Principle.* Every nonempty set X can be totally ordered and every nonempty subset of X has a minimal element.

(iv) *The Axiom of Choice.* If $\{X_\lambda\}_{\lambda \in \Lambda}$ is a nonempty collection of nonempty sets, then the Cartesian product $\Pi_{\lambda \in \Lambda} X_\lambda$ is nonempty.

The operations of addition and multiplication can be thought of as functions from $\mathbb{R} \times \mathbb{R}$ to $\mathbb{R}$. Addition assigns $(x, y) \in \mathbb{R}$ to an element in $\mathbb{R}$ denoted $x + y$

and multiplication assigns an element of $\mathbb{R}$ denoted xy. The familiar properties of these functions can be listed as follows:

Axioms of Addition:

    A1. $(x + y) + z = x + (y + z)$, for any $x, y, z \in \mathbb{R}$.
    A2. $x + y = y + x$, for any $x, y \in \mathbb{R}$.
    A3. There is an element $0 \in \mathbb{R}$ such that $x + 0 = x$ for every $x \in \mathbb{R}$.
    A4. For each $x \in \mathbb{R}$, there is an element $-x \in \mathbb{R}$ such that $x + (-x) = 0$.

Axioms of Multiplication:

    M1. $(xy)z = x(yz)$, for any $x, y, z \in \mathbb{R}$.
    M2. $xy = yx$, for any $x, y \in \mathbb{R}$.
    M3. There is an element $1 \in \mathbb{R}$ such that $x1 = x$ for any $x \in \mathbb{R}$.
    M4. For every nonzero $x \in \mathbb{R}$, there is an element $x^{-1} \in \mathbb{R}$ such that $xx^{-1} = 1$.

For the addition operation, the zero element 0 is unique in $\mathbb{R}$. Also, for every $x \in \mathbb{R}$, the additive inverse element $-x$ is unique. Therefore, we have $-(-x) = x$. For the multiplication operation, the unit element 1 is unique and also for every $x \in \mathbb{R}$, the inverse element x^{-1} is unique.

Concerning the interaction of addition and multiplication we have:

Distributive Law: For any $x, y, z \in \mathbb{R}$, we have $x(y + z) = xy + xz$.

We also hypothesize a subset P of $\mathbb{R}$, called the set of "positive" elements satisfying the following:

Axioms of Order:

    O1. If $x \in \mathbb{R}$, then exactly one of the following holds: $x \in P$, $-x \in P$, or $x = 0$.
    O2. If $x, y \in P$, then $x + y \in P$.
    O3. If $x, y \in P$, then $xy \in P$.
    O4. If we define $x < y$ by $y - x \in P$, then $x \in P$ if and only if $x > 0$. In terms of this, we have the *Archimedean Axiom*: if $x, y > 0$, then there is a positive integer n such that $nx > y$.

We complete our list of axioms describing $\mathbb{R}$ with the following:

Axiom of Completeness: If A is a nonempty subset of $\mathbb{R}$ which is bounded above (i.e., there is $y \in R$ such that for any $x \in A$, either $x < y$ or $x = y$, y is called an upper bound for A), then A has a least upper bound. ($x \in \mathbb{R}$ is said to be a *least upper bound* for a nonempty set $A \subset \mathbb{R}$, denoted by lub A, if x is an upper bound, and any upper bound x' of A satisfies $x' \geq x$. The *greatest lower bound* for A, glb A, is defined similarly.)

It can be shown that any set with addition, multiplication, and a subset of positive elements, which satisfies all the axioms above, is just a copy of $\mathbb{R}$ (see, for example, [15]).

Starting from $\mathbb{R}$, we can construct the set $\mathbb{C}$ of complex numbers. As usual, we write the elements of $\mathbb{C}$ by $z = x + iy$ and $x = Re(z) \in \mathbb{R}$ and $y = Im(z) \in \mathbb{R}$, which are called the *real part* and *imaginary part* of z, respectively. The *complex conjugate* of $\bar{z} = x + iy$, denoted by $\bar{z}$, is defined by $\bar{z} = x - iy$. It is easy to see that $z\bar{z} = x^2 + y^2$. We define the *modulus* of z as

$$|z| = (z\bar{z})^{1/2} = \sqrt{x^2 + y^2}.$$

Exercises

1. Let $f : X \mapsto Y$ with $A_1, A_2 \subset X$ and $B_1, B_2 \subset Y$. Show that
   (a) $f(A_1 \cup A_2) = f(A_1) \cup f(A_2)$.
   (b) $f(A_1 \cap A_2) \subset f(A_1) \cap f(A_2)$.
   (c) $f^{-1}(B_1 \cup B_2) = f^{-1}(B_1) \cup f^{-1}(B_2)$.
   (d) $f^{-1}(B_1 \cap B_2) = f^{-1}(B_1) \cap f^{-1}(B_2)$.
   (e) Give an example where $f(A_1 \cap A_2) \neq f(A_1) \cap f(A_2)$.
2. Prove that the following are generally proper inclusions
   (a) $f[f^{-1}(B)] \subset B$;
   (c) $f^{-1}[f(A)] \supset A$.
3. Let ψ^* be the set of all finite strings of a's, b's, and c's, and Σ^* the set of all finite strings of a's and b's. Define $f : \psi^* \mapsto \Sigma^*$ by f of a string $s \in \psi^*$ is the string in Σ^* obtained by omitting all the c's in s (so that $f(abccbca) = abba$, for example).
   (a) Is f one-to-one? Prove this or give a counterexample.
   (b) Is f onto? Prove this or give a counterexample.
4. (a) Give an example of a function $f : \mathbb{N} \mapsto \mathbb{N}$ that is onto but not one-to-one. Be sure to prove that your example has these properties.
   (b) Give an example of a function $g : \mathbb{N} \mapsto \mathbb{N}$ that is one-to-one but not onto. Be sure to prove that your example has these properties.
5. Let $f : A \mapsto B$, $g : B \mapsto C$. Prove that
   (a) if $g \circ f : A \mapsto C$ is onto, then $g : B \mapsto C$ is also onto;
   (b) if $g \circ f : A \mapsto C$ is one-to-one, $f : A \mapsto B$ is also one-to-one.
6. Prove that if $f : A \mapsto B$, $g : B \mapsto C$, and $h : C \mapsto D$, then $(h \circ g) \circ f = h \circ (g \circ f)$.

7. If $f : A \mapsto B$ is bijective, then prove that there is a unique function $f^{-1} : B \mapsto A$ such that $f^{-1} \circ f(x) = x$, for all $x \in A$, and $f \circ f^{-1}(y) = y$ for all $y \in B$.

8. Let $(f_n(x))$ be a sequence of real-valued functions defined on $\mathbb{R}$. Then for real-valued function $f(x)$ over $\mathbb{R}$, prove that we have

$$\{x \mid \lim_{n \to \infty} f_n(x) \neq f(x)\} = \bigcup_{k=1}^{\infty} \bigcap_{N=1}^{\infty} \bigcup_{n=N}^{\infty} \left\{ x \mid |f_n(x) - f(x)| \geq \frac{1}{k} \right\}.$$

9. Prove that, for the addition operation satisfying the axioms of addition, the zero element 0 is unique and also for any $x \in \mathbb{R}$, the additive inverse element $-x$ is unique.

10. Prove that, for the multiplication operation satisfying the axioms of multiplication, the unit element 1 is unique and also for any $x \in \mathbb{R}$, the inverse element x^{-1} is unique.

11. Under the axioms of addition and multiplication together with the distribution law, prove that $0x = 0$, $(-1)x = -x$, and $(-x)y = -xy$.

12. Define addition and multiplication over a set of two elements so that the axioms of addition and multiplication and the distribution law are satisfied.

13. Show that $\mathbb{Q}$ does not satisfy the completeness axiom.

14. Show that $Re(z + w) = Re(z) + Re(w)$ and $Im(z + w) = Im(z) + Im(w)$ for $z, w \in \mathbb{C}$.

15. Show that $|z + w| \leq |z| + |w|$ for any $z, w \in \mathbb{C}$.

16. Define a relation R on $\mathbb{N}$ by nRm if n and m are relatively prime. Is R reflexive? Symmetric? Transitive? Why or why not?

17. Define a relation $\sim$ on the set of positive reals by $x \sim y$ if either $x = y$ or $xy = 1$ (so that $3 \sim 3$ and $3 \sim \frac{1}{3}$). Is $\sim$ an equivalence relation? Why or why not?

18. Let R be an equivalence relation on a set A. For each $x \in A$, the *equivalence class* of x, denoted by $[x]$, is defined by $[x] = \{y \in A \mid yRx\}$. Show that for $x, y \in A$, (a) $x \in [x]$; (b) xRy if and only if $[x] = [y]$, and (c) either $[x] = [y]$ or $[x] \cap [y] = \emptyset$. Therefore, any equivalence relation R on A induces a partition of A.

19. Give an example of a partial ordering on a set X where neither the maximal nor the minimal element of X is unique.

20. Show that if f is surjective from $A \mapsto B$, then $f[f^{-1}(B)] = B$.

21. For any set X and its power set $\mathcal{P}(X)$, prove that there is no surjection $f : X \mapsto \mathcal{P}(X)$. (*Hint.* Consider the set $\{x \in X \mid x \notin f(x)\}$.)

22. For set $A \subset X$, we define the *characteristic function* of A as

$$\chi_A(x) = \begin{cases} 1, & x \in A, \\ 0, & x \in A^c. \end{cases}$$

Prove

(a) $\chi_{A \cup B}(x) = \chi_A(x) + \chi_B(x) - \chi_{A \cap B}(x)$;

(b) $\chi_{A \cap B}(x) = \chi_a(x) \cdot \chi_B(x)$;

(c) $\chi_{A \setminus B}(x) = \chi_A(x)[1 - \chi_B(x)]$;

(d) $\chi_{A \Delta B}(x) = |\chi_A(x) - \chi_B(x)|$.

23. Prove that if $f : X \mapsto Y$ is surjective, then the following are equivalent:

(a) f is injective;

(b) For any sets $A, B \subset X, f(A \cap B) = f(A) \cap f(B)$;

(c) For any sets $A, B \subset X$ with $A \cap B = \emptyset, f(A) \cap f(B) = \emptyset$.

24. A proof by *mathematical induction* that a proposition $P(n)$ is true for every positive integer n consists of two steps:

(i) *Basic Step.* $P(1)$ is shown to be true.

(ii) *Inductive Step.* The implication $P(k) \to P(k+1)$ is shown to be true for every positive integer k.

Show that the well ordering principle of $\mathbb{N}$ is equivalent to the principle of mathematical induction.

3 Cardinality and Countability

The *cardinality* of a finite set is merely the number of elements that the set possesses. That is, for a set E with finitely many elements the *cardinal number* of E, denoted by $|E|$ or card(E), is the number of elements in E. It is easy to see that for finite sets A and B, if $f : A \mapsto B$ is one-to-one and onto, or bijective, then we have $|A| \leq |B|$, $|A| \geq |B|$, $|A| = |B|$, respectively. In general, if X and Y are nonempty sets, we define the formulas

$$|X| \leq |Y|, \ |X| = |Y|, \ |X| \geq |Y|$$

to mean that there exists $f : X \mapsto Y$ that is injective, bijective, or surjective, respectively. We say X and Y have *the same cardinality*, denoted by $X \sim Y$, if $|X| = |Y|$ (i.e., if there is a bijection from X to Y). If $|X| \leq |Y|$, then there is a subset Y_1 of Y such that $X \sim Y_1$.

The reader can easily prove the following.

Theorem 1.3.1 For sets A, B, and C, we have

    (i) $A \sim A$.

    (ii) If $A \sim B$, then $B \sim A$.

    (iii) If $A \sim B$ and $B \sim C$, then $A \sim C$.

Therefore, $\sim$ is an equivalence relation on any collection of sets.

> **EXAMPLE 1.3.1:** Let $O = \{2k - 1 \mid k \in \mathbb{N}\}$ and $E = \{2k \mid k \in \mathbb{N}\}$. Then $O \sim E \sim \mathbb{N}$ since $f_1(k) = 2k - 1 : \mathbb{N} \mapsto O$ and $f_2(k) = 2k : \mathbb{N} \mapsto E$ are bijective.

> **EXAMPLE 1.3.2:** Let $A = \{x_1, \cdots, x_m\}$ and $B = \{a_1, \cdots, a_n\}$. To define a function $f : A \mapsto B$, there are n choices from B for the value of $f(x_i)$, $i = 1, \cdots, m$. Therefore, there are $n^m = |B|^{|A|}$ functions from A to B. For a given set A with $|A| = n$, we consider a set F of functions from A to $\{0, 1\}$, i.e., the set of *binary functions* on A. Clearly, $|F| = |\mathcal{P}(A)|$. Thus, $|F| = 2^n$. Therefore, we denote by 2^A the set of all binary functions on A. For given two sets A and B, B^A denotes the set of all functions $A \mapsto B$.

Let $\{A_\lambda \mid \lambda \in \Lambda\}$ and $\{B_\lambda \mid \lambda \in \Lambda\}$ be two families of mutually disjoint sets. That is, for $\lambda, \mu \in \Lambda$ and $\lambda \neq \mu$, $A_\lambda \cap A_\mu = \emptyset$ and $B_\lambda \cap B_\mu = \emptyset$. If for every $\lambda \in \Lambda$, $A_\lambda \sim B_\lambda$, then $\bigcup_{\lambda \in \Lambda} A_\lambda \sim \bigcup_{\lambda \in \Lambda} B_\lambda$.

The following theorem is useful to determine if two sets have the same cardinality.

Theorem 1.3.2 [Bernstein's Theorem] If $\mathrm{card}(X) \leq \mathrm{card}(Y)$ and $\mathrm{card}(Y) \leq \mathrm{card}(X)$, then $\mathrm{card}(X) = \mathrm{card}(Y)$.

Proof: Let f_1 be a bijection from X to a subset Y_1 of Y and f_2 a bijection from Y to a subset X_1 of X. Let $X_2 = f_2(Y_1)$. Then f_2 is a bijection from Y_1 to X_2. That is,

$$X \overset{f_1}{\sim} Y_1 \overset{f_2}{\sim} X_2, \ (X_2 \subset X_1).$$

Therefore, $f = f_2 \circ f_1$ is a bijection from X to X_2. Let $X_3 = f(X_1)$. Then $X_1 \overset{f}{\sim} X_3$, $X_3 \subset X_2$ since $X_1 \subset X$ and $X_3 = f(X_1) \subset X_2$. Continuing this process, we obtain a sequence of decreasing sets:

$$X \supset X_1 \supset X_2 \supset \cdots \supset X_n \supset \cdots$$

and under the same bijection f,

$$X \sim X_2 \sim X_4 \sim \cdots$$

$$X_1 \sim X_3 \sim X_5 \sim \cdots.$$

Let $X_0 = X$ and $D = \cap_{n=1}^{\infty} X_n$. Then X can be written as

$$X = \cup_{n=0}^{\infty}(X_n \setminus X_{n+1}) \cup D.$$

Similarly, we have a decomposition of X_1:

$$X_1 = \cup_{n=1}^{\infty}(X_n \setminus X_{n+1}) \cup D.$$

Since f is a bijection, we have

$$(X \setminus X_1) \overset{f}{\sim} (X_2 \setminus X_3), \cdots, (X_n \setminus X_{n+1}) \overset{f}{\sim} (X_{n+2} \setminus X_{n+3}), \cdots.$$

Rewrite the decompositions of X and X_1 as the following:

$$X = D \cup (X \setminus X_1) \cup (X_1 \setminus X_2) \cup (X_2 \setminus X_3) \cup (X_3 \setminus X_4) \cup \cdots,$$

$$X_1 = D \cup (X_2 \setminus X_1) \cup (X_1 \setminus X_2) \cup (X_4 \setminus X_5) \cup (X_3 \setminus X_4) \cup \cdots.$$

Notice that $X_{2n} \setminus X_{2n+1} \overset{f}{\sim} X_{2n+2} \setminus X_{2n+3}$ for $n = 0, 1, \cdots$ and D, $X_{2n+1} \setminus X_{2n+2}$ for $n = 1, 2, \cdots$ are identical to themselves in X and X_1; we obtain that $X \sim X_1$. Therefore, $X \sim Y$ since $X_1 \sim Y$. ∎

We know that $|\mathcal{P}(A)| = 2^n$ if $|A| = n$. In general we have the following.

Theorem 1.3.3 For any set X, $\text{card}(X) < \text{card}(\mathcal{P}(X))$.

Proof: It is clear that $\text{card}(X) \leq \text{card}(\mathcal{P}(X))$, since X can be injectively mapped to the set of one-element sets of X, which is a subset of $\mathcal{P}(X)$. We need to show there is no onto map between X and $\mathcal{P}(X)$. Consider any map $f : X \mapsto \mathcal{P}(X)$. Let $E = \{x \mid x \notin f(x)\}$. Then $E \in \mathcal{P}(X)$. E has no pre-image from f. In fact, if there exists $x_0 \in X$ such that $f(x_0) = E$, then there are two cases:

Case 1: $x_0 \in E$, then $x_0 \notin f(x) = E$, which is a contradiction.

Case 2: $x_0 \notin E$, then $x_0 \in f(x) = E$, which is again a contradiction.

Therefore, $\text{card}(X) < \text{card}(\mathcal{P}(X))$. ∎

Definition 1.3.1 A set E is called *countable* if $\text{card}(E) \leq \text{card}(\mathbb{N})$. A set E is said to be *countable but not finite*, also called an *infinite countable set*, if $\text{card}(E) = \text{card}(\mathbb{N})$. We denote $\text{Card}(\mathbb{N}) = \aleph_0$ (aleph zero).

Clearly, every subset of a countable set is still countable. In addition, we have the following. (Notice the use of the axiom of choice and the principle of mathematical induction in the proof.)

Theorem 1.3.4 For any infinite set E, there is an infinite countable subset.

Proof: Choose $x_1 \in E$. Then $E - \{x_1\}$ is not empty. Next choose $x_2 \in E - \{x_1\}$. If we have taken $\{x_1, x_2, \ldots, x_n\}$ from E, then $E - \{x_1, x_2, \ldots, x_n\}$ is not empty. Then we can choose $x_{n+1} \in E - \{x_1, x_2, \ldots, x_n\}$. By induction we have a sequence $A = \{x_1, \ldots, x_n, \ldots\}$. Then A is an infinite countable subset of E. ∎

From this theorem, we see that the smallest cardinality among infinite sets is $\aleph_0$. From Theorem 1.3.3, we have $\aleph_0 < 2^{\aleph_0}$ for $X = \mathbb{N}$. This means that an uncountable set does exist.

EXAMPLE 1.3.3: The interval $[0, 1] = \{x \in \mathbb{R} \mid 0 \leq x \leq 1\}$ is uncountable!

Proof: It suffices to show that $(0, 1]$ is uncountable. Suppose that $(0, 1]$ is countable; then we can list the points in $(0, 1]$ as a sequence $\{x_1, x_2, \cdots, x_n, \cdots\}$. Notice that each real number in $(0, 1]$ can be uniquely written as a decimal number with infinitely many nonzero digits. For the numbers with a finite representation, we can write them with infinitely many nines. For example, $0.1 = 0.0999999 \cdots$ and $1 = 0.999999 \cdots$. Express each $x_n \in (0, 1]$ in this form:

$$x_1 = 0.x_{11}x_{12}x_{13}x_{14} \cdots$$
$$x_2 = 0.x_{21}x_{22}x_{23}x_{24} \cdots$$
$$x_3 = 0.x_{31}x_{32}x_{33}x_{34} \cdots$$
$$x_4 = 0.x_{41}x_{42}x_{43}x_{44} \cdots$$
$$\vdots$$

Define $x_0 = 0.a_1a_2a_3 \cdots$ by $0 \neq a_k \neq x_{kk}$ for $k = 1, 2, \cdots$. Then x_0 is not in the sequence, for by the definition, the kth digits of x_0 and x_k are different for any $k \in \mathbb{N}$. The contradiction completes the proof. ∎

The method just used is called the *Cantor diagonalization argument.*

We can prove that $[0, 1]$ and the set $\mathbb{R}$ of real numbers have the same cardinality (exercise). The cardinality of $\mathbb{R}$ is denoted by c and is called the *cardinality of the continuum.* Since no bijection between $\mathbb{N}$ and $\mathbb{R}$ exists and $\mathbb{N} \subset \mathbb{R}$, we can write $\aleph_0 < c$.

For a long time mathematicians did not know if there was a set with cardinality strictly between $\aleph_0$ and c. The *Continuum Hypothesis* states that there is no such set. In 1939 Kurt Gödel showed that the Continuum Hypothesis does not contradict the axioms of set theory, and in 1964 Paul Cohen showed that it also does not follow from them. Therefore, the existence (or, as the Continuum Hypothesis states, the nonexistence) of the set A such that $\aleph_0 < |A| < c$ can be taken as a new and independent axiom of set theory.

We now present some facts concerning cardinality:

1. A union of a finite set and a countable set is still countable.

2. If $X_k, k \in \mathbb{N}$ are countable sets, then $\bigcup_{k=1}^{\infty} X_k$ is also countable.

   In fact, without loss of generality, we assume that $X_k, k \in \mathbb{N}$ are mutually disjoint. Since $X_k, k = 1, 2, \cdots$, are countable, we can list them as $X_1 = \{x_{11}, x_{12}, \ldots, x_{1n}, \ldots\}$, $X_2 = \{x_{21}, x_{22}, \cdots, x_{2n}, \cdots\}$, and so on. Then $\bigcup_{k=1}^{\infty} X_k$ can be listed as $\{x_{11}, \{x_{12}, x_{21}\}(x_{ij}, i + j = 3), \{x_{13}, x_{22}, x_{31}\}(x_{ij}, i + j = 4), \cdots \}$. Therefore, it is countable.

3. If X_1 and X_2 are countable, then $X_1 \times X_2$ is countable.
   To show this, let $X_k \times X_2 = \{(x_k, x) \mid x \in X_2\}$ for any $x_k \in X_1$. Notice that $X_1 \times X_2 = \bigcup_{k=1}^{\infty} X_k \times X_2 = \{(x, y) \mid x \in X_1, y \in X_2\}$, and the conclusion follows from Fact 2 above. Inductively, we can also obtain the following.

4. If $X_k, k = 1, \cdots, n$ are countable, then $X_1 \times X_2 \times \cdots \times X_n = \Pi_{k=1}^{n} X_k$ is countable.

Theorem 1.3.5 Let A be a countable set. Then the set of all finite-length strings from A is also countable.

Proof: Let $A = \{a_1, a_2, \ldots, a_n, \cdots\}$. There are a countable number of 1-strings, "a_1", "a_2", $\cdots$, "a_n", $\cdots$. There are also a countable number of 2-sequences, "a_1, a_1", "a_1, a_2", "a_1, a_3", $\cdots$, "a_2, a_1", "a_2, a_2", "a_2, a_3", $\cdots$, and so on. Consider an n-string: "$a_1, a_2, \cdots, a_n$" for each $a_k \in A$. The set of n-sequences is the same as

$$A^n = \Pi_{i=1}^{n} A.$$

This is countable by Fact 4 above. The set of all finite strings is the union of the sets of n-string, $n = 1, 2, \cdots$. Therefore, it is countable by Fact 2 above. ∎

EXAMPLE 1.3.4: The set $\mathbb{Q}$ of rational numbers is countable.

EXAMPLE 1.3.5:   The set of disjoint intervals on $\mathbb{R}$ is countable.

Proof: Let E be the set of disjoint intervals. Define a map f from E to $\mathbb{Q}$ so that for any interval $I \in E$, it maps to a rational number contained in I. Then f is one-to-one since intervals in E are disjoint. Thus, E is countable. ∎

Recall that $\text{card}(\mathbb{N}) = \aleph_0$ and $\text{card}(\mathbb{N}) < \text{card}(\mathcal{P}(\mathbb{N}))$, we write $\text{card}(\mathcal{P}(\mathbb{N})) = 2^{\aleph_0}$. We can prove that $c = 2^{\aleph_0}$. From the next theorem and the fact that $\text{card}([0,1]) = \text{card}(\mathbb{R})$.

Theorem 1.3.6 $\text{Card}([0,1]) = 2^{\aleph_0}$.

Proof: Let $\{0,1\}^{\mathbb{N}}$ be the set of maps from $\mathbb{N}$ to $\{0,1\}$; i.e., $\{0,1\}^{\mathbb{N}} = \{(a_1, a_2, \cdots, a_n, \cdots) \mid a_n = 0 \text{ or } 1 \text{ for all } n \in \mathbb{N}\}$. We claim that $\text{card}(\{0,1\}^{\mathbb{N}}) = 2^{\aleph_0}$. In fact the map $f : \mathcal{P}(\mathbb{N}) \mapsto \{0,1\}^{\mathbb{N}}$ defined on $E \subset \mathbb{N}$ as $f(E) = (a_1, a_2, \cdots, a_n, \cdots)$, where

$$a_n = \begin{cases} 0, & \text{if } n \notin E; \\ 1, & \text{if } n \in E, \end{cases}$$

is a one-to-one correspondence. So, it is sufficient to show that $\text{card}([0,1]) = \text{card}(\{0,1\}^{\mathbb{N}})$. Let

$$A = \{(a_1, a_2, \cdots, a_n, \cdots) \mid a_n = 0 \text{ or } 1 \text{ for all } n \in \mathbb{N},$$
$$\text{and only finitely many } a_n\text{'s are } 1\}$$

and

$$B = \{(a_1, a_2, \cdots, a_n, \cdots) \mid a_n = 0 \text{ or } 1 \text{ for all } n \in \mathbb{N},$$
$$\text{and infinitely many } a_n\text{'s are } 1\}.$$

Then $\{0,1\}^{\mathbb{N}} = A \cup B$ and $A \cap B = \emptyset$. Note that $\text{card}(A) = \aleph_0$ and $\text{card}(\{0,1\}^{\mathbb{N}}) = 2^{\aleph_0}$. Then we have $\text{card}(B) = 2^{\aleph_0}$.

Consider a map g from B to $(0,1]$ defined as $g : \psi \in B \mapsto \sum_{n=1}^{\infty} \dfrac{\psi(n)}{2^n}$, where $\psi = \{\psi(n)\}_{n=1}^{\infty} \in B$. Then g is a one-to-one correspondence. Therefore, $\text{card}((0,1]) = \text{card}(B)$ and so,

$$\text{card}([0,1]) = \text{card}([0,1]) = \text{card}(A \cup B) = \text{card}(\{0,1\}^{\mathbb{N}}) = 2^{\aleph_0}.$$

Exercises

1. For any sets X and Y, prove that either $\text{card}(X) \leq \text{card}(Y)$ or $\text{card}(X) \geq \text{card}(Y)$. [Hint: consider the set J of all injections from subsets of X

into Y. The members of J can be regarded as subsets of $X \times Y$, and so J is partially ordered by inclusion. Then Zorn's Lemma can be applied.]

2. For sets A and B, if $|A| = m$ and $|B| = n$, then
   (a) how many relations are there from A to B?
   (b) how many functions are there from A to B?
   (c) how many one-to-one functions are there from A to B?

3. Let $\{A_\lambda \mid \lambda \in \Lambda\}$ and $\{B_\lambda \mid \lambda \in \Lambda\}$ be two families of mutually disjoint sets. If for every $\lambda \in \Lambda$, $A_\lambda \sim B_\lambda$, then prove $\cup_{\lambda \in \Lambda} A_\lambda \sim \cup_{\lambda \in \Lambda} B_\lambda$.

4. Prove that a union of two countable sets is countable.

5. Prove that the set $\mathbb{Q}$ of rational numbers is countable.

6. Show that the set of intervals with rational numbers as endpoints is countable.

7. Show that the function $f(x) = \frac{1}{x}$ is a one-to-one function from the finite segment $(0, 1]$ onto the infinite segment $[1, \infty)$.

8. Prove that $[0, 1] \sim (0, 1) \sim \mathbb{R}$.

9. Prove that a nonempty set is countable if and only if it is the range of an infinite sequence.

10. Prove that the intervals (a, b) and $[a, b]$ have the same cardinality for $b > a$.

11. Prove that A is an infinite set if and only if there is a proper subset $A_1 \subset A$ such that $A_1 \sim A$.

12. Show that $[0, 1] \times [0, 1]$ and $[0, 1]$ have the same cardinality.

13. If f is a monotone function on $\mathbb{R}$, then prove the set of discontinuity points of f is countable.

14. Let $R[a, b]$ be the set of all real valued functions on $\mathbb{R}$. Prove $\text{card}(R[a, b]) > c$.

15. Let $C[a, b]$ denote the space of continuous functions over interval $[a, b]$ with $a < b$. Prove $\text{card}(C[a, b]) = 2^{\aleph_0}$.

16. Let E be the set of real numbers so that any two elements in E, say x and y, satisfy $\text{dist}(x, y) = |x - y| > 1$. Prove E is countable.

4 The Topology of $\mathbb{R}^n$

Without any doubt, the most important set for real analysis is the set $\mathbb{R}$ of real numbers, also called the real line $\mathbb{R} = (-\infty, \infty)$. In many situations, functions of interest depend on several variables. This leads us into the realm of multivariate calculus, which is naturally set in $\mathbb{R}^n$.

Let $\mathbb{R}^n$ be the set of n-vectors $x = (x_1, x_2, \cdots, x_n)$ with $x_i \in \mathbb{R}$ for $i = 1, 2, \cdots, n$. x_i is called the ith coordinate of the vector x. We define the following operations on $\mathbb{R}^n$:

(i) *Addition.* For $x = (x_1, x_2, \cdots, x_n)$, $y = (y_1, y_2, \cdots, y_n) \in \mathbb{R}^n$, we define

$$x + y = (x_1 + y_1, \cdots, x_n + y_n).$$

(ii) *Scalar Multiplication.* For $\lambda \in \mathbb{R}$ and $x = (x_1, x_2, \cdots, x_n) \in \mathbb{R}^n$, we define $\lambda x = \lambda(x_1, x_2, \cdots, x_n) = (\lambda x_1, \lambda x_2, \cdots, \lambda x_n)$.

Then $\mathbb{R}^n$ becomes a vector space, called the *Euclidean space.* If $x_i, i = 1, \cdots, n$ are rationals, then $x = (x_1, x_2, \cdots, x_n)$ is called a *rational point* in $\mathbb{R}^n$.

Definition 1.4.1 For $x = (x_1, x_2, \cdots, x_n) \in \mathbb{R}^n$, we define the *length* of x as $|x| = (x_1{}^2 + x_2{}^2 + \cdots + x_n{}^2)^{\frac{1}{2}}$. $|x|$ is also called the norm of x in general and denoted by $\|x\|$.

Theorem 1.4.1 The length $|x|$ satisfies:

(i) $|x| \geq 0$ and $|x| = 0$ if and only if $x = 0 = (0, \cdots, 0)$.
(ii) For all $\lambda \in \mathbb{R}$, $|\lambda x| = |\lambda| \, |x|$.
(iii) For all $x, y \in \mathbb{R}^n$, we have $|x + y| \leq |x| + |y|$. This is called the *triangle inequality*.

Proof: (i) and (ii) are obvious. Showing (iii) is equivalent to showing

$$|x + y|^2 \leq (|x| + |y|)^2 = |x|^2 + 2|x||y| + |y|^2,$$

i.e.,

$$\sum_{i=1}^{n} (x_i + y_i)^2 \leq \sum_{i=1}^{n} x_i^2 + 2\left(\sum_{i=1}^{n} x_i\right)^{\frac{1}{2}} \left(\sum_{i=1}^{n} y_i\right)^{\frac{1}{2}} + \sum_{i=1}^{n} y_i^2.$$

Thus, we need only to show

$$\sum_{i=1}^{n} x_i y_i \leq \left(\sum_{i=1}^{n} (x_i)^2\right)^{\frac{1}{2}} \left(\sum_{i=1}^{n} (y_i)^2\right)^{\frac{1}{2}}.$$

Or equivalently,

$$\left(\sum_{i=1}^{n} x_i y_i\right)^2 \leq \sum_{i=1}^{n} x_i^2 \sum_{i=1}^{n} y_i^2.$$

This is the *Schwarz inequality*! To prove it, we consider a quadratic function of λ as follows.

$$f(\lambda) = |x + \lambda y|^2 = \sum_{i=1}^{n}(x_i + \lambda y_i)^2 = \sum_{i=1}^{n}x_i^2 + 2\lambda\sum_{i=1}^{n}x_i y_i + \lambda^2\sum_{i=1}^{n}y_i^2.$$

Since $f(\lambda) \geq 0$ for all λ, we have its discriminant

$$\Delta = 4\left(\sum_{i=1}^{n}x_i y_i\right)^2 - 4\left(\sum_{i=1}^{n}x_i^2\right)\left(\sum_{i=1}^{n}y_i^2\right) \leq 0.$$

This gives the Cauchy–Schwarz Inequality. ∎

The *inner product* of two vectors $x, y \in \mathbb{R}^n$ is defined by

$$\langle x, y\rangle = \sum_{i=1}^{n}x_i y_i.$$

Clearly, $\langle x, x\rangle = |x|^2$. From the proof of the above property (iii), we see that for any $x, y \in \mathbb{R}^n$,

$$\langle x, y\rangle \leq |x|\,|y|$$

and equality holds if and only if x and y are collinear. If $\langle x, y\rangle = 0$, then the vectors x and y are said to orthogonal, denoted $x \perp y$

Definition 1.4.2 For any $x, y \in \mathbb{R}^n$, we define the *distance*, $d(x, y)$ between x and y by $|x - y|$. Then $d(x, y) = |x - y|$ satisfies
   (i) $d(x, y) \geq 0$ and $d(x, y) = 0$ if and only if $x = y$.
   (ii) $d(x, y) = d(y, x)$.
   (iii) $d(x, y) \leq d(x, z) + d(z, y)$.

Definition 1.4.3 Let $E \subset \mathbb{R}^n$. We define $\mathrm{diam}(E) = \sup\{|x - y| \mid x, y \in E\}$ and call it the *diameter* of E. If $diam(E) < \infty$, then E is said to be a *bounded set*.

EXAMPLE 1.4.1: Consider the set of positive integers $\mathbb{N}$. For any two elements say $x = n, y = m$, we have $\mathrm{dist}(x, y) = |n - m| < \infty$, but $\mathrm{diam}(\mathbb{N}) = \infty$. Clearly, if E is bounded then there exist $M > 0$, such that $|x| \leq M$ for any $x \in E$.

Definition 1.4.4
   (1) Let $x_0 \in \mathbb{R}^n$, and $\delta > 0$, we define the *δ-neighborhood* of x_0 as the *open ball* $B(x_0, \delta) = \{x \in \mathbb{R}^n, |x - x_0| < \delta\}$.

(2) Let $a_i, b_i \in \mathbb{R}$ and $a_i < b_i$ for $i = 1, 2, \cdots, n$. We define the *open rectangle* as $I = \{\mathbf{x} \in \mathbb{R}^n; \mathbf{x} = (x_1, x_2, \cdots, x_n), a_i < x_i < b_i, i = 1, 2, \cdots, n\}$. That is, $I = (a_1, b_1) \times (a_2, b_2) \times \cdots \times (a_n, b_n)$.

The notion of distance for points in $\mathbb{R}^n$ immediately allows us to discuss convergence of sequences in $\mathbb{R}^n$.

Definition 1.4.5 For $\mathbf{x}_k \in \mathbb{R}^n$, $k = 1, 2, \cdots$, if $\mathbf{x} \in \mathbb{R}^n$ satisfies that for any ϵ-ball of $\mathbf{x}$, $B(\mathbf{x}, \epsilon)$, there exists $N \in \mathbb{N}$ and for any $k > N$, we have $\mathbf{x}_k \in B(\mathbf{x}, \epsilon)$, i.e., $\lim_{k \to \infty} |\mathbf{x}_k - \mathbf{x}| = 0$, then we say the sequence $(\mathbf{x}_k)$ *converges* to $\mathbf{x}$. $\mathbf{x}$ is called the *limit of the sequence* $(\mathbf{x}_k)$, denoted by $\lim_{k \to \infty} \mathbf{x}_k = \mathbf{x}$.

The proof of the following result is left to the reader as an exercise.

Theorem 1.4.2 Suppose $\mathbf{x}_k \in \mathbb{R}^n$ and $\mathbf{x}_k = (x_{1,k}, x_{2,k}, \cdots, x_{n,k})$ then $\lim_{k \to \infty} \mathbf{x}_k = \mathbf{x}$ if and only if $\lim_{k \to \infty} x_{i,k} = x_i$ for $i = 1, 2, \cdots, n$ where $\mathbf{x} = (x_1, x_2, \cdots, x_n)$.

Definition 1.4.6 A sequence $(\mathbf{x}_k)$ in $\mathbb{R}^n$ is said to be a *Cauchy sequence* if for $\epsilon > 0$, there exists $N \in \mathbb{N}$ such that $d(\mathbf{x}_k, \mathbf{x}_\ell) < \epsilon$ for all $\ell, k > N$.

Using Theorem 1.4.2 and the Cauchy criterion for the real line (see the exercises), we obtain the following.

Theorem 1.4.3 A sequence $(\mathbf{x}_k)$ in $\mathbb{R}^n$ converges if and only if it is a Cauchy sequence.

Definition 1.4.7 For a sequence $(\mathbf{x}_k)$ in $\mathbb{R}^n$ and a subsequence (k_i) of possible integers so that $k_1 < k_2 < \cdots$, the sequence $(\mathbf{x}_{k_i})$ is called a *subsequence* of $(\mathbf{x}_k)$.

Theorem 1.4.4 The sequence $(\mathbf{x}_k)$ converges if and only if any subsequence of $(\mathbf{x}_k)$ converges.

Proof: "$\Longrightarrow$" Suppose $(\mathbf{x}_k)$ converges to $\mathbf{x}$. Then for all $\epsilon > 0$, there exists $N \in \mathbb{N}$ such that $\text{dist}(\mathbf{x}_k, \mathbf{x}) < \epsilon$ for $k > N$. For any subsequence of $(\mathbf{x}_k)$, say $(\mathbf{x}_{k_i})$, $\text{dist}(\mathbf{x}_{k_i}, \mathbf{x}) < \epsilon$ for $k_i > N$. So $(\mathbf{x}_{k_i})$ converges to $\mathbf{x}$.

"$\Longleftarrow$" (Proof of contrapositive). Suppose $(\mathbf{x}_k)$ diverges; then we need to show that there exists a subsequence $(\mathbf{x}_{k_i})$ that diverges. From the definition of $(\mathbf{x}_k)$ divergent, we have that there exists $\epsilon_0 > 0$ such that for any $N \in \mathbb{N}$, there exists

$k_2 > k_1 > N$ such that $\text{dist}(x_{k_2} - x_{k_1}) \geq \epsilon_0$. For $N_1 = 1$, we have $k_2 > k_1 > 1$ such that $\text{dist}(x_{k_2} - x_{k_1}) \geq \epsilon_0$. For $N_2 = \max\{k_1, k_2\}$, we have $k_4 > k_3 > N_2$ such that $\text{dist}(x_{k_4} - x_{k_3}) \geq \epsilon_0$, and so on. In such a way, we obtained a subsequence (x_{k_i}) satisfying $\text{dist}(x_{k_{i+1}} - x_{k_i}) \geq \epsilon$. Clearly, (x_{k_i}) is not a Cauchy sequence and thus diverges. ∎

Similar to the Bolzano–Weierstrass Theorem on $\mathbb{R}$, we have the following.

Theorem 1.4.5 Every bounded sequence (x_k) in $\mathbb{R}^n$ has a convergent subsequence.

For a sequence (x_n) of real numbers, (x_n) is said to be nondecreasing if $x_1 \leq x_2 \leq \cdots$. (x_n) is said to be nonincreasing if $x_1 \geq x_2 \geq \cdots$. In either case, (x_n) is called a *monotone sequence*.

Corollary 1.4.1 *Any bounded monotone sequence has a limit point.*

Definition 1.4.8 Let $E \subset \mathbb{R}^n$ and $x \in \mathbb{R}^n$. If there is a sequence (x_k) of distinct points in E such that $\lim_{k \to \infty} |x_k - x| = 0$, (i.e., $\lim_{k \to \infty} x_k = x$) then x is said to be an *accumulation point* of E. The set of all accumulation points of E is denoted E'. The set $E \cup E'$ is called the *closure* of E and denoted by $\overline{E}$.

    EXAMPLE 1.4.2:
(a) Let $E = (a, b)$ then $E' = [a, b]$. So $\overline{E} = [a, b]$.
(b) Let $E = \{1, \frac{1}{2}, \cdots, \frac{1}{n}, \cdots\}$. Then $E' = \{0\}$.

Theorem 1.4.6 Some basic facts of E' include:
(i) If E is a finite set, then E' is empty.
(ii) If $E \subset \mathbb{R}^n$, then $x \in E'$ if and only if for $\delta > 0$, $(B(x, \delta) \setminus \{x\}) \cap E \neq \emptyset$.
(iii) If $E \subset \mathbb{R}^n$ then $x \notin E'$ if and only if there exists $\delta > 0$ such that $(B(x, \delta) \setminus \{x\}) \cap E) = \emptyset$. Such points are called *isolated points* of E.
(iv) $(E \cup F)' = E' \cup F'$.

    EXAMPLE 1.4.3: If $E = \{\sqrt{n} - \sqrt{m} \mid m, n \in \mathbb{N}\}$, then $E' = \mathbb{R}$. In fact, let $x_n = \sqrt{\lfloor (x + n)^2 \rfloor} - \sqrt{n^2}$. where $\lfloor \cdot \rfloor$ denotes the greatest integer $\leq x$. Then $\sqrt{(x + n)^2 - 1} - \sqrt{n^2} \leq x_n \leq x + n - n = x$. Noticing that $\sqrt{(x + n)^2 - 1} - \sqrt{n^2} = \frac{x^2 + 2nx - 1}{\sqrt{(x+n)^2 - 1} + \sqrt{n^2}} \to x$ as $n \to \infty$, we obtain that $\lim_{n \to \infty} x_n = x$.

Definition 1.4.9 Let $E \subset \mathbb{R}^n$. If $E' \subset E$, then E is called a *closed set*.

Similar to the definition of the real-valued continuous function f on $\mathbb{R}$, we say that $f : \mathbb{R}^n \mapsto \mathbb{R}^m$ is *continuous* at $a \in \mathbb{R}^n$ if for any $\epsilon > 0$, there exists $\delta > 0$ such that $f(x) \in B(f(a, \epsilon)$ whenever $x \in B(a, \delta)$.

   EXAMPLE 1.4.4:

   (i) $[0, 1]$ is a closed set of $\mathbb{R}$ and $[0, 1]^n$ is closed in $\mathbb{R}^n$.
   (ii) $\mathbb{R}^n$ and $\emptyset$ are closed sets.
   (iii) Let f be a continuous function on $\mathbb{R}^n$. Then $E_1 = \{x \mid f(x) \leq a\}$ and $E_2 = \{x \mid f(x) \geq a\}$ are closed sets of $\mathbb{R}^n$.

Proof: We show that E_1 is a closed set. If $E_1' = \emptyset$, then the conclusion holds. Assume that $x_0 \in E'$. Then there exists $(x_k) \subset E$ such that $x_k \to x_0$ as $k \to \infty$. Notice that $x_k \in E$ implies that $f(x_k) \leq a$ and thus, $f(x_0) = \lim_{k \to \infty} f(x_k) \leq a$. Therefore, $x_0 \in E_1$. This means that $E_1' \subset E_1$ and so, E_1 is closed. ∎

Theorem 1.4.7
   (i) If F_1 and F_2 are closed sets of $\mathbb{R}^n$, then $F_1 \cup F_2$ is closed. Furthermore, the union of any finitely many closed sets is closed.
   (ii) If $\{F_\lambda\}$ is a collection of closed sets of $\mathbb{R}^n$, then the intersection $F = \cap_{\lambda \in \Lambda} F_\lambda$ is closed.

Proof: (i) Suppose that $\bar{F}_1 = F_1, \bar{F}_2 = F_2$. Then $\overline{F_1 \cup F_2} = (F_1 \cup F_2) \cup (F_1 \cup F_2)' = (F_1 \cup F_2) \cup (F_1' \cup F_2') = (F_1 \cup F_1') \cup (F_2 \cup F_2') = \bar{F}_1 \cup \bar{F}_2 = F_1 \cup F_2$. Therefore, $F_1 \cup F_2$ is closed. Inductively, we can have that the finite union of closed sets is again closed.

(ii) If $F_\lambda, \lambda \in \Lambda$ are closed, then $\bar{F}_\lambda = F_\lambda, \lambda \in \Lambda$. Let $F = \cap_{\lambda \in \Lambda} F_\lambda$. Then $F \subset F_\lambda$, $\lambda \in \Lambda$. Thus, $\bar{F} \subset \bar{F}_\lambda, \lambda \in \Lambda$. So, $\bar{F} \subset \cap_{\lambda \in \Lambda} \bar{F}_\lambda = \cap_{\lambda \in \Lambda} F_\lambda = F$. Therefore, F is a closed set. ∎

   EXAMPLE 1.4.5: The union of infinitely many closed sets may not be a closed set. For instance, $F_k = [\frac{1}{k+1}, \frac{1}{k}], k = 1, \cdots$, then $\cup_{k \in \mathbb{N}} F_k = (0, 1]$.

Definition 1.4.10 $G \subset \mathbb{R}^n$ is said to be *open* if $G^c = \mathbb{R}^n \setminus G$ is closed.

   EXAMPLE 1.4.6:

   (i) For any real numbers $a < b$, (a, b) is an open set of $\mathbb{R}$ and $(a, b)^n$ is open in $\mathbb{R}^n$.

(ii) $\mathbb{R}^n$ and $\emptyset$ are open sets.

(iii) Let f be a continuous function on $\mathbb{R}^n$. Then $G_1 = \{x \mid f(x) < c\}$ and $G_2 = \{x \mid f(x) > c\}$ are open sets of $\mathbb{R}^n$ for any $c \in \mathbb{R}$.

Theorem 1.4.8

(i) If G_1 and G_2 are open sets on $\mathbb{R}^n$, then $G_1 \cap G_2$ is open. Furthermore, the intersection of finitely many open sets is open.

(ii) If $\{G_\lambda\}_{\lambda \in \Lambda}$ is a collection of open sets of $\mathbb{R}^n$, then $\cup_{\lambda \in \Lambda} G_\lambda$ is open.

As we have seen in the above example, a countable union of closed sets is not necessarily a closed set, and it is easy to see that a countable intersection of open sets needs not to be an open set. A set is called an F_σ set if it is the union of countably many closed sets. Similarly, a set is said to be a G_δ set if it is the intersection of countably many open sets. Clearly, a set is an F_σ set if and only if its complement is a G_δ set.

Theorem 1.4.9 If (F_k) is a sequence of nonempty bounded closed sets of $\mathbb{R}^n$ and

$$F_1 \supset F_2 \supset \cdots \supset F_k \supset \cdots,$$

then $\cap_{k=1}^\infty F_k \neq \emptyset$.

Proof: If there are only finitely many different sets in (F_k), then there exists $k_0 \in \mathbb{N}$ such that $F_k = F_{k_0}$ for $k \geq k_0$. In this case, $\cap F_k = F_{k_0} \neq \emptyset$.

Now, without loss of generality, we assume that for any $k \in \mathbb{N}$, $F_k \setminus F_{k+1} \neq \emptyset$. Then we can choose $x_k \in F_k \setminus F_{k+1}$, $k = 1, \cdots$, and thus (x_k) consists of distinct points in $\mathbb{R}^n$ and (x_k) is bounded. By the Bolzano–Weierstrass Theorem, (x_k) has a convergent subsequence, say (x_{k_i}) and $\lim_{i \to \infty} x_{k_i} = x_0$. Since F_k is closed for any $k \in \mathbb{N}$, we have $x_0 \in F_k$ for $k = 1, 2, \cdots$, i.e., $x_0 \in \cap_{k=1}^\infty F_k$. ∎

Definition 1.4.11 For $E \subset \mathbb{R}^n$, let $\mathcal{C}$ be a collection of open sets satisfying for all $x \in E$, there exists $O \in \mathcal{C}$ such that $x \in O$. Then $\mathcal{C}$ is called an *open covering* of E. If there exists $\mathcal{C}_1 \subset \mathcal{C}$ and $\mathcal{C}_1$ covers E, then $\mathcal{C}_1$ is called a *subcovering*. In particular, if $\mathcal{C}_1$ is finite, then it is called a *finite covering*.

Theorem 1.4.10 Let $\mathcal{C}$ be a collection of open sets of $\mathbb{R}^n$. Then there exists a countable subcollection $\{O_i\}_{i \in I}$ such that $\cup_{O \in \mathcal{C}} O = \cup_{i \in I} O_i$, where $I \subset \mathbb{N}$.

Corollary 1.4.2 *Any open covering $\mathcal{C}$ of E contains a countable subcovering.*

Theorem 1.4.11 [Heine–Borel] Let F be a closed and bounded set of $\mathbb{R}^n$. Then each open covering $\mathcal{C}$ of F contains a finite subcovering.

Proof: By the above corollary, we can assume $\mathcal{C}$ is a countable covering. Let $H_k = \cup_{i=1}^k O_i$, and $L_k = F \setminus H_k$, $k = 1, \cdots$. Then for each k, H_k is open and $\cup_{k=1}^\infty H_k = \cup_{i=1}^\infty O_i$. For each k, L_k is closed and $L_k \supset L_{k+1}$ for $k = 1, 2, \cdots$. If there exists $k_0 \in \mathbb{N}$ such that $L_{k_0} = \emptyset$, then $F \subset H_{k_0} = \cup_{i=1}^{k_0} O_i$. The conclusion holds.

If for all $k \in \mathbb{N}$, $L_k \neq \emptyset$, then since L_k is closed and bounded ($L_k \subset F$ and F is bounded), by Theorem 1.4.9, we have that $\cap_{k=1}^\infty L_k \neq \emptyset$. That is, there is $x \in F \setminus H_k$ for $k = 1, 2 \cdots$, and thus, $x \in F$ but $x \notin H_k$ for any $k \in \mathbb{N}$. This is a contradiction since (H_k) is an open covering of F.   ∎

The boundedness condition and closed condition are important in the Heine–Borel Theorem, as the next example shows.

EXAMPLE 1.4.7:

(1) $F = \mathbb{N}$ is closed but not bounded. The set of $O_k = (k - \frac{1}{k}, k + \frac{1}{k})$, $k = 1, 2 \cdots$, is an open covering of F, but there is no finite subcovering.

(2) $F = \{1, \frac{1}{2}, \frac{1}{3}, \cdots\}$ is a bounded set but F is not closed. The set of $O_k = (\frac{1}{k} - \frac{1}{2^k}, \frac{1}{k} + \frac{1}{2^k})$, $k = 1, 2, \cdots$, is an open covering of F, but there is no finite subcovering.

Theorem 1.4.12 Let $E \subset \mathbb{R}^n$. If any open covering of E contains a finite subcovering, then E is a bounded and closed set.

Proof: Consider any given point y in E^c. For all $x \in E$ we can have a $\delta_x > 0$ such that $B(y, \delta_x) \cap B(x, \delta_x) = \emptyset$. The collection of all such δ-balls is an open covering of E. By the assumption, there is a finite subcovering, say, $B(x_1, \delta_1) \cup \cdots \cup B(x_n, \delta_n) \supset E$. Therefore, E must be bounded. Let $\delta = \min\{\delta_1, \cdots, \delta_n\}$. Then $B(y, \delta) \cap E = \emptyset$. Thus, E^c is open and so, E is closed.   ∎

Definition 1.4.12 A set E is called a *compact set* if any open covering of E contains a finite subcovering.

Corollary 1.4.3 *In $\mathbb{R}^n$, a set E is compact if and only if E is bounded and closed.*

Theorem 1.4.13 Let $\mathcal{C}$ be a collection of closed sets of $\mathbb{R}^n$ with the property that every finite subcollection of $\mathcal{C}$ has a nonempty intersection and suppose there is at least one set in $\mathcal{C}$ which is bounded. Then $\cap_{F \in \mathcal{C}} F \neq \emptyset$.

Proof: Assume that F_1 is bounded and $C = \{F_\lambda \mid \lambda \in \Lambda\}$. Let $G_\lambda = F_\lambda^c$, $\lambda \in \Lambda$. If $\cap_{\lambda \in \Lambda} F_\lambda = \emptyset$, then $(\cap_{\lambda \in \Lambda} F_\lambda)^c = \mathbb{R}^n$, i.e., $\cup_{\lambda \in \Lambda} F_\lambda^c = \cup_{\lambda \in \Lambda} G_\lambda = \mathbb{R}^n \supset F_1$.

Since F_1 is closed and bounded, there is a finite subcovering from $\{G_\lambda \mid \lambda \in \Lambda\}$, say $G_{\lambda_1} \cup \cdots \cup G_{\lambda_m} \supset F_1$. Therefore, $F_1 \cap G_{\lambda_1}^c \cap \cdots \cap G_{\lambda_m}^c = \emptyset$, i.e., $F_1 \cap F_{\lambda_1} \cap \cdots \cap F_{\lambda_m} = \emptyset$. This is a contradiction. ∎

Next, we discuss the open set structure on $\mathbb{R}$.

Theorem 1.4.14 Any open set on $\mathbb{R}$ is a countable union of disjoint open intervals.

Proof: Let O be an open set on $\mathbb{R}$. Then for all $x \in O$, there exists $\delta > 0$ such that $B(x, \delta) \subset O$. For a given $x \in O$, let $b_x = \sup\{b \mid (x, b) \subset O\}$ and $a_x = \inf\{a \mid (a, x) \subset O\}$. Here, b_x could be ∞ and a_x could be $-\infty$.

Let $I_x = (a_x, b_x)$ for $x \in O$. Then $\cup_{x \in O} I_x = O$. In fact, for all $x \in O$, $x \in I_x$ and thus $\cup_{x \in O} I_x \supset O$. On the other hand, $I_x \subset O$: for any $y \in I_x$, if $y = x$, then clearly $x \in O$; if $y \in (x, b_x)$, then by the definition of b_x, there is b such that $y \in (x, b) \subset O$. Thus, $[x, b_x) \subset O$ and similarly we have $(a_x, x] \subset O$.

Now, we show that for any $x, y \in O$ and $x \neq y$, we have either $I_x \cap I_y = \emptyset$ or $I_x = I_y$. Assume that $x < y$ and $I_x \cap I_y \neq \emptyset$. Then $a_y < b_x$. Since $a_y \notin O$ by the definition of a_y, we must have that $a_x = a_y$. Similarly, we have $b_x = b_y$. That is $I_x = I_y$.

Notice that the collection $\{I_x \mid x \in O\}$ is a set of disjoint open intervals and thus it is countable. Therefore, $\cup_{x \in O} I_x = O$ shows that the open set O is a union of countably many disjoint open intervals. ∎

Definition 1.4.13 A set E is called a *perfect set* if $E' = E$.

Notice that every point in a perfect set E is a limit point of E. Any closed interval is a perfect set.

Next, we discuss a very special subset of $[0, 1]$ called the *Cantor set*. Let $K_0 = [0, 1]$. First, we remove the segment $(\frac{1}{3}, \frac{2}{3})$ from K_0. The remaining part is $[0, \frac{1}{3}] \cup [\frac{2}{3}, 1]$, denoted by K_1. Then, remove the middle thirds of these intervals and let K_2 be the union of the remaining intervals. So, $K_2 = [0, \frac{1}{9}] \cup [\frac{2}{9}, \frac{3}{9}] \cup [\frac{6}{9}, \frac{7}{9}] \cup [\frac{8}{9}, 1]$. Inductively continuing this process, we obtain a sequence of compact sets K_n satisfying (i) $K_0 \supset K_1 \supset K_2 \supset \cdots$ and (ii) K_n is the union of 2^n closed intervals with the length $\frac{1}{3^n}$ of each interval. The limit set $K = \cap_{n=0}^{\infty} K_n$ is called the Cantor set.

Theorem 1.4.15 The Cantor set K is nonempty but K contains no segment. K is compact, perfect, and card$(K) = c$.

Proof: Obviously K is nonempty. K is closed and thus compact as it is an intersection of closed sets and bounded. From the construction, it is also clear that K does not contain any interval and thus it has empty interior. Next, for any $x \in K$, we need to show $x \in K'$. For each $n \in \mathbb{N}$ we have $x \in K_n$. Let I_{n,i_0} be the interval which contains x where $K_n = \cup_{i=1}^{2^n} I_{n,i}$. For any $\delta > 0$, we can choose n large enough so that $I_{n,i_0} \subset B(x, \delta)$. Let x_n be one of the end points of I_{n,i_0}. Then $x_n \in B(x, \delta) \cap K$. Hence, x is an accumulation point of K or $x \in K'$. This concludes that K is perfect.

For any point $x \in [0, 1]$, we choose the ternary (base three) representation of x: $x = \sum_{k=1}^{\infty} \frac{p_k}{3^k}$, where each p_k is 0, 1, or 2. Then $x \in [0, 1] \setminus K$ if and only if there exists k such that $p_k = 1$. So, $x \in K$ if and only if $x = 2 \sum_{k=1}^{\infty} \frac{a_k}{3^k}$ for each $a_k = 0$ or 1. Define $\phi : K \mapsto [0, 1]$ as $\phi(x) = \phi(2 \sum_{k=1}^{\infty} \frac{a_k}{3^k}) = \sum_{k=1}^{\infty} \frac{a_k}{2^k}$. Then ϕ is a one-to-one correspondence. So, card$(K) = $ card$2^{\{0,1\}} = $ card$([0, 1]) = c$. ∎

The concept of distance in $\mathbb{R}^n$ can be extended on any nonempty set X, usually called a *metric* on X. More precisely, a metric (or a distance) d on a nonempty set X is a function $d : X \times X \mapsto \mathbb{R}$ satisfying (i) $d(x, y) \geq 0$ for all $x, y \in X$ and $d(x, y) = 0$ if and only if $x = y$; (ii) $d(x, y) = d(y, x)$ for all $x, y \in X$; (iii) $d(x, y) \leq d(x, z) + d(z, y)$ for all $x, y, z \in X$. The set X with the distance d is called a *metric space*. For a metric space X with distance d, let $x \in X$. The *open ball* of x with radius $r > 0$ is the set $B(x, r) = \{y \in X \mid d(x, y) < r\}$. A set $O \subset X$ is said to be an *open set* if for every $x \in O$, there is some $r > 0$ such that $B(x, r) \subset O$. A point x is called an *interior point* of a set E if there exists an open ball $B(x, r)$ such that $B(x, r) \subset E$. Similar to the setting of $\mathbb{R}^n$, we can define the convergence of a sequence in a metric space X, the accumulation points, and the closure set $\overline{E}$ of a set E. A subset E of a metric space X is called *dense* in X if $\overline{E} = X$. A set $E \subset X$ is said to be *nowhere dense* if its closure has an empty interior set. Cantor set K is nowhere dense in $[0, 1]$. A set $E \subset X$ is said to be *first category* if there exists a sequence (E_n) of nowhere dense subsets such that $E = \cup_{n=1}^{\infty} E_n$. A metric space is called a *Baire space* if every nonempty open set is not first category. A set that is not first category is said to be *second category*.

The Baire Theorem in set theory states that if E is an F_σ set, $E = \cup_k F_k$, and each F_k has no interior points, then E has no interior points as well. As an application of the Baire Theorem, we can show that the set $\mathbb{Q}$ of rationals is not a G_δ set (though it is easy to see that $\mathbb{Q}$ is an F_σ set).

Exercises

1. For $x, y \in \mathbb{R}^n$, we say x and y are orthogonal if $\langle x, y \rangle = 0$. Prove that for two orthogonal vectors x and y in $\mathbb{R}^n$,

$$|x + y| = (|x|^2 + |y|^2)^{\frac{1}{2}}.$$

2. Show that $|x + y|^2 + |x - y|^2 = 2(|x|^2 + |y|^2)$ for any two vectors x and y in $\mathbb{R}^n$. This is called the *parallelogram law*.

3. Prove the *Bolzano–Weierstrass Theorem* on $\mathbb{R}$: Every bounded sequence in $\mathbb{R}$ contains a convergent subsequence.

4. Establish the Cauchy Criterion of the real line:
   (a) Show that a sequence $\langle x_n \rangle$ in $\mathbb{R}$ which converges to a real number l is a Cauchy sequence.
   (b) Show that each Cauchy sequence in $\mathbb{R}$ is bounded.
   (c) Show that if a Cauchy sequence in $\mathbb{R}$ has a subsequence that converges to l, then the original sequence converges to l.
   (d) There is a real number l to which the sequence $\langle x_n \rangle$ converges if and only if $\langle x_n \rangle$ is a Cauchy sequence.

5. Consider the infinite continued fraction

$$\cfrac{1}{2 + \cfrac{1}{2 + \cfrac{1}{2 + \cdots}}}.$$

We have a recursion formula $a_{n+1} = \frac{1}{2 + a_n}, n \geq 1$, and $a_1 = \frac{1}{2}$ for the following approximation sequence of the continued fraction:

$$a_1 = \frac{1}{2}, a_2 = \frac{1}{2 + \frac{1}{2}}, a_3 = \frac{1}{2 + \frac{1}{2 + \frac{1}{2}}}, \cdots.$$

Show that (a_n) is a convergent sequence and evaluate $\lim_{n \to \infty} a_n$, the value of the continued fraction.

6. Construct a bounded set of real numbers with exactly two accumulation points.

7. Show that every bounded monotone sequence has a unique limit point.

8. If a function g is defined on a set E and $g(p) = p$ for some $p \in E$, then g is said to have a *fixed point* p in E. Show that if $g \in C[a, b]$ and $g(x) \in [a, b]$ for all $x \in [a, b]$, then g has a fixed point in $[a, b]$. Further, suppose $g'(x)$ exists on (a, b) and $|g'(x)| \leq \lambda < 1$ for all $x \in (a, b)$. Then g has a unique fixed point p in $[a, b]$.

9. Let $g \in C[a, b]$ and suppose that $g(x) \in [a, b]$ for all $x \in [a, b]$ and g' exists on (a, b) with

$$|g'(x)| \leq \lambda < 1$$

for all $x \in (a, b)$. If p_0 is any number in $[a, b]$, then the sequence defined by

$$p_n = g(p_{n-1}),$$

$n \geq 1$ will converge to the unique fixed point in $[a, b]$.

10. *Newton's method* is one of the most powerful and well-known numerical methods for finding a root of $f(x) = 0$: Let $f \in C^2[a, b]$. If $p \in (a, b)$ is such that $f(p) = 0$ and $f'(p) \neq 0$, then prove that there exists $\delta > 0$ such that the sequence

$$p_n = p_{n-1} - \frac{f(p_{n-1})}{f'(p_{n-1})}, \ n \geq 1$$

converges to p for any initial approximation $p_0 \in [p - \delta, p + \delta]$.

11. Give an example of an open covering of $(0, 1)$ that has no finite subcovering.

12. If X is the space of all rational numbers, and E is the set of all rational p such that $2 < p^2 < 3$, then prove E is closed and bounded, but not compact.

13. For a set A of real numbers, show that the following are equivalent:
    (a) A is closed and bounded.
    (b) Every sequence of points in A has a limit point.
    (c) Every open cover of A has a finite subcover.

14. Suppose that A and B are closed (open). Show that $A \times B = \{(a, b) \mid a \in A, b \in B\}$ is also closed (open).

15. Show that $\mathbb{Q}$ is an F_σ set and therefore $\mathbb{Q}^c$ is a G_δ set.

16. Show that compact sets are closed under arbitrary intersections and finite unions.

17. Show that if $f : \mathbb{R}^n \mapsto \mathbb{R}^m$ is continuous, then $U \subset \mathbb{R}^m$ is open implies that $f^{-1}(U)$ is also open in $\mathbb{R}^n$.

18. Give an example of a continuous function f and open set U such that $f(U)$ is not open.

19. Show that a function $f : \mathbb{R}^n \mapsto \mathbb{R}^n$ is continuous if and only if $f^{-1}(E)$ is closed for every closed set $E \subset \mathbb{R}^m$.

20. Let C be a compact set of $\mathbb{R}^n$ and $f : \mathbb{R}^n \mapsto \mathbb{R}^m$ a continuous function. Then the image set $f(C)$ is compact in $\mathbb{R}^m$.

21. A function $f : E \mapsto \mathbb{R}^m$ is called a *Lipschitz function* on $E \subset \mathbb{R}^n$ if there is a constant C such that $|f(x) - f(y)| \leq C|x - y|$ for all $x, y \in E$. Show that every Lipschitz function is continuous.

22. A function $f : S \mapsto \mathbb{R}^m$ on $S \subset \mathbb{R}^n$ is said to be *uniformly continuous* if for every $\epsilon > 0$, there is a positive real number $\delta > 0$ such that $|f(x) - f(y)| < \epsilon$ whenever $|x - y| < \delta$ for $x, y \in S$. Show that every Lipschitz function is uniformly continuous.
23. Let f be a differentiable real-valued function on $[a, b]$ with bounded derivative. Then prove f is uniformly continuous on $[a, b]$.
24. Suppose that $E \subset \mathbb{R}^n$ is compact and $f : E \mapsto \mathbb{R}^m$ is continuous. Then prove f is uniformly continuous on E.
25. Show that the function f defined by $f(x) = \frac{1}{x} \sin x$ for $x \neq 0$ and $f(0) = 1$ is uniformly continuous on $\mathbb{R}$.
26. A sequence $(f_k(x))$ of real-valued functions on $D \subset \mathbb{R}^n$ is said to *converge uniformly* to a function $f(x)$ if for any $\epsilon > 0$, there is $N \in \mathbb{N}$ such that $f_k(x) \in B(f(x), \epsilon)$ for all $x \in D$ whenever $k > N$.
    (a) (Cauchy criterion) Show that $(f_k(x))$ converges to $f(x)$ uniformly if and only if for any $\epsilon > 0$, there is $N \in \mathbb{N}$ such that $|f_k(x) - f_\ell(x)| < \epsilon$ for all $x \in D$ whenever $k, \ell > N$.
    (b) Show that if $(f_k(x))$ converges to $f(x)$ uniformly on D and all the $f_k(x)$ are continuous on D, then $f(x)$ is continuous on D.
    (c) Give an example of a sequence of continuous functions on a compact domain converging pointwise but not uniformly.
27. Prove that every continuous function on $[a, b]$ is the uniform limit of linear splines.
28. Define a *step function* to be a function that is piecewise constant, $f(x) = \sum_{k=1}^m \alpha_k \chi_{[a_k, b_k)}(x)$, where $[a_k, b_k)$ are disjoint intervals. Show that every continuous function on $[a, b]$ is a uniform limit of step functions.
29. For a given function f defined on $[0, 1]$, define its *Bernstein polynomial*
$$B_n(f, x) = \sum_{k=0}^n f\left(\frac{k}{n}\right)\binom{n}{k} x^k(1 - x)^{n-k}.$$
Show that
    (a) If f is linear, then $B_n(f, x) = f(x)$.
    (b) If $f(x) = x^2$, then $B_n(f, x) = x^2 + \frac{1}{n}(x - x^2)$.
    (c) If $f \in C[0, 1]$, then $B_n(f, x)$ converges to f uniformly on $[0, 1]$.
    (d) Let $f \in C[a, b]$ and $p_n(x) = B_n\left(f, \frac{x-a}{b-a}\right)$. Then $p_n(x)$ converges to f uniformly on $[a, b]$.
30. A metric space is called *separable* if it contains a countable dense subset. Show that $\mathbb{R}^n$ is separable.

31. A sequence (x_n) of a metric space X with distance d is called a *Cauchy sequence* if for every $\epsilon > 0$, there is $N \in \mathbb{N}$ such that $d(x_n, x_m) < \epsilon$ for all integers $n, m > N$. A metric space is called a *complete metric space* if all of its Cauchy sequences converge in the space. Show that if the monotone decreasing sequence (E_n) of closed, nonempty subsets of a complete metric space X satisfies that the diameter of E_n, $d(E_n) \to 0$ as $n \to \infty$, then $\cap_{n=1}^{\infty} E_n$ consists of only one element.

32. Show that every complete metric space is a Baire space.

33. Let $B[a, b]$ be the set of all real-valued, bounded functions defined on $[a, b]$. For $f, g \in B[a, b]$, we define $d(f, g) = \sup\{|f(y) - g(y)| \mid y \in [a, b]\}$. Then prove d is a distance function on $B[a, b]$ and $B[a, b]$ becomes a complete metric space with this distance.

34. Prove that for a metric space X, the following statements are equivalent: (i) X is a Baire space; (ii) every countable intersection of open dense sets is also dense; (iii) if $X = \cup_{n=1}^{\infty} F_n$ for closed sets F_n, then the open set $\cup_{n=1}^{\infty} (F_n)^{\circ}$ is dense. Here F° denotes the set of interior points of F.

Chapter 2

Measure Theory

The geometric idea to define the Riemann integral of a function $f \geq 0$ defined on an interval $[a, b]$ is the following: To find the area under the curve f, let's chop up the interval $[a, b]$ into a partition consisting of a number of small subintervals $[x_{k-1}, x_k]$. Approximate f by functions ϕ that are constant $f(\xi_k)$ on each subinterval $[x_{k-1}, x_k]$. The union of rectangle areas $f(\xi_k) \cdot (x_k - x_{k-1})$ approximates the region bounded by f. As the number of rectangles is increased, one gets "better" approximation and the required area is obtained by a limiting process. This intuitive idea can be expressed precisely as follows. For a bounded function $f : [a, b] \mapsto \mathbb{R}$, the Riemann integral of f over $[a, b]$ is the limit

$$(R) \int_a^b f(x) \, dx = \lim_{|\Delta| \to 0} \sum_{k=1}^n f(\xi_k) \Delta x_k,$$

where Δ is a partition: $a = x_0 < x_1 < \cdots < x_n = b$ and the meshsize of the partition $|\Delta| = \max_k \Delta x_k$, $\Delta x_k = x_k - x_{k-1}$ and $\xi_k \in [x_{k-1}, x_k]$ for $k = 1, \cdots, n$, if the limit exists independently to the choices of ξ_k in $[x_{k-1}, x_k]$.

There are many cases in which we need to extend the Riemann integral. One of the drawbacks of Riemann integral is the tough restriction for the interchange of the limit and integration. As a simple example, let us consider the *Dirichlet function*

$$D(x) = \begin{cases} 0, & \text{if } x \text{ is an irrational in } [0, 1]; \\ 1, & \text{if } x \text{ is a rational in } [0, 1]. \end{cases}$$

Let $\{r_k\}_{k=1}^{\infty}$ denote the set of all rational numbers in $[0,1]$. Define

$$\varphi_{r_k}(x) = \begin{cases} 0, & \text{if } x \neq r_k; \\ 1, & \text{if } x = r_k. \end{cases}$$

Then $D(x) = \sum_{k=1}^{\infty} \varphi_{r_k}(x)$. Clearly, $(R) \int_0^1 \varphi_{r_k}(x)\,dx = 0$, where $(R) \int$ means Riemann integral. Therefore, $\sum_{k=1}^{\infty}(R) \int_0^1 \varphi_{r_k}(x)\,dx = 0$. On the other hand, $(R) \int_0^1 \sum_{k=1}^{\infty} \varphi_{r_k}\,dx = (R) \int_0^1 D(x)dx$ does not exist. We would like to have a new integral so that

$$\int_0^1 \sum_{k=1}^{\infty} \varphi_{r_k}(x)dx = \sum_{k=1}^{\infty} \int_0^1 \varphi_{r_k}(x)dx.$$

H. Lebesgue (1875–1941) first established such an integral, which is well known as the *Lebesgue integral*.

We notice that $\omega_k = 1$ on any small subinterval for the function $D(x)$. For a bounded function f ($A \leq f(x) \leq B$) over $[a,b]$, rather than using the sum $\sum_{k=1}^{n} f(x_k)\Delta x_k$, the Lebesgue integral considers the sum $\sum_{k=1}^{n} y_k m(E_k)$, where $A = y_0 < y_1 < \cdots < y_n = B$, $E_k = \{x \in [a,b] \mid y_{k-1} \leq f(x) < y_k\}$, and $m(E_k)$ is the measure of the set E_k. Lebesgue measure, to be defined later, is an extension of the notion of interval length.

For this application, we need to define the measure of sets first.

The concept of measure is an extension of the concept of length. E. Borel (1871–1956) in 1898 formulated several postulates which outline essential properties of the length of an interval by defining the measure of sets as follows:

- (Nonnegativity) A measure is always nonnegative.
- (Countable additivity) The measure of the union of a countable number of nonoverlapping sets equals the sum of their measures.
- (Monotonicity) The measure of the difference of a set and a subset is equal to the difference of their measures.
- Every set whose measure is not zero is uncountable.

H. Lebesgue presented a mathematically rigorous description of the class of sets for which he defined a measure satisfying Borel's postulates. This measure is well known as Lebesgue measure and is perhaps the most important and useful concept of a measure found on $\mathbb{R}$ to date. Our approach to Lebesgue's theory follows the elegant treatment given by a Greek mathematician, C. Carathéodory (1873–1950).

This chapter presents measure theory roughly as follows. A general procedure is developed for constructing measures by first defining the measure on a ring of subsets of an arbitrary abstract nonempty set, then obtaining a measure on a σ-ring using an outer measure. Throughout this chapter, we apply this procedure to construct Lebesgue measure on $\mathbb{R}$ as a concrete example.

I Classes of Sets

In this section, we introduce certain classes of sets. Elements of these classes will be in the domains of measures.

Let X be a nonempty set. There are five types of classes of subsets of X we wish to examine: rings, algebras, σ-rings, σ-algebras, and monotone classes.

Definition 2.1.1 A nonempty collection $\mathcal{R}$ of subsets of X is called a *ring* on X if for any $E_1, E_2 \in \mathcal{R}$, we have $E_1 \cup E_2 \in \mathcal{R}$ and $E_1 \setminus E_2 \in \mathcal{R}$. A ring on X is called a *σ-ring* if it is closed under countable unions, that is, if for $E_k \in \mathcal{R}, k \in \mathbb{N}$, then $\cup_{k=1}^{\infty} E_k \in \mathcal{R}$.

Some basic facts of a ring/σ-ring include the following:

- If $\mathcal{R}$ is a ring, then $\emptyset \in \mathcal{R}$.
- If $\mathcal{R}$ is a ring and $A, B \in \mathcal{R}$, then $A \setminus B \in \mathcal{R}$ and $A \Delta B \in \mathcal{R}$.
- If $\mathcal{R}$ is a ring and $E_k \in \mathcal{R}$ for $k = 1, \cdots, n$, then $\cup_{k=1}^{n} E_k \in \mathcal{R}$ and $\cap_{k=1}^{n} E_k \in \mathcal{R}$.
- If $\mathcal{R}$ is a σ-ring and $E_k \in \mathcal{R}$ for $k \in \mathbb{N}$, then $\cap_{k=1}^{\infty} E_k \in \mathcal{R}$ since $\cap_{k=1}^{\infty} E_k = E \setminus \cup_{k=1}^{\infty}(E \setminus E_k)$ for $E = \cup_{k=1}^{\infty} E_k$. That is, a σ-ring is also closed under countable intersections.
- If $\mathcal{R}$ is a σ-ring and $E_k \in \mathcal{R}$ for $k \in \mathbb{N}$, then $\underline{\lim}_{k \to \infty} E_k \in \mathcal{R}$ and $\overline{\lim}_{k \to \infty} E_k \in \mathcal{R}$.
- Let $\mathcal{C} \subset 2^X$. Then there exists a unique ring $\mathcal{R}(\mathcal{C})$ such that $\mathcal{C} \subset \mathcal{R}(\mathcal{C})$ and for any ring $\mathcal{R}$ containing $\mathcal{C}$, we have $\mathcal{R}(\mathcal{C}) \subset \mathcal{R}$. That is, $\mathcal{R}(\mathcal{C})$ is the smallest ring that contains $\mathcal{C}$.
- Let $\mathcal{C} \subset 2^X$. Then there exists a unique σ-ring $\sigma_r(\mathcal{C})$ such that $\mathcal{C} \subset \sigma_r(\mathcal{C})$ and that any σ-ring σ_r containing $\mathcal{C}$, we have $\sigma_r(\mathcal{C}) \subset \sigma_r$. That is, $\sigma_r(\mathcal{C})$ is the smallest σ-ring that contains $\mathcal{C}$.

We leave the proof of these facts as exercises.

Definition 2.1.2 An *algebra* of sets on X is a nonempty collection $\mathcal{A}$ of subsets of X which is closed under finite unions and complements. That is, $\mathcal{A} \subset 2^X$ is called an algebra if $E_1, \cdots, E_n \in \mathcal{A}$ implies that $\cup_{k=1}^{n} E_k \in \mathcal{A}$ and $E^c \in \mathcal{A}$ for all

$E \in \mathcal{A}$. An algebra is called a *σ-algebra* if it is closed under countable unions. That is, if $A_k \in \mathcal{A}$ for $k \in \mathbb{N}$, then $\cup_{k=1}^{\infty} A_k \in \mathcal{A}$.

Similar to the facts of a ring/σ-ring, the following properties of an algebra/σ-algebra are basic. We provide the verification of some facts and leave the others as exercises:

- If $\mathcal{A}$ is an algebra of sets on X, then $X \in \mathcal{A}$ and $\emptyset \in \mathcal{A}$.
- If $\mathcal{A}$ is an algebra, then $\mathcal{A}$ is closed for finite intersections. That is, if $A_1, \cdots, A_n \in \mathcal{A}$, then $\cap_{k=1}^{n} A_k \in \mathcal{A}$. Therefore, any algebra is a ring.

In fact, since $\mathcal{A}$ is an algebra, we have that $A_1, A_2, \cdots, A_n \in \mathcal{A}$ implies $A_1^c, A_2^c, \cdots, A_n^c \in \mathcal{A}$, so $\cup_{k=1}^{n} A_k^c \in \mathcal{A}$ and then $\left(\cup_{k=1}^{n} A_k^c \right)^c \in \mathcal{A}$. By De Morgan's Law, we have $\cap_{k=1}^{n} A_k \in \mathcal{A}$. Since $E_1 \setminus E_2 = E_1 \Delta E_2$, we see that any algebra is a ring.

- If $\mathcal{A}$ is a σ-algebra then $\mathcal{A}$ is closed under countable intersections.
- Let $\{\mathcal{A}_\lambda\}_{\lambda \in \Lambda}$ be an arbitrary collection of algebras in X. Then $\cap_{\lambda \in \Lambda} \mathcal{A}_\lambda$ is again an algebra in X.

In fact, let $\mathcal{A}_\lambda$, where $\lambda \in \Lambda$, be an algebra. For any $A \in \cap_{\lambda \in \Lambda} \mathcal{A}_\lambda$ we have $A \in \mathcal{A}_\lambda, \lambda \in \Lambda$, and then $A^c \in \mathcal{A}_\lambda$ for each $\lambda \in \Lambda$. Thus, $A^c \in \cap_{\lambda \in \Lambda} \mathcal{A}_\lambda$, i.e., $\cap_{\lambda \in \Lambda} \mathcal{A}_\lambda$ is closed for complements. Also, for $A, B \in \cap_{\lambda \in \Lambda} \mathcal{A}_\lambda$, we have $A, B \in \mathcal{A}_\lambda$, for each $\lambda \in \Lambda$ and so $A \cup B \in \mathcal{A}_\lambda$, for each $\lambda \in \Lambda$. Therefore, $A \cup B \in \cap_{\lambda \in \Lambda} \mathcal{A}_\lambda$ i.e., $\cap_{\lambda \in \Lambda} \mathcal{A}_\lambda$ is an algebra.

- For any collection $\mathcal{C}$ of subsets of X there exists a smallest algebra (σ-algebra) containing $\mathcal{C}$. Such an algebra (σ-algebra) is called the algebra (σ-algebra) generated by $\mathcal{C}$ and is denoted by $\mathcal{A}(\mathcal{C})$ $(\sigma_a(\mathcal{C}))$.
- If $\mathcal{A}$ is an algebra on X and (A_n) is a sequence of elements in $\mathcal{A}$, then there exists a disjoint sequence (B_n) in $\mathcal{A}$ such that $\cup_{n=1}^{\infty} A_n = \cup_{n=1}^{\infty} B_n$.

Clearly, let $B_1 = A_1$ and proceed by induction so that $B_{n+1} = A_{n+1} \setminus (\cup_{k=1}^{n} A_k)$.

- For a ring $\mathcal{C}$ on X, if $X \in \mathcal{C}$, then $\mathcal{C}$ is an algebra.
- For a σ-ring $\mathcal{C}$ on X, if $X \in \mathcal{C}$, then $\mathcal{C}$ is a σ-algebra.
- For any collection $\mathcal{C}$ of sets in X, we have $\sigma_a(\mathcal{C}) = \sigma_a(\mathcal{R}(\mathcal{C}))$.

EXAMPLE 2.1.1:

(a) Let $X = \mathbb{R}$ and $\mathcal{R} = \{E \mid E \in 2^X, |E| < \infty\}$. Then $\mathcal{R}$ is a ring on $\mathbb{R}$, but $\mathcal{R}$ is not a σ-ring nor an algebra.

(b) $X = \mathbb{R}$, $\mathcal{R} = \{E \mid E \in 2^X, E \text{ is countable}\}$, then $\mathcal{R}$ is a σ-ring, but $\mathcal{R}$ is not an algebra.

EXAMPLE 2.1.2:

(a) Let X be a set of infinitely many elements. $\mathcal{C}$ denotes the collection of countable subsets of X. Then $\mathcal{C}$ is a σ-ring. If X is countable, then $\mathcal{C}$ becomes a σ-algebra.
(b) For any set X, the power set 2^X is an algebra and it is a σ-algebra.
(c) Let $X \neq \emptyset$ and $\mathcal{C}$ denote the set of all single-element subsets of X. Then $\mathcal{R}(\mathcal{C})$ is the set of all finite subsets of X (including $\emptyset$).

The following example is important to our later discussion since we will use the ring $\mathcal{R}_0$ defined below to construct Lebesgue measure on $\mathbb{R}$.

EXAMPLE 2.1.3: Let $X = \mathbb{R}$. Then the class $\mathcal{R}_0$ of all finite unions of "half-open" bounded intervals $(a, b]$ is a ring. In fact, it is obvious that $\mathcal{R}_0$ is closed under the union operation. Let E_1 and E_2 be two elements in $\mathcal{R}_0$. That is, let them be finite unions of "half-open" bounded intervals. Noticing that the difference of any two "half-open" bounded intervals satisfies

$$(a, b] \setminus (c, d] = \begin{cases} \emptyset, & \text{if } (a, b] \subset (c, d]; \\ (a, b], & \text{if } [a, b) \cap (c, d) = \emptyset; \\ (a, c], & \text{if } a < c \leq b \leq d; \\ (a, c] \cup (d, b], & \text{if } a \leq c < d \leq b; \\ (d, b], & \text{if } c \leq a < d \leq b, \end{cases}$$

we have $(a, b] \setminus (c, d] \in \mathcal{R}$. Let $E_1 = \cup_{i=1}^m (a_i, b_i]$ and $E_2 = \cup_{j=1}^n (c_j, d_j]$. Then $E_1 \setminus (c, d] = \cup_{i=1}^m (a_i, b_i] \setminus (c, d] = \cup_{i=1}^m ((a_i, b_i] \setminus (c, d]) \in \mathcal{R}_0$. Therefore, $E_1 \setminus E_2 \in \mathcal{R}$ followed by $(E_1 \setminus (c_1, d_1]) \setminus \cup_{j=2}^n (c_j, d_j]$ and math induction.

Definition 2.1.3 The σ-algebra generated from $\mathcal{R}_0$ is called the σ-algebra of *Borel sets* of $\mathbb{R}$ and denoted by $\mathcal{B}_{\mathbb{R}}$.

Clearly intervals, open sets, closed sets, and thus F_σ and σ_δ sets, are Borel sets. Let $\mathcal{I}$ denote the set of intervals (open, closed, or half-open and half-closed) of $\mathbb{R}$. For an interval $I \in \mathcal{I}$ with endpoints a and b ($a \leq b$), we denote I by $I(a, b)$ (here, a could be $-\infty$ and b could be ∞). By convention, the open interval $(a, a) = \emptyset$ for any $a \in \mathbb{R}$. Then we have $\sigma_r(\mathcal{I}) = \sigma_a(\mathcal{I}) = \mathcal{B}_{\mathbb{R}}$.

Definition 2.1.4 For a nonempty class C of subsets of X, if for any monotone sequence (E_n) of sets, $\lim_{n \to \infty} E_n \in C$, then C is called a *monotone class*.

EXAMPLE 2.1.4:

(a) For any set X, $\{\emptyset, X\}$ and 2^X are monotone classes.
(b) On the real line, $C = \{[0,1], [3,4]\}$ is a monotone class. Thus, it is not necessary to be closed under the operations of unions and set differences.

Some basic facts of monotone classes include the following:

- The intersection of any collection of monotone classes is again a monotone class.
- Any σ-ring is a monotone class. A monotone class is a σ-ring if and only if it is a ring.
- For any collection C of subsets of X there exists a *smallest monotone* class containing C. Such a monotone class is called the *monotone class generated by C* and is denoted by $\mathcal{M}(C)$.

Theorem 2.1.1 If $\mathcal{R}$ is a ring on X, then $\mathcal{M}(\mathcal{R}) = \sigma_r(\mathcal{R})$.

Proof: Since σ-ring $\sigma_r(\mathcal{R})$ is a monotone class and $\mathcal{M}(\mathcal{R})$ is the smallest monotone class containing $\mathcal{R}$, we have $\mathcal{M}(\mathcal{R}) \subset \sigma_r(\mathcal{R})$.

On the other hand, we can verify that $\mathcal{M}(\mathcal{R})$ is a ring. Therefore, it is a σ-ring. Thus, $\mathcal{M}(\mathcal{R}) \supset \sigma_r(\mathcal{R})$. ∎

Exercises

1. Show the following properties on a ring:
   (a) If $\mathcal{R}$ is a ring and $A, B \in \mathcal{R}$, then $A \setminus B \in \mathcal{R}$ and $A \triangle B \in \mathcal{R}$.
   (b) If $\mathcal{R}$ is a ring and $E_k \in \mathcal{R}$ for $k = 1, \cdots, n$, then $\cup_{k=1}^n E_k \in \mathcal{R}$ and $\cap_{k=1}^n E_k \in \mathcal{R}$.
   (c) Let $C \subset 2^X$. Then there exists a unique ring $\mathcal{R}(C)$ generated by C.
2. Prove the properties of a σ-ring:
   (a) If $\mathcal{R}$ is a σ-ring and $E_k \in \mathcal{R}$ for $k \in \mathbb{N}$, then $\underline{\lim}_{k \to \infty} E_k \in \mathcal{R}$ and $\overline{\lim}_{k \to \infty} E_k \in \mathcal{R}$.
   (b) Let $C \subset 2^X$. Then there exists a unique σ-ring $\sigma_r(C)$ generated by C.
3. Show each of the following:
   (a) If $\mathcal{A}$ is an algebra of sets on X, then $X \in \mathcal{A}$ and $\emptyset \in \mathcal{A}$.

(b) If $\mathcal{A}$ is a σ-algebra then $\mathcal{A}$ is closed for countable intersections.

(c) For any collection $\mathcal{C}$ of subsets of X, there exists a unique algebra (σ-algebra) generated by $\mathcal{C}$.

(d) Show that a ring is not necessarily an algebra.

4. Let $\mathcal{F}$ be any collection of subsets of a set X. Show that $\mathcal{F}$ is an algebra if and only if the following hold:

   (i) $\emptyset, X \in \mathcal{F}$.

  (ii) $A^c \in \mathcal{F}$ whenever $A \in \mathcal{F}$.

 (iii) $A \cup B \in \mathcal{F}$ whenever $A, B \in \mathcal{F}$.

5. Let $\mathcal{F}$ be an algebra of subsets of X. Show that

  (a) If $A, B \in \mathcal{F}$, then $A \triangle B \in \mathcal{F}$.

  (b) If $E_1, \cdots, E_n \in \mathcal{F}$, then there exist $F_1, \cdots, F_n \in \mathcal{F}$ such that $F_i \subset E_i$ for each i, $F_i \cap F_j = \emptyset$ for $i \neq j$ and $\cup_{i=1}^n E_i = \cup_{j=1}^n F_j$.

6. Let $\mathcal{C}$ be a collection of subsets of X and $A \subset X$. Denote

$$A \cap \mathcal{C} = \{A \cap B \mid B \in \mathcal{C}\}.$$

Then $\sigma_a(A \cap \mathcal{C}) = A \cap \sigma_a(\mathcal{C})$.

7. For $\mathcal{C} = \{E_1, \cdots, E_n\}$, find $\mathcal{R}(\mathcal{C})$ and $\mathcal{A}(\mathcal{C})$.

8. Show that $\sigma_r(\mathcal{I}) = \sigma_a(\mathcal{I}) = \mathcal{B}_{\mathbb{R}}$.

9. Let $\mathcal{I}_r$ denote the class of all open intervals of $\mathbb{R}$ with rational endpoints. Show that $\sigma_a(\mathcal{I}_r) = \mathcal{B}_{\mathbb{R}}$.

10. (a) The intersection of any collection of monotone classes is again a monotone class.

  (b) Any σ-ring is a monotone class. A monotone class is a σ-ring if and only if it is a ring.

  (c) For any collection $\mathcal{C}$ of subsets of X there exists a unique monotone class generated by $\mathcal{C}$.

  (d) If $\mathcal{R}$ is a ring, then $\mathcal{M}(\mathcal{R})$ is also a ring.

11. For any collection $\mathcal{C}$ of sets on X, we have

  (a) $\mathcal{M}(\mathcal{C}) \subset \sigma_a(\mathcal{C})$.

  (b) $\mathcal{M}(\mathcal{C}) = \sigma_a(\mathcal{C})$ if and only if

$A \in \mathcal{C}$ implies $A^c \in \mathcal{M}(\mathcal{C})$ and $A, B \in \mathcal{C}$ implies $A \cap B \in \mathcal{M}(\mathcal{C})$.

2 Measures on a Ring

In the following discussion, we would like to define μ, a measure on a ring $\mathcal{R}$, and then to extend it to a σ-ring and even a larger collection of sets using an outer measure in the next section. As a concrete example, we first define the Lebesgue measure on the ring $\mathcal{R}_0$ over the real line, and then extend it to a σ-ring and an even larger collection of sets on $\mathbb{R}$. As usual, we denote by $\overline{\mathbb{R}} = \mathbb{R} \cup \{-\infty\} \cup \{\infty\}$ the set of *extended real numbers*.

Let us recall some properties of the length function of intervals first. For an interval $I \in \mathcal{I}$ with end points a and b ($a \leq b$), the length function on $\mathcal{I}$ is defined by $\lambda(I(a,b)) = b - a$ if $a, b \in \mathbb{R}$ and $\lambda(I(a,b)) = \infty$ if either $a = -\infty$, or $b = \infty$, or both.

Theorem 2.2.1 We have the following properties on the length function $\lambda(I)$ on $\mathcal{I}$:

(a) $\lambda(\emptyset) = 0$.

(b) (Monotonicity) $\lambda(I) \leq \lambda(J)$ for $I \subset J$ and $I, J \in \mathcal{I}$.

(c) (Finite additivity) If $I \in \mathcal{I}$ and $I = \cup_{k=1}^{n} J_k$ for mutually disjoint subintervals $J_k \in \mathcal{I}, k = 1, \cdots, n$, then $\lambda(I) = \sum_{k=1}^{n} \lambda(J_k)$.

(d) (Countable subadditivity) Let $I \in \mathcal{I}$ satisfy $I \subset \cup_{k=1}^{\infty} I_k$ for $I_k \in \mathcal{I}$, $k \in \mathbb{N}$. Then $\lambda(I) \leq \sum_{k=1}^{\infty} \lambda(I_k)$.

(e) (Countable additivity) Let $I \in \mathcal{I}$ satisfy $I = \cup_{k=1}^{\infty} I_k$ for mutually disjoint $I_k \in \mathcal{I}, k \in \mathbb{N}$. Then $\lambda(I) = \sum_{k=1}^{\infty} \lambda(I_k)$.

(f) (Translation invariance) $\lambda(I) = \lambda(I + y)$ for every $I \in \mathcal{I}$ and $y \in \mathbb{R}$, where $I + y = \{x + y \mid x \in I\}$.

To extend the length of intervals to a larger class of subsets of $\mathbb{R}$, our first choice would be to measure all sets. However, this turns out to be impossible if we want to retain the properties of the length function. Let us define a measure on a ring as follows.

Definition 2.2.1 Let $\mathcal{R} \subset 2^X$ be a ring. The extended real-valued set function $\mu : \mathcal{R} \mapsto \overline{\mathbb{R}}$ is called a *measure* on ring $\mathcal{R}$ if μ satisfies

(i) $\mu(\emptyset) = 0$,

(ii) for any $E \in \mathcal{R}, \mu(E) \geq 0$, and

(iii) for $E_k \in \mathcal{R}, k \in \mathbb{N}$ and $E_k \cap E_\ell = \emptyset$ for $k \neq \ell$, if $\cup_{k=1}^{\infty} E_k \in \mathcal{R}$, then $\mu(\cup_{k=1}^{\infty} E_k) = \sum_{k=1}^{\infty} \mu(E_k)$.

The extended real value $\mu(E)$ is called the measure of the set E.

Except $\mu \equiv \infty$, any set function satisfying (ii) and (iii) implies (i). In fact, if $\mu(E) < \infty$, for some $E \in \mathcal{C}$, we choose $E_1 = E, E_k = \emptyset, k = 2, \cdots$, then $\cup_{k=1}^{\infty} E_k = E$; by (iii), we have $\mu(E) = \mu(E_1) + \sum_{k=2}^{\infty} \mu(E_k)$. If $\mu(E) < \infty$, then $\sum_{k=2}^{\infty} \mu(E_k) = 0$. By (ii), we have $\mu(\emptyset) = 0$. Therefore, a finite measure μ on a ring $\mathcal{R}$ is a nonnegative and countably additive set function.

If $\mathcal{R}$ is an algebra, the extended real-valued set function μ in this definition defines a measure on an algebra.

EXAMPLE 2.2.1:

(a) For a set X, let $\mathcal{R}$ be a ring of finite subsets of X. Define μ on $\mathcal{R}$ as $\mu(E) = |E|$, $E \in \mathcal{R}$. Then μ is a measure on $\mathcal{R}$.

(b) Let $X \neq \emptyset$ and $\mathcal{A} = 2^X$. For a fixed element $a \in X$ and any set $E \in \mathcal{A}$, we define

$$\mu(E) = \begin{cases} 0, & \text{if } a \notin E; \\ 1, & \text{if } a \in E. \end{cases}$$

Then μ is a measure on $\mathcal{A}$.

(c) Let $\mathcal{A} = 2^X$ and μ be defined on $\mathcal{A}$ by

$$\mu(E) = \begin{cases} |E|, & \text{if } E \text{ is finite}; \\ \infty, & \text{if } E \text{ is infinite}. \end{cases}$$

Then μ is a measure. This is usually called the *counting measure* in X.

(d) Let X be an uncountable set and

$$\mathcal{A} = \{E \subset X \mid \text{either } E \text{ or } X \setminus E \text{ is countable}\}.$$

Then $\mathcal{A}$ is a σ-algebra. Define μ on $\mathcal{A}$ by

$$\mu(E) = \begin{cases} 0, & \text{if } E \text{ is countable}; \\ 1, & \text{if } X \setminus E \text{ is countable}. \end{cases}$$

Then μ is a measure on $\mathcal{A}$.

Theorem 2.2.2 If μ is the measure on a ring $\mathcal{R}$, then it has the following properties:

(i) (Finite additivity) If $E_k \in \mathcal{R}, k = 1, \cdots, n$, and $E_k \cap E_\ell = \emptyset$ for $k \neq \ell$, then $\mu(\cup_{k=1}^n E_k) = \sum_{k=1}^n \mu(E_k)$.

(ii) (Monotonicity) If $E, F \in \mathcal{R}$ and $E \subset F$, then $\mu(E) \leq \mu(F)$.

(iii) (Countable subadditivity) If $E_k \in \mathcal{R}$ for $k = 1, 2, \cdots$, then $\mu(\cup_{k=1}^\infty E_k) \leq \sum_{k=1}^\infty \mu(E_k)$.

(iv) (Continuity from below) If $E_k \in \mathcal{R}$ and $E_1 \subset E_2 \subset \cdots \subset E_n \subset \cdots$, then $\mu(\cup_{k=1}^\infty E_k) = \lim_{n \to \infty} \mu(E_n)$.

(v) (Continuity from above) If $E_k \in \mathcal{R}$ for $k \in \mathbb{N}$, $E_1 \supset E_2 \supset E_3 \supset \cdots \supset E_n \supset \cdots$, and $\mu(E_n) < \infty$ for some n, then $\mu(\cap_{k=1}^\infty E_k) = \lim_{k \to \infty} \mu(E_k)$.

Proof:

(i) Let $E_{n+k} = \emptyset$ for $k \in \mathbb{N}$. Then $E_k \cap E_\ell = \emptyset$ for any $k \neq \ell$. By the countable additivity of μ, we have $\mu(\cup_{k=1}^\infty E_k) = \sum_{k=1}^\infty \mu(E_k)$. Since $\mu(E_{n+\ell}) = \mu(\emptyset) = 0$ for $\ell \geq 1$, we have $\mu(\cup_{k=1}^n E_k) = \sum_{k=1}^n \mu(E_k)$.

(ii) Since $F = E \cup (F \setminus E)$ and $E \cap (F \setminus E) = \emptyset$, we have $\mu(F) = \mu(E) + \mu(F \setminus E)$ by (i). So, $\mu(F) \geq \mu(E)$ since $\mu(F \setminus E) \geq 0$.

(iii) Let $F_1 = E_1, F_k = E_k \setminus \cup_{\ell=1}^{k-1} E_\ell$ for $k \geq 2$. Then $F_k \cap F_{k'} = \emptyset$ if $k \neq k'$ and $\cup_{k=1}^\infty F_k = \cup_{k=1}^\infty E_k$. Thus $\mu(\cup_{k=1}^\infty E_k) = \mu(\cup_{k=1}^\infty F_k) = \sum_{k=1}^\infty \mu(F_k)$. $F_k \subset E_k$, so by (ii) we have $\mu(F_k) \leq \mu(E_k)$. Therefore, $\mu(\cup_{k=1}^\infty E_k) = \sum_{k=1}^\infty \mu(F_k) \leq \sum_{k=1}^\infty \mu(E_k)$.

(iv) Let $F_1 = E_1, F_k = E_k \setminus E_{k-1}$ for $k \geq 2$; then $F_\ell \cap F_k = \emptyset$ when $\ell \neq k$ and $\cup_{k=1}^\infty F_k = \cup_{k=1}^\infty E_k$. Thus $\mu(\cup_{k=1}^\infty E_k) = \mu(\cup_{k=1}^\infty F_k) = \sum_{k=1}^\infty \mu(F_k) = \lim_{n \to \infty} \sum_{k=1}^n \mu(F_k) = \lim_{n \to \infty} \mu(\cup_{k=1}^n F_k) = \lim_{n \to \infty} \mu(E_n)$.

(v) Without loss of generality, we assume $\mu(E_1) < \infty$. Let $F_k = E_1 \setminus E_k$ for $k \in \mathbb{N}$; then (F_k) is increasing. By (iv), we have $\mu(\cup_{k=1}^\infty F_k) = \lim_{n \to \infty} \mu(F_n)$. Noticing that $E_1 = E_k \cup F_k$ and then $\mu(F_k) = \mu(E_1) - \mu(E_k)$ and $\cap_{k=1}^n E_k = E_1 \setminus (\cup_{k=1}^n F_k)$, $\mu(\cap_{k=1}^n E_k) = \mu(E_n) = \mu(E_1) - \mu(\cup_{k=1}^n F_k) = \mu(E_1) - \mu(F_n)$. Therefore $\lim_{n \to \infty} \mu(E_n) = \mu(E_1) - \lim_{n \to \infty} \mu(F_n) = \mu(E_1) - \mu(\cup_{k=1}^\infty F_k) = \mu(E_1 \setminus \cup_{k=1}^\infty F_k) = \mu(\cap_{k=1}^\infty E_k)$.

Let $\mathcal{R}_0 = \{E \subset \mathbb{R} \mid E = \cup_{i=1}^n (a_i, b_i] \text{ for some } n \in \mathbb{N} \text{ and } a_i, b_i \in \mathbb{R}\}$ be the ring discussed in Example 2.1.3. Then for any set $E \in \mathcal{R}_0$, we have always a disjoint decomposition $E = \cup_{i=1}^m (a_i, b_i]$ for some $m \in \mathbb{N}$.

Definition 2.2.2 For interval $E = (a, b]$, we define the *measure* $m(E) = b - a$. If a general element $E = \cup_{i=1}^n (a_i, b_i] \in \mathcal{R}_0$ and $(a_i, b_i] \cap (a_j, b_j] = \emptyset$ for $i \neq j$, then $m(E) = \sum_{i=1}^n (b_i - a_i)$.

The following result and Theorem 2.2.4 show that m is an extension of the interval length function λ to $\mathcal{R}_0$.

Theorem 2.2.3 m is a well-defined set function over $\mathcal{R}_0$.

Proof: Noting that for a set $E \in \mathcal{R}_0$, there are different decompositions: $E = \cup_{i=1}^n (a_i, b_i], E = \cup_{j=1}^m (c_j, d_j]$, it suffices to show $m(E)$ is independent of the decompositions. First, for any decomposition of $(a, b]$, $(a, b] = \cup_{i=1}^n (a_i, b_i]$, assume $a = a_1 \leq b_1 \leq a_2 \leq b_2 \leq \cdots \leq a_n \leq b_n = b$. Since $(a_i, b_i] \cap (a_j, b_j] = \emptyset$ if $i \neq j$, we have $m((a, b]) = \sum_{i=1}^n (b_i - a_i) = \sum_{i=1}^n m((a_i, b_i]) = b - a$. Now consider $E \in \mathcal{R}_0$: say $E = \cup_{i=1}^n (a_i, b_i] = \cup_{j=1}^m (c_j, d_j]$ and $(a_i, b_i] \cap (a_{i'}, b_{i'}] = \emptyset$ for $i \neq i'$ and $(c_j, d_j] \cap (c_{j'}, d_{j'}] = \emptyset$ for $j \neq j'$. Since $(a_i, b_i] = (a_i, b_i] \cap E = (a_i, b_i] \cap \cup_{j=1}^m (c_j, d_j] = \cup_{j=1}^m (a_i, b_i] \cap (c_j, d_j] = \cup_{j=1}^m (e_{ij}, f_{ij}]$ where $e_{ij} \leq f_{ij}$ for $j = 1, 2, \cdots, m$. We have $m((a_i, b_i]) = \sum_{j=1}^{m}(f_{ij} - e_{ij})$ and $m(E) = m(\cup_{i=1}^{n}(a_i, b_i]) = \sum_{i=1}^{n}\sum_{j=1}^{m}(f_{ij} - e_{ij})$. Similarly, $m(E) = m(\cup_{j=1}^{m}(c_j, d_j]) = \sum_{j=1}^{m}\sum_{i=1}^{n}m(e_{ij}, f_{ij}) = \sum_{j=1}^{m}\sum_{i=1}^{n}(f_{ij} - e_{ij})$. Thus $m(E)$ is well defined on $\mathcal{R}_0$. ∎

Some facts of set function m on $\mathcal{R}_0$ include the following:

- m is finitely additive.

Proof: We need to show $m(\cup_{i=1}^{n}E_i) = \sum_{i=1}^{n}m(E_i)$ for disjoint sets $E_i, i = 1, \cdots n$. Let $E = \cup_{i=1}^{n}E_i$ and $E_i = \cup_{j=1}^{n_i}(a_{ij}, b_{ij}]$ be a disjoint decomposition of E_i in $\mathcal{R}_0$. Then $E = \cup_{i=1}^{n}\cup_{j=1}^{n_i}(a_{ij}, b_{ij}]$ is a disjoint decomposition of E since $E_i, i = 1, \cdots, n$ are disjoint. Thus, $m(E) = \sum_{i=1}^{n}\sum_{j=1}^{n_i}(b_{ij} - a_{ij}) = \sum_{i=1}^{n}m(E_i)$. ∎

- For $E_1, \cdots, E_n, E \in \mathcal{R}_0$, if $E_i \cap E_j = \emptyset$ for $i \neq j$ and $\cup_{i=1}^{n}E_i \subset E$, then $\sum_{i=1}^{n}m(E_i) \leq m(E)$.

Proof: Let $E_{n+1} = E \setminus \cup_{i=1}^{n}E_i$, then $E_i, i = 1, \cdots, (n+1)$ are disjoint sets in $\mathcal{R}_0$, and $\cup_{i=1}^{n+1}E_i = E$. By (1), we have $m(E) = \sum_{i=1}^{n+1}m(E_i) \geq \sum_{i=1}^{n}m(E_i)$ since $m(E_{n+1}) \geq 0$. ∎

- If $E_1, \cdots, E_n, E \in \mathcal{R}_0$ and $E \subset \cup_{i=1}^{n}E_i$ then $m(E) \leq \sum_{i=1}^{n}m(E_i)$.

Proof: Let $F_1 = E_1, F_i = E_i \setminus \cup_{j=1}^{i-1}E_j$, then $\cup_{i=1}^{n}F_i = \cup_{i=1}^{n}E_i$ and $F_i, i = 1, \cdots, n$ are disjoint. Since $E \subset \cup_{i=1}^{n}E_i$, we have $E = E \cap \cup_{i=1}^{n}E_i = E \cap \cup_{i=1}^{n}F_i = \cup_{i=1}^{n}(E \cap F_i)$. Here $E \cap F_i, i = 1, \cdots, n$ are disjoint sets in $\mathcal{R}_0$. Thus $m(E) = \sum_{i=1}^{n}m(E \cap F_i) \leq \sum_{i=1}^{n}m(F_i) \leq \sum_{i=1}^{n}m(E_i)$. ∎

Theorem 2.2.4 m is a measure on $\mathcal{R}_0$.

Proof: It suffices to prove m is countably additive. Let $E_i, i \in \mathbb{N}$ be a collection of disjoint sets in $\mathcal{R}_0$ and $E = \cup_{i=1}^{\infty}E_i \in \mathcal{R}_0$. Then $\sum_{i=1}^{\infty}m(E_i) \leq m(E)$ since $\cup_{i=1}^{n}E_i \subset E$ for any $n \in \mathbb{N}$ (monotonicity of m). On the other hand, let $E = \cup_{k=1}^{n}(a_k, b_k]$ be a disjoint decomposition of E in $\mathcal{R}_0$ and $E_i = \cup_{j=1}^{n_i}(\alpha_{ij}, \beta_{ij}]$ be the disjoint decomposition of E_i in $\mathcal{R}_0, i \in \mathbb{N}$. Then $\cup_{i=1}^{\infty}E_i = \cup_{i=1}^{\infty}\cup_{j=1}^{n_i}(\alpha_{ij}, \beta_{ij}]$. For convenience, we relabel $(\alpha_{ij}, \beta_{ij}], j = 1, \cdots, n_i, i = 1, 2, \cdots$ as $(\alpha_\ell, \beta_\ell], \ell = 1, 2, \cdots$. For any $\epsilon > 0$ (assume $\epsilon < n(b_k - a_k), k = 1, 2, \cdots, n$), we construct closed intervals $[a_k + \frac{\epsilon}{n}, b_k], k = 1, 2, \cdots, n$ and open intervals $(\alpha_\ell, \beta_\ell + \frac{\epsilon}{2^\ell}), \ell = 1, 2, \cdots$. Notice that $(\alpha_\ell, \beta_\ell + \frac{\epsilon}{2^\ell}), \ell = 1, 2, \cdots$, form an open covering of E, and so it covers every $[a_k + \frac{\epsilon}{n}, b_k]$. There exists a finite open covering for each of $[a_k + \frac{\epsilon}{n}, b_k], k = 1, \cdots, n$. Putting all these finite subcoverings together, we obtain a set of open intervals

$(\alpha_{n_1}, \beta_{n_1} + \frac{\epsilon}{2^{n_1}}), \cdots, (\alpha_{n_s}, \beta_{n_s} + \frac{\epsilon}{2^{n_s}})$ such that $\cup_{k=1}^n [a_k + \frac{\epsilon}{n}, b_k] \subset \cup_{i=1}^s (\alpha_{n_i}, \beta_{n_i} + \frac{\epsilon}{2^{n_i}})$. So, $\cup_{k=1}^n (a_k + \frac{\epsilon}{n}, b_k] \subset \cup_{i=1}^s (\alpha_{n_i}, \beta_{n_i} + \frac{\epsilon}{2^{n_i}}]$. Since $[a_k + \frac{\epsilon}{n}, b_k], k = 1, 2, \cdots, n$ are disjoint, we have $m(\cup_{k=1}^n (a_k + \frac{\epsilon}{n}, b_k]) \leq m(\cup_{i=1}^s (\alpha_{n_i}, \beta_{n_i} + \frac{\epsilon}{2^{n_i}}])$. That is, $\sum_{k=1}^n (b_k - a_k - \frac{\epsilon}{n}) \leq \sum_{i=1}^s (\beta_{n_i} + \frac{\epsilon}{2^{n_i}} - \alpha_{n_i})$, and so, $\sum_{k=1}^n (b_k - a_k) - \epsilon \leq \sum_{i=1}^\infty (\beta_i + \frac{\epsilon}{2^i} - \alpha_i)$. Therefore,

$$\sum_{k=1}^n (b_k - a_k) - \epsilon \leq \sum_{i=1}^\infty (\beta_i - \alpha_i) + \epsilon.$$

Since ϵ is arbitrary, we have $\sum_{k=1}^n (b_k - a_k) \leq \sum_{i=1}^\infty (\beta_i - \alpha_i)$. That is, $m(E) = \sum_{i=1}^\infty m(E_i)$. ∎

Exercises

1. Directly verify the properties of the length function λ listed in Theorem 2.2.1.
2. If μ is a measure on a ring $\mathcal{R}$, then prove that for $E_1, E_2 \in \mathcal{R}$ and $E_1 \subset E_2$, we have $\mu(E_2 - E_1) = \mu(E_2) - \mu(E_1)$.
3. If μ is a measure on a ring Σ, then prove that for $E_1, E_2, \cdots \in \Sigma$, we have
   (a) $\mu(\underline{\lim}_{n \to \infty} E_n) \leq \underline{\lim}_{n \to \infty} \mu(E_n)$, and
   (b) if there is some $k \in \mathbb{N}$ such that $\mu(\cup_{n=k}^\infty E_n) < \infty$, then $\mu(\overline{\lim}_{n \to \infty} E_n) \geq \overline{\lim}_{n \to \infty} \mu(E_n)$;
   (c) if there is some $k \in \mathbb{N}$ such that $\sum_{n=k}^\infty \mu(E_n) < \infty$, then $\mu(\overline{\lim}_{n \to \infty} E_n) = 0$.
4. Show by a counterexample that the condition $\mu(E_n) < \infty$ for some $n \in \mathbb{N}$ in Theorem 2.2.2 (v) is necessary.
5. Let $\mu_i, c = 1, 2 \cdots, n$ be measures on a σ-algebra Σ and F a set in Σ. Define the restriction of μ to F, denoted by $\mu|_F$, by $\mu|_F(A) = \mu(A \cap F)$. Prove that $\mu|_F$ is a measure on Σ.
6. Let μ be measure on a σ-algebra Σ and $a_1, a_2, \cdots, a_n \in [0, \infty)$. Then prove $\mu = \sum_{i=1}^n a_i \mu_i$ is a measure on Σ.
7. Let μ_n be a sequence of measures on a ring $\mathcal{R}$ satisfying the condition that for any $E \in \mathcal{R}$, the limit $\lim_{n \to \infty} \mu_n(E)$ exists, denoted by $\mu(E)$. Show that μ is a nonnegative and finite additive set function satisfying $\mu(\emptyset) = 0$. Give a counterexample to show that μ may not be a measure on $\mathcal{R}$.
8. Let (μ_n) be a sequence of measures on a ring $\mathcal{R}$ such that for all $E \in \mathcal{R}$ and any $n \in \mathbb{N}$, $\mu_n(E) \leq 1$. Show that $\mu(E) = \sum_{n=1}^\infty \frac{1}{2^n} \mu_n(E)$, $\mu \in \mathcal{R}$ is also a measure on $\mathcal{R}$ satisfying $\mu(E) \leq 1$ for $E \in \mathcal{R}$.

9. Let X be any infinite set and (x_n) a sequence of distinct elements in X. Let (p_n) be a sequence of nonnegative real numbers. For $E \subset X$, define
$$\mu(E) = \sum_{\{k \mid x_k \in E\}} p_k.$$

   (a) Prove μ is a measure on 2^X. μ is called a *discrete measure* with "mass" p_k at x_k.
   (b) Prove the measure μ is finite (i.e., $\mu(X) < \infty$) if and only if $\sum_{k=1}^{\infty} p_k < \infty$. If $\sum_{k=1}^{\infty} p_k = 1$, the measure μ is called a *discrete probability measure*.

10. Let C_0 be the collection of rectangles of the form $(a, b] \times (c, d] = \{(x, y) \mid a < x \le b, c < y \le d\}$ and define a set function m on C_0 by
$$m((a, b] \times (c, d]) = (b - a)(d - c).$$

   Show that m can be uniquely extended to be a measure on $\mathcal{R}(C_0)$.

11. Let $g(x)$ be a monotone increasing function on $\mathbb{R}$ satisfying for each $x \in \mathbb{R}$ $g(x) = \lim y \to x^- g(y)$. For interval $(\alpha, \beta]$, define its length by
$$g((\alpha, \beta]) = g(\beta) - g(\alpha).$$

   Show that this set function g can be uniquely extended to be a measure on $\mathcal{R}_0$.

3 Outer Measures and Lebesgue Measure

In the previous section, we have defined a measure m on the ring $\mathcal{R}_0$ on $\mathbb{R}$ as
$$m(E) = \sum_{i=1}^{n} (b_i - a_i)$$

if $E = \cup_{i=1}^{n} (a_i, b_i]$ for some $n \in \mathbb{N}$ and $(a_i, b_i] \cap (a_j, b_j] = \emptyset$, for $i \ne j$. Next, we would like to extend m to a larger class of subsets of $\mathbb{R}$ (e.g., at least on a $\sigma-$ring such as $\mathcal{B}_\mathbb{R}$). This is usually done through the introduction of outer measures, a concept due to C. Carathéodory. Let us briefly explain the idea as follows.

Suppose that a set E is covered by a countable union of elements of $\mathcal{R}_0$, $E \subset \cup_{k=1}^{\infty} E_k$ for $E_k \in \mathcal{R}_0$. Then if E would be measurable, the countable subadditivity would yield $m(E) \le \sum_{k=1}^{\infty} m(E_k)$. Therefore, any countable covering of a set gives some information of the measure of the set. Observe that the information from countable coverings is much more precise than that obtained from finite coverings. For example, let E be the set of rationals on $[0, 1]$. Then any finite covering from $\mathcal{R}_0$ should cover $[0, 1]$ entirely. Thus, the measures of these finite sets of $\mathcal{R}_0$ should add up to at least one. However, using the countable covering, we can show that $m(E)$ can be as small as any given $\epsilon > 0$. In fact, let $E = \{r_1, \cdots, r_n, \cdots\}$ be the set of all rational numbers on $[0, 1]$. Then $\{(r_k - \frac{\epsilon}{2^{k+1}}, r_k + \frac{\epsilon}{2^{k+1}}] \mid k \in \mathbb{N}\}$ will be a countable covering of E from $\mathcal{R}_0$. Therefore, $m(E) \leq \sum_{k=1}^{\infty} \frac{\epsilon}{2^k} = \epsilon$. Since this is true for any $\epsilon > 0$, we have $m(E) = 0$. Based on this discussion, for any "measurable" set E, it is natural to define

$$m(E) = \inf\left\{ \sum_{k=1}^{\infty} m(E_k) \;\middle|\; E_k \in \mathcal{R}_0 \text{ and } E \subset \cup_{k=1}^{\infty} E_k \right\}.$$

In general, let $\mathcal{R}$ be a ring on X and μ a measure on $\mathcal{R}$. We introduce a σ-ring $S(\mathcal{R})$ containing $\mathcal{R}$ defined by

$$S(\mathcal{R}) = \{E \in 2^X \mid E \subset \cup_{i=1}^{\infty} E_i \text{ and } E_i \in \mathcal{R}, i \in \mathbb{N}\}.$$

Lemma 2.3.1 For any ring $\mathcal{R}$, $S(\mathcal{R}) \supset \mathcal{R}$ is a σ-ring. Also, for any $E \in \mathcal{R}$, its power set $2^E \subset S(\mathcal{R})$.

Proof: We leave the proof as an exercise.   ∎

It is clear that on the real line $\mathbb{R}$, $S(\mathcal{R}_0) = \mathcal{P}(\mathbb{R})$, the entire power set of $\mathbb{R}$.

Definition 2.3.1 For any set $E \in S(\mathcal{R})$, we define

$$\mu^*(E) = \inf\left\{ \sum_{i=1}^{\infty} \mu(E_i) \;\middle|\; E \subset \cup_{i=1}^{\infty} E_i, \; E_i \in \mathcal{R} \right\}.$$

$\mu^*(E)$ is called an *outer measure* of E.

Clearly, if we restrict μ^* on $\mathcal{R}$, we have $\mu^*|_{\mathcal{R}} = \mu$. That is, $\mu^*(E) = \mu(E)$ for $E \in \mathcal{R}$. The following properties can be easily verified.

Lemma 2.3.2 μ^* on $S(\mathcal{R})$ has the following properties.
   (1) $\mu^*(\emptyset) = 0$.
   (2) $\mu^*(E) \geq 0$ for any $E \in S(\mathcal{R})$.
   (3) $\mu^*(E_1) \leq \mu^*(E_2)$ if $E_1 \subset E_2$.
   (4) $\mu^*(E) = \mu(E)$ if $E \in \mathcal{R}$.

In addition, we have the countable subadditivity property of μ^*:

Theorem 2.3.1 Let (E_i) be a sequence of sets in $S(\mathcal{R})$. Then $\mu^*(\cup_{i=1}^{\infty} E_i) \leq \sum_{i=1}^{\infty} \mu^*(E_i)$.

Proof: Clearly, the theorem holds if $\mu^*(E_i) = \infty$ for some i. We assume that $\mu^*(E_i) < \infty$ for all $i \in \mathbb{N}$.

For a fixed i and for all $\epsilon > 0$, by the definition of μ^*, we have $(E_i^{(j)})$ such that

$$E_i^{(j)} \in \mathcal{R}_0, \ \cup_{j=1}^{\infty} E_i^{(j)} \supset E_i$$

with

$$\sum_{j=1}^{\infty} \mu(E_i^{(j)}) < \mu^*(E_i) + \frac{\epsilon}{2^i}.$$

Thus,

$$\sum_{i=1}^{\infty} \sum_{j=1}^{\infty} \mu(E_i^{(j)}) < \sum_{i=1}^{\infty} \mu^*(E_i) + \epsilon.$$

Noticing that $\cup_{i=1}^{\infty} \cup_{j=1}^{\infty} E_i^{(j)} \supset \cup_{i=1}^{\infty} E_i$, we have

$$\mu^*(\cup_{i=1}^{\infty} E_i) \leq \sum_{i,j} \mu(E_i^{(j)}) < \sum_{j=1}^{\infty} \mu^*(E_i) + \epsilon.$$

The conclusion follows by letting $\epsilon \to 0$. ∎

So far we have seen that μ^* is a set function extended to $S(\mathcal{R})$ from $\mathcal{R}$. A question arises naturally: is μ^* a measure on $S(\mathcal{R})$? The following example shows that μ^* is not, in general, a measure on $S(\mathcal{R})$.

EXAMPLE 2.3.1: Let $X = (0,1]$, $\mathcal{R} = \{\emptyset, X\}$. Define μ on $\mathcal{R}$ by $\mu(\emptyset) = 0$ and $\mu(X) = 1$. Then $S(\mathcal{R})$ is the power set of X. Obviously, for any nonempty set $E \in S(\mathcal{R})$, $\mu^*(E) = 1$. Therefore, μ^* is not finite additive, and so μ^* is not a measure on $S(\mathcal{R})$.

It seems that we have to adjust our goal and hope to find a class, say $\mathcal{R}^*$, in $S(\mathcal{R})$, which is a σ-ring containing $\mathcal{R}$, and μ^* will be a measure on $\mathcal{R}^*$. The following theorem provides a characteristic to identify the sets in the class $\mathcal{R}^*$ and is usually call the *splitting condition* of a measurable set.

Theorem 2.3.2 Let $E \in \mathcal{R}$ and F be any set of $S(\mathcal{R})$, then

$$\mu^*(F) = \mu^*(F \cap E) + \mu^*(F \setminus E).$$

Proof: Since $F = (F \cap E) \cup (F \setminus E)$, we have

$$\mu^*(F) \leq \mu^*(F \cap E) + \mu^*(F \setminus E).$$

If $\mu^*(F) = \infty$, then we are done.

Assume $\mu^*(F) < \infty$. For all $\epsilon > 0$, there exists (E_i) such that $E_i \in \mathcal{R}$, $\cup E_i \supset F$ and

$$\sum_{i=1}^{\infty} \mu(E_i) < \mu^*(F) + \epsilon.$$

Let $E_i^{(1)} = E \cap E_i$, $E_i^{(2)} = E_i \setminus E$. Then $E_i^{(1)}, E_i^{(2)} \in \mathcal{R}$, and $E_i^{(1)} \cap E_i^{(2)} = \emptyset$. So, $\mu(E_i) = \mu(E_i^{(1)} \cup E_i^{(2)}) = \mu(E_i^{(1)}) + \mu(E_i^{(2)})$. Thus, $\sum_{i=1}^{\infty} \mu(E_i) = \sum_{i=1}^{\infty} \mu(E_i^{(1)}) + \sum_{i=1}^{\infty} \mu(E_i^{(2)})$.

Noticing that

$$\cup_{i=1}^{\infty}(E_i^{(1)}) = \cup_{i=1}^{\infty}(E_i \cap E) = (\cup_{i=1}^{\infty} E_i) \cap E \supset F \cap E$$

and

$$\cup_{i=1}^{\infty}(E_i^{(2)}) = \cup_{i=1}^{\infty}(E_i \setminus E) = (\cup_{i=1}^{\infty} E_i)) \setminus E \supset F \setminus E,$$

we have,

$$\sum_{i=1}^{\infty} \mu(E_i) = \sum_{i=1}^{\infty} \mu(E_i \cap E) + \sum_{i=1}^{\infty} \mu(E_i \setminus E) \geq \mu^*(F \cap E) + \mu^*(F \setminus E).$$

Therefore, $\mu^*(F) + \epsilon > \mu^*(F \cap E) + \mu^*(F \setminus E)$. Letting $\epsilon \to 0$, we obtain $\mu^*(F) \geq \mu^*(F \cap E) + \mu^*(F \setminus E)$. ∎

The equation

$$\mu^*(F) = \mu^*(F \cap E) + \mu^*(F \setminus E)$$

is also called the *Carathéodory condition*. It is used to characterize measurable sets.

Definition 2.3.2 For $E \subset S(\mathcal{R})$, if for any set $F \in S(\mathcal{R})$, we have $\mu^*(F) = \mu^*(F \cap E) + \mu^*(F \setminus E)$, then E is said to be μ^*-*measurable*. The set of all μ^*-measurable sets is denoted by $\mathcal{R}^*$.

It can be proved that such an extension of μ-measure on $\mathcal{R}$ to μ^*-measure on $\mathcal{R}^*$ is unique. The following result meets our expectation.

Theorem 2.3.3 The class $\mathcal{R}^*$ of all μ^*-measurable sets is a σ-ring.

Proof: We leave the proof as an exercise. ∎

In the following, we restrict our study to Lebesgue measure on the real line.

Definition 2.3.3 Based on the ring $\mathcal{R}_0$ on the real line $\mathbb{R}$ and the measure m on $\mathcal{R}_0$, the extended measure m^* on $\mathcal{R}_0^*$ is called the *Lebesgue measure* and is still denoted by m. The collection of all Lebesgue measurable sets is denoted by $\mathcal{M}$.

Though most of the following properties are also true for a general measure μ, we state them only for Lebesgue measure m.

Theorem 2.3.4 The set E is Lebesgue measurable if and only if for any $A \subset E$ and $B \subset E^c$,

$$m^*(A \cup B) = m^*(A) + m^*(B).$$

Proof: Let $F = A \cup B$. Then $F \cap E = A$ and $F \setminus E = B$. E is measurable implies that $m^*(A \cup B) = m^*(F) = m^*(F \cap E) + m^*(F \setminus E) = m^*(A) + m^*(B)$.

On the other hand, for any $F \subset \mathbb{R}$, let $A = F \cap E$ and $B = F \setminus E$. Then $A \subset E$ and $B \subset E^c$. Therefore, $m^*(F) = m^*(A \cup B) = m^*(A) + m^*(B) = m^*(F \cap E) + m^*(F \setminus E)$. ∎

Theorem 2.3.5 On $\mathcal{M}$, we have the following properties.
  (1) $E \in \mathcal{M}$ if and only if $E^c \in \mathcal{M}$.
  (2) If $m^*(E) = 0$, then $E \in \mathcal{M}$.
  (3) If E_1 and E_2 are in $\mathcal{M}$, then $E_1 \cup E_2 \in \mathcal{M}$. In particular, if $E_1 \cap E_2 = \emptyset$, then for any set $F \subset \mathbb{R}$,

$$m^*(F \cap (E_1 \cup E_2)) = m^*(F \cap E_1) + m^*(F \cap E_2).$$

  (4) If E_1 and E_2 are measurable, so are $E_1 \cap E_2$ and $E_1 \setminus E_2$.
  (5) If $E_k \in \mathcal{M}$ for $k \in \mathbb{N}$, then $\cup_{k=1}^{\infty} E_k \in \mathcal{M}$.

Proof:
  (1) This is obvious from the definition.
  (2) For any $F \subset \mathbb{R}$, we have $0 \leq m^*(E \cap F) \leq m^*(E) = 0$. Therefore, $m^*(F) \geq m^*(F \setminus E) = m^*(F \cap E) + m^*(F \setminus E)$. On the other hand, we always have $m^*(F) \leq m^*(F \cap E) + m^*(F \setminus E)$ and thus, E is measurable.
  (3) For any set F, we have the decomposition

$$F = (F \cap E_1 \setminus E_2) \cup (F \cap E_2 \setminus E_1) \cup (F \cap E_1 \cap E_2) \cup (F \setminus E_1 \setminus E_2) := A \cup B \cup C \cup D.$$

Noticing that $A \cup C \subset E_1$, $B \cup D \subset E_1^c$, and E_1 is measurable, by Theorem 2.3.4 we have

$$m^*(A \cup C) + m^*(B \cup D) = m^*(F).$$

Similarly, we have

$$m^*(B) + m^*(A \cup C) = m^*(A \cup B \cup C).$$

Since E_2 is measurable, we obtain

$$m^*(B \cup D) = m^*(B) + m^*(D).$$

Combining the above three equalities, we have

$$m^*(F) = m^*(A \cup B \cup C) + m^*(D) = m^*\big(F \cap (E_1 \cup E_2)\big) + m^*\big(F \setminus (E_1 \cup E_2)\big).$$

Thus, $E_1 \cup E_2$ is measurable.

If $E_1 \cap E_2 = \emptyset$ and E_1 is measurable, then for any set F, $F \cap E_1 \subset E_1$ and $F \cap E_2 \subset E_1^c$. The equality

$$m^*(F \cap (E_1 \cup E_2)) = m^*(F \cap E_1) + m^*(F \cap E_2)$$

follows from Theorem 2.3.4.

(4) Noticing that $E_1 \cap E_2 = (E_1^c \cup E_2^c)^c$, we know that $E_1 \cap E_2$ is measurable from facts (1) and (3). The fact that $E_1 \setminus E_2$ is measurable can be seen from $E_1 \setminus E_2 = E_1 \cap E_2^c$.

(5) Without loss of any generality, we can assume that (E_k) is the sequence of pairwise disjoint measurable sets because of the decomposition formula

$$\cup_{k=1}^\infty E_k = E_1 \cup (E_1 \setminus E_2) \cup (E_1 \setminus E_2 \setminus E_3) \cup \cdots$$

and fact (3).

Let $S = \cup_{k=1}^\infty E_k$ and $S_n = \cup_{k=1}^n E_k$. From fact (3) and math induction, we know that S_n is measurable for any $n \in \mathbb{N}$ and also that for any $F \subset \mathbb{R}$,

$$m^*(F \cap S_n) = \sum_{k=1}^n m^*(F \cap E_k).$$

Therefore,

$$
\begin{aligned}
m^*(F) &= m^*(F \cap S_n) + m^*(F \setminus S_n) \geq m^*(F \cap S_n) + m^*(F \setminus S) \\
&= \sum_{k=1}^n m^*(F \cap E_k) + m^*(F \setminus S).
\end{aligned}
$$

Letting $n \to \infty$, we obtain

$$m^*(F) \geq \sum_{k=1}^{\infty} m^*(F \cap E_k) + m^*(F \setminus S) \geq m^*(F \cap S) + m^*(F \setminus S)$$

by the countable subadditivity of m^*, and thus S is measurable. ∎

Corollary 2.3.1

(1) If $E_k, k = 1, 2, \cdots , n$ are measurable, then so are $\cup_{k=1}^{n} E_k$ and $\cap_{k=1}^{n} E_k$.

(2) If (E_k) is a sequence of measurable sets, then

$$m(\cup_{k=1}^{\infty} E_k) \leq \sum_{k=1}^{\infty} m(E_k).$$

If (E_k) is a sequence of pairwise disjoint measurable sets, then

$$m(\cup_{k=1}^{\infty} E_k) = \sum_{k=1}^{\infty} m(E_k).$$

(3) If E_k is measurable for every $k \in \mathbb{N}$, then so is $\cap_{k=1}^{\infty} E_k$.

Theorem 2.3.6 Let (E_k) be an increasing sequence of measurable sets. That is, $E_k \subset E_{k+1}$ for each $k \in \mathbb{N}$ and $E = \cup_{k=1}^{\infty} E_k = \lim_{k \to \infty} E_k$. Then for any set $F \subset \mathbb{R}$, we have

$$m^*(F \cap E) = \lim_{k \to \infty} m^*(F \cap E_k).$$

In particular, we have

$$m(E) = \lim_{k \to \infty} m(E_k).$$

Proof: We leave the proof as an exercise. ∎

Theorem 2.3.7 Let (E_k) be a decreasing sequence of measurable sets. That is, $E_k \supset E_{k+1}$ for each $k \in \mathbb{N}$ and $E = \cup_{k=1}^{\infty} E_k = \lim_{k \to \infty} E_k$. Then for any set $F \subset \mathbb{R}$ with $m^*(F) < \infty$, we have

$$m^*(F \cap E) = \lim_{k \to \infty} m^*(F \cap E_k).$$

In particular, if $m(E_n) < \infty$ for some $n \in \mathbb{N}$, we have

$$m(E) = \lim_{k \to \infty} m(E_k).$$

Proof: We leave the proof as an exercise.   ∎

Recall that $\mathcal{B}_\mathbb{R}$ denotes the collection of Borel sets, which is the smallest σ-algebra generated by $\mathcal{R}_0$. Thus, $\mathcal{B}_\mathbb{R} \subset \mathcal{M}$. Therefore, all open sets and closed sets are in $\mathcal{M}$. In fact, we can see that $\mathcal{M}$ is a σ-algebra. Noticing that the Cantor set K has cardinality c and measure zero, we see that $P(K) \subset \mathcal{M}$. On the other hand, obviously, we have $\mathcal{M} \subset P(\mathbb{R})$. Therefore, the cardinality of $\mathcal{M}$ is 2^c.

Noticing that $\mathcal{B}_\mathbb{R} = \sigma_a(\mathcal{R}_0) = \sigma_a(\mathcal{I})$, the following result is not unexpected.

Lemma 2.3.3 For any $E \in 2^\mathbb{R}$, the outer measure m^* can also be defined as $m^*(E) = \inf\{m(O) \mid O \supset E, O \text{ is open}\}$.

Proof: By the monotonicity of m^*, we have $m^*(E) \leq m^*(O)$ if $O \supset E$. Noticing that $O \in \mathcal{M}$, we have $m^*(E) \leq \inf\{m(O) \mid O \supset E, O \text{ is open}\}$. If $m^*(E) = \infty$, then we are done. Therefore, we assume $m^*(E) < \infty$. For any $\epsilon > 0$, by definition there exists a sequence (E_i) of sets in $\mathcal{R}_0$, such that $E \subset \cup_{i=1}^\infty E_i$ and $m^*(E) + \epsilon > \sum_{i=1}^\infty m(E_i)$. Let $E_i = \cup_{j=1}^{n_i} E_i^{(j)}$ be a disjoint decomposition of E_i in $\mathcal{R}_0$. $E_i^{(j)} = (a_i^{(j)}, b_i^{(j)}]$, $E \subset \cup_{i=1}^\infty \cup_{j=1}^{n_i} E_i^{(j)}$ and $m(E_i) = \sum_{j=1}^{n_i} m(E_i^{(j)}), i = 1, 2, \cdots$.

Since $\{E_i^{(j)} \mid j = 1, 2, \cdots, n_i, \ i = 1, 2, \cdots\}$ is countable, we can relabel the collection as $\{(a_n, b_n] \mid n = 1, 2, \cdots\}$. Then $E \subset \cup_{i,j} E_i^{(j)} = \cup_{n=1}^\infty (a_n, b_n]$.

Define $O_n = (a_n, b_n + \frac{\epsilon}{2^n})$. Then the open set $O = \cup_{n=1}^\infty O_n \supset E$ and

$$m(O) = m^*(O) \leq \sum_{n=1}^\infty m^*(O_n) = \sum_{n=1}^\infty (b_n + \frac{\epsilon}{2^n} - a_n)$$

$$= \sum_{n=1}^\infty (b_n - a_n) + \epsilon < m^*(E) + 2\epsilon.$$

So, $\inf\{m(O) \mid O \supset E, O \text{ is open}\} \leq m^*(E) + 2\epsilon$.

Let $\epsilon \to 0$. We have $\inf\{m(O) \mid E \subset O, O \text{ is open}\} \leq m^*(E)$.   ∎

We also have the following results on approximating Lebesgue measurable sets using open sets or closed sets.

Theorem 2.3.8 The following are equivalent:
  (1) $E \in \mathcal{M}$.
  (2) For all $\epsilon > 0$, there exists an open set $O \supset E$, such that $m^*(O \setminus E) < \epsilon$.
  (3) For all $\epsilon > 0$, there exists a closed set $F \subset E$, such that $m^*(E \setminus F) < \epsilon$.

(4) For all $\epsilon > 0$, there exists an open set O and a closed set F such that $F \subset E \subset O$, and $m^*(O \setminus F) < \epsilon$.

Proof: We leave the proof as an exercise. ∎

If E can be expressed as an intersection of countably many open sets, then E is called a G_δ set. Similarly, if F is a union of countably many closed sets, then F is called an F_σ set. Therefore, we obtain the following.

Corollary 2.3.2 If $E \in \mathcal{M}$ then there exists a G_δ set G and an F_σ set F such that $F \subset E \subset G$, and $m(G \setminus E) = m(E \setminus F) = m(G \setminus F) = 0$.

Proof: Let E be a measurable set. Then by Theorem 2.3.8, for all $\epsilon_n = \frac{1}{n}$, there exists an open set $O_n \supset E$ and a closed set $F_n \subset E$, such that $m^*(O_n \setminus E) < \epsilon$ and $m^*(E \setminus F_n) < \epsilon$.

Now let $G = \cap_{n=1}^\infty O_n$ and $F = \cup_{n=1}^\infty F_n$. Then G is a G_δ set and F is an F_σ set.

Now $m(G \setminus E) \leq m(O_n \setminus E) \leq \frac{1}{n}$ for all n and thus $m(G \setminus E) = 0$. Similarly, $m(E \setminus F) = 0$. Also, $m(G \setminus F) \leq m(G \setminus E) + m(E \setminus F) = 0$. ∎

Since any subset of a measure zero set is measurable, we obtain the following.

Corollary 2.3.3 For any $E \in \mathcal{M}$, E is a union of a Borel set and a set of Lebesgue measure zero.

For $E \subset \mathbb{R}$ and $y \in \mathbb{R}$, we define $E_y = \{x + y \mid x \in E\}$. E_y is called the y-translation of E. For Lebesgue measure m, we can show that $m(E) = m(E_y)$ if E is measurable. That is, m is a translation-invariant measure. Using this property, we can show that there is a nonmeasurable set $P \subset [0,1)$ and thus a nonmeasurable set in any interval of positive length. See for example, the book by Royden [23] for details.

Are there Lebesgue measurable sets that are not Borel sets? The answer to this question is "yes." In fact, we can use the Cantor function to construct such a set. As a reference, see [23], Problem 2.48, and Problem 3.28.

Exercises

1. Show that the class $\mathcal{R}^*$ is a σ-ring.
2. If $m^*(E) = 0$, then prove any subset of E is measurable.

3. Show that the Cantor set K on $[0,1]$ has measure zero.

4. Define the distance between two sets as $d(E_1, E_2) = \inf\{|x_1 - x_2| \mid x_1 \in E_1, x_2 \in E_2\}$. Prove that if $d(E_1, E_2) > 0$, then $m^*(E_1 \cup E_2) = m^*(E_1) + m^*(E_2)$.

5. For $0 < m^*(E) < \infty$ and $f(x) = (-\infty, x) \cap E$, show that $f(x)$ is a continuous function on $\mathbb{R}$.

6. If both $A \cup B$ and A are measurable, can you conclude that B is also measurable? Discuss the cases when $m(A) = 0$ and $A \cap B = \emptyset$, respectively.

7. For a set $A \subset \mathbb{R}$, m^*A is the outer measure of A. Prove that if $m^*A = 0$, then $m^*(A \cup B) = m^*B$.

8. Let A and E be subsets of $\mathbb{R}$ and let E be measurable. Prove that

$$m^*(A) + m^*(E) = m^*(A \cup E) + m^*(A \cap E).$$

9. Let $E \subset \mathbb{R}$ and $m^*(E) > 0$. Prove that for any $0 < a < m^*(E)$, there is a subset A of E such that $m^*(A) = a$.

10. If $A \in \mathcal{M}$ and $m(A \Delta B) = 0$, then prove B is also measurable.

11. Prove Theorem 2.3.6 and Theorem 2.3.7.

12. Let $m^*(E) < \infty$. Show that there is a G_δ set H such that $H \supset E$ and $m^*(E) = m(H)$.

13. If $A \cup B \in \mathcal{M}$ and $m(A \cup B) = m^*(A) + m^*(B) < \infty$, then both A and B are measurable.

14. Prove Theorem 2.3.8.

15. For $E \in \mathcal{M}$, if $m(E) < \infty$, then prove that for all $\epsilon > 0$, there are open intervals $I_\ell, \ell = 1, \cdots, k$ with rational endpoints such that $m(E \Delta G) < \epsilon$, where $G = \cup_{\ell=1}^{k} I_\ell$.

4 Measurable Functions

In the previous sections, we have extended the concept of length of intervals to measure of Lebesgue measurable sets. Consequently, we are now able to generalize Riemann integrals to Lebesgue integrals. For this purpose, we first define Lebesgue measurable functions, which play a similar role in Lebesgue integrals as the almost continuous functions in Riemann integrals. In the following, we always consider real-valued functions defined over measurable sets of $\mathbb{R}$.

Definition 2.4.1 A function f is said to be *measurable* on a measurable set E if for any $\alpha \in \mathbb{R}$, the set $E(f \leq \alpha) = E \cap \{x \mid f(x) \leq \alpha\}$ is measurable.

Therefore, by saying a function f is measurable on E, we mean its domain is a measurable set E and that for any $\alpha \in \mathbb{R}$, the set $E(f \leq \alpha)$ is measurable. Similarly, we define $E(f < \alpha) = E \cap \{x \mid f(x) < \alpha\}$, $E(f > \alpha) = E \cap \{x \mid f(x) > \alpha\}$, and $E(f \geq \alpha) = E \cap \{x \mid f(x) \geq \alpha\}$.

Theorem 2.4.1 Let f be a real-valued function defined on a measurable set E. Then the following are equivalent:
    (1) f is measurable on E.
    (2) $E(f < \alpha)$ is measurable for any $\alpha \in \mathbb{R}$.
    (3) $E(f > \alpha)$ is measurable for any $\alpha \in \mathbb{R}$.
    (4) $E(f \geq \alpha)$ is measurable for any $\alpha \in \mathbb{R}$.

Proof: The equivalence follows immediately from the set equations

$$E(x \mid f(x) < \alpha) = \bigcup_{k=1}^{\infty} E\left(x \mid f(x) \leq \alpha - \frac{1}{k}\right),$$

$$E(x \mid f(x) \leq \alpha) = \bigcap_{k=1}^{\infty} E\left(x \mid f(x) < \alpha + \frac{1}{k}\right),$$

and

$$E(f > \alpha) = E \setminus E(f \leq \alpha), E(f \geq \alpha) = E \setminus (f < \alpha).$$

EXAMPLE 2.4.1:

(a) If $f \in C[a, b]$, then $E(f \leq \alpha)$ is a closed set. Thus, f is a measurable function on $E = [a, b]$.
(b) For the Dirichlet function on $E = [0, 1]$,

$$E(D \leq \alpha) = \begin{cases} E \setminus \mathbb{Q}, & \text{if } 0 \leq \alpha < 1; \\ \emptyset, & \text{if } \alpha < 0; \\ E, & \text{if } \alpha \geq 1. \end{cases}$$

Thus, $D(x)$ is measurable.
(c) A step function $\phi(x) = \sum_{k=1}^{n} c_k \chi_{(k,k+1]}$ for $k \in \mathbb{Z}$ is measurable, since $E(x : \phi(x) \leq \alpha)$ is a union of half-open and half-closed intervals.

Theorem 2.4.2 Let D be a dense set in $\mathbb{R}$ and E a measurable set. Then f is measurable on E if and only if for any number $r \in D$, the set $E(f \leq r)$ is measurable.

Proof: The necessary condition is obvious. For the sufficient condition, notice that for any $\alpha \in \mathbb{R}$, since set D is dense we can choose a sequence (r_k) such that $r_k \geq r$ and $\lim_{k \to \infty} r_k = r$. From the fact that

$$E(f \leq r) = \cap_{k=1}^{\infty} E(f \leq r_k),$$

we see that $E(f \leq r)$ is measurable from Corollary 2.3.1 since each $E(f \leq r_k)$ is measurable. ∎

We would like to choose measurable functions as the Lebesgue integrable functions. For this, we need to consider the measurability of functions after some operations on measurable functions.

Lemma 2.4.1 If f is measurable, then f^2 is measurable.

Proof: In fact, the conclusion follows from the fact that f is measurable and that if $\alpha > 0, E(f^2 < \alpha) = [E(f < \sqrt{\alpha}) \cap E(f \geq 0)] \cup [E(f < 0) \cap E(f > -\sqrt{\alpha})]$. If $\alpha < 0, E(f^2 < \alpha) = \emptyset$. ∎

Theorem 2.4.3 If f and g are measurable functions on $E \subset \mathbb{R}$, then $cf, f + g, fg, \frac{f}{g}$ $(g \neq 0)$, $\max(f, g)$, and $\min(f, g)$ are measurable.

Proof: cf is measurable: If $c = 0$, then cf is measurable since

$$E(x : cf < \alpha) = \begin{cases} E, & \text{if } \alpha > 0; \\ \emptyset, & \text{if } \alpha \leq 0. \end{cases}$$

If $c > 0$, then $E(x : cf < \alpha) = E\left(x : f < \dfrac{\alpha}{c}\right)$ is measurable for any $\alpha \in \mathbb{R}$. Similarly, if $c < 0, E(x : cf < \alpha)$ is measurable. Thus, cf is measurable.

$f + g$ is measurable since

$$E(f + g < \alpha) = \bigcup_{r \in \mathbb{Q}} E(f < r) \cap E(g < \alpha - r).$$

We leave this claim as an exercise.

fg is measurable if f and g are measurable: This follows from Lemma 2.4.1 since $fg = \frac{1}{4}[(f + g)^2 - (f - g)^2]$.

$\dfrac{f}{g}$ is measurable because $\dfrac{1}{g}$ is measurable if $g \neq 0$:

$$E\left(\frac{1}{g} < \alpha\right) = \begin{cases} E(g < 0) \bigcup [E(g > 0) \bigcap E(g > \frac{1}{\alpha})], & \text{if } \alpha > 0; \\ E(g < 0), & \text{if } \alpha = 0; \\ E(g < 0) \bigcap E(g > \frac{1}{\alpha}), & \text{if } \alpha < 0. \end{cases}$$

$\max(f, g)$ is measurable since $E(\max(f, g) > \alpha) = E(f > \alpha) \bigcup E(g > \alpha)$. $\min(f, g)$ is measurable since $\min(f, g) = -\max(-f, -g)$. ∎

Noticing that $|f| = \max(f, -f)$, we obtain the following.

Corollary 2.4.1 If f is measurable, then $|f|$ is measurable.

Next, we need to consider the measurability of a limit of a sequence of measurable functions.

Theorem 2.4.4 Let (f_n) be a sequence of measurable functions on E. Then $\underline{\lim}_{n \to \infty} f_n$ and $\overline{\lim}_{n \to \infty} f_n$ are also measurable. In particular, if $\lim_{n \to \infty} f_n$ exists, then $f = \lim_{n \to \infty} f_n$ is also measurable on E.

Proof: We leave the proof as an exercise. ∎

The following results are easy to see.

Theorem 2.4.5
(1) If f is defined on $E_1 \cup E_2$ and f is measurable on E_1 and E_2, respectively, then f is measurable on $E_1 \cup E_2$.
(2) If f is a measurable function on E and $A \subset E$ is measurable, then $f|_A$, the function restricted on A, is measurable on A.

Definition 2.4.2 For a statement $P(x)$ on E, if $P(x)$ is true on E except a measure zero subset of E, then we say $P(x)$ is true *almost everywhere* on E, denoted by $P(x)$ a.e. on E.

EXAMPLE 2.4.2: Let f and g be functions defined on a measurable set E. If $m\big(E(x \mid f(x) \neq g(x))\big) = 0$, then f equals g almost everywhere, denoted by $f = g$ a.e. on E. For instance, $f(x) = 0$ for $x \in [0, 1]$ and the Dirichlet function $D(x)$ are equal almost everywhere, since $m(\mathbb{Q} \cap [0, 1]) = 0$, so we have $D(x) = 0$ a.e. on $[0, 1]$.

Theorem 2.4.6 Let f and g be two functions on E and f measurable. If $f = g$ a.e. on E, then g is measurable.

Proof: Let $A = \{ x \in E \mid f(x) \neq g(x)\}$. $E(x : g(x) < \alpha) = \{x \in E \setminus A \mid g(x) < \alpha\} \cup \{x \in A \mid g(x) < \alpha\} = \{x \in E \setminus A \mid f(x) < \alpha\} \cup \{x \in A \mid g(x) < \alpha\}$. ∎

Definition 2.4.3 A real-valued function ϕ on E is called a *simple function* if it is measurable and assumes only a finite number of values.

Thus, if f is a simple function on E, then the set $\{y \mid y = f(x), x \in E\}$ is a finite set. A step function is of course a simple function.

> **EXAMPLE 2.4.3:** The *greatest integer function* $\lfloor x \rfloor$ (floor function) on $[a, b]$ is a simple function. Also, the Dirichlet function, $D(x)$ is a simple function.

Let f be a simple function on E. Then the range of f, $R(f) = \{c_1, c_2, \cdots, c_n\}$ for some $n \in \mathbb{N}$. Let $E_i = E\{ x : f(x) = c_i\}$. Then $E = \bigcup_{i=1}^{n} E_i$. Therefore, $f = \sum_{i=1}^{n} c_i \chi_{E_i}$ where χ_{E_i} is the characteristic function of E_i. Recall that any continuous function on $[a, b]$ can be approximated by step functions. For a bounded measurable function, we obtain the following.

Theorem 2.4.7 For any real-valued measurable bounded function f on E and for all $\epsilon > 0$, there exists a sequence (φ_n) of simple functions on E such that $\varphi_n \to f$ on E as $n \to \infty$.

Proof: For $n \in \mathbb{N}$, let

$$E_j^{(n)} = E \left(\frac{j}{n} \leq f < \frac{j+1}{n} \right), j = -n^2, -n^2 + 1, \cdots, 0, 1, \cdots, n^2 - 1.$$

Define

$$\varphi_n = \sum_{j=-n^2}^{n^2-1} \frac{j}{n} \chi_{E_j^{(n)}}.$$

Then φ_n's are simple functions. For any $x_0 \in E$, there exists $N \in \mathbb{N}$ such that $|f(x_0)| \leq N$ since f is finite. Thus, for $n > N$, there exists $j \in \mathbb{N}$ such that

$-n^2 \leq j \leq n^2 - 1$, and $\frac{i}{n} \leq f(x_0) \leq \frac{i+1}{n}$. So, $x_0 \in E_j^{(n)}$. By the definition of φ_n, we see that $\varphi_n(x_0) = \frac{i}{n}$, therefore, when $n > N$,

$$|\varphi_n(x_0) - f(x_0)| < \frac{1}{n}.$$

That is, $\varphi_n(x_0) \to f(x_0)$ as $n \to \infty$. $x_0 \in E$ is arbitrarily chosen and so (φ_n) converges to f on E. ∎

More generally, we can prove the following.

Theorem 2.4.8 If f is measurable on E, then there is a sequence (φ_n) such that $\varphi_n \to f$ on E and
    (i) if $f \geq 0$, then for every $x \in E$, (φ_n) is an increasing sequence of nonnegative simple functions convergent to f on E;
    (ii) if f is bounded, then (φ_n) converges to f uniformly.

Proof: We leave the proof as an exercise. ∎

Finally, we end this section by discussing the relationship between measurable functions and continuous functions.

Theorem 2.4.9 Let f be a measurable function on E. For all $\delta > 0$, there exists a closed set $F \subset E$ such that $m(E \setminus F) < \delta$ and $f|_F$ is continuous.

Proof: We leave the proof as an exercise. ∎

Lemma 2.4.2 If $F \subset \mathbb{R}$ is closed and f is continuous on F, then f can be extended to a continuous function on $\mathbb{R}$. That is, there exists $h \in C(\mathbb{R})$ such that $h|_F = f$.

Proof: F^c is open, and thus by Theorem 1.4.14 we have $F^c = \cup(a_k, b_k)$ and the open intervals (a_k, b_k) are mutually disjoint. Define $h(x)$ by $h(x) = f(x)$ if $x \in F$, $h(x) = f(a_k)\frac{b-x}{b_k-a_k} + f(b_k)\frac{x-a_k}{b_k-a_k}$ if (a_k, b_k) is a finite interval; otherwise, $h(x) = f(b_k)$ if $a_k = -\infty$ and $h(x) = f(a_k)$ if $b_k = \infty$. It is easy to see now that $h(x)$ is continuous on $\mathbb{R}$ and $h|_F = f$. ∎

Theorem 2.4.10 [Lusin] If f is measurable on E, then for any $\delta > 0$, there exists a continuous function h on $\mathbb{R}$, such that $m(\{x \mid f(x) \neq h(x)\}) < \delta$.

Proof: For $\delta > 0$, there is a closed set F_δ such that $F_\delta \subset E$ and $m(E \setminus F_\delta) < \delta$. Using the above lemma, we can have a continuous function h such that $E(f \neq h) \subset E \setminus F_\delta$. Therefore, $m(f \neq h) < \delta$. ∎

Exercises

1. Let f be a measurable function on E. Prove that for any $a \in \mathbb{R}$, the set $E(f = a)$ is measurable.

2. Let f be a measurable function on $\mathbb{R}$. Then prove
   (a) For any open set O on the real line, $f^{-1}(O)$ is measurable.
   (b) For any closed set F on the real line, $f^{-1}(F)$ is measurable.
   (c) For any G_δ or F_σ set E on the real line, $f^{-1}(E)$ is measurable.
   (d) For any Borel set B on the real line, $f^{-1}(B)$ is measurable. [Hint: The class of sets for which $f^{-1}(E)$ is measurable is a σ-algebra.]

3. Show that $E \in \mathcal{M}$ if and only if its characteristic function $\chi_E(x)$ is a measurable function.

4. Show that $E(f + g < \alpha) = \cup_{r \in \mathbb{Q}} E(f < r) \cap E(g < \alpha - r)$.

5. Construct a function f satisfying f^2 is measurable but f is nonmeasurable.

6. If f and g are measurable on E, then prove
   (a) for any $\alpha, \beta \in \mathbb{R}$, $E(f = \alpha)$ and $E(\alpha < f < \beta)$ are measurable and so are $E(\alpha \leq f \leq \beta)$, $E(\alpha \leq f < \beta)$, and $E(\alpha < f \leq \beta)$;
   (b) $E(f > g)$ is measurable.

7. Show that if f is a measurable function and g a continuous function defined on $\mathbb{R}$, then $g \circ f$ is measurable.

8. Prove Theorem 2.4.4.

9. Let f be a bounded measurable function on E. Prove there is a sequence (f_n) of simple functions such that (f_n) converges to f uniformly on E and $|f_n(x)| \leq \sup_{x \in E} |f(x)|, n = 1, 2, \cdots$.

10. Prove that if $f \in C(\mathbb{R})$ (that is, f is continuous on $\mathbb{R}$), then f is measurable. Furthermore, if f is differentiable, then prove f' is measurable.

11. Let $\varphi(x) = \sum c_k \cdot \chi_{F_k}$ and $F_k, k = 1, 2, \cdots, n$ be disjoint closed sets. Then prove $\varphi(x)$ is continuous when restricted to $E = \cup_{k=1}^n F_k$.

12. Prove Lusin's Theorem: Let f be a measurable function on E. For all $\delta > 0$, there exists a closed set $F \subset E$ such that $m(E \setminus F) < \delta$ and $f|_F$ is continuous.

5 Convergence of Measurable Functions

In this section, we discuss the convergence properties of measurable function sequences by only stating results without giving proof. First, we give the following definitions.

Definition 2.5.1

(1) (*Convergence a.e.*) (f_n) *converges to f almost everywhere* on E means that $\lim_{n\to\infty} f_n(x) = f(x)$ for all $x \in E$ except a measure zero subset of E. $f_n \to f$ almost everywhere on E is denoted by $f_n \to f$ a.e. on E.

(2) (*Uniform convergence a.e.*) (f_n) is said to be *almost everywhere uniformly convergent* to f on E if there is a measure zero subset A of E such that (f_n) uniformly converges to f on $E \setminus A$. (f_n) almost everywhere uniformly converges to f on E is denoted by $f_n \rightrightarrows f$ a.e. on E.

(3) (*Convergence in measure*) A sequence (f_n) of measurable functions on E is said to *converge to f in measure* if for any $\epsilon > 0$, there exists an positive integer N such that $m\{x \mid |f_n(x) - f(x)| \geq \epsilon\} < \epsilon$ for all integers $n \geq N$. $f_n \to f$ in measure on E is denoted by $f_n \xrightarrow{m} f$ on E.

In the following, we give characterizations of these convergences.

Theorem 2.5.1 Let (f_n) and f be measurable on E. Then

(1) $f_n \to f$ a.e. on E if and only if for any $\epsilon > 0$,

$$m\left(\cap_{n=1}^{\infty} \cup_{k=n}^{\infty} E(|f_k - f| \geq \epsilon) \right) = 0.$$

(2) $f_n \rightrightarrows f$ a.e. on E if and only if for any $\epsilon > 0$,

$$\lim_{n\to\infty} m\left(\cup_{k=n}^{\infty} E(|f_k - f| \geq \epsilon) \right) = 0.$$

(3) $f_n \xrightarrow{m} f$ on E if and only if for any subsequence (f_{n_k}) of (f_n), there is a subsequence $(f_{n_k'})$ such that $f_{n_k'} \to f$ a.e. on E.

From the characterization of $f_n \to f$ a.e. on E in the above theorem and Theorem 2.3.7, we obtain the following.

Theorem 2.5.2 If $m(E) < \infty$ and $f_n \to f$ a.e. on E, then $f_n \xrightarrow{m} f$ on E.

If $m(E) = \infty$, then $f_n \to f$ a.e. on E may not imply $f_n \xrightarrow{m} f$ on E.

EXAMPLE 2.5.1: Define

$$f_n(x) = \begin{cases} x, & \text{if } x \in [0, n]; \\ 0, & \text{if } x \in (n, \infty). \end{cases}$$

$E = [0, \infty)$. Then $\lim_{n\to\infty} f_n(x) = x$ and so $f(x) = x$. $\lim_{n\to\infty} f_n = f$ almost everywhere. Noticing that $m\{x \mid |f_n(x) - f(x)| \geq \epsilon\} = \infty$ for any fixed n, we have $f_n \to f$ in measure on $[0, \infty)$.

The following example shows that $f_n \xrightarrow{m} f$ on E does not imply $f_n \to f$ a.e. on E.

EXAMPLE 2.5.2: Let $E = (0, 1]$ and define (f_n) as follows. First, divide E into two even subintervals and define

$$f_1^{(1)} = \begin{cases} 1, & x \in (0, \frac{1}{2}]; \\ 0, & x \in (\frac{1}{2}, 1]. \end{cases}$$

$$f_2^{(1)} = \begin{cases} 0, & x \in (0, \frac{1}{2}]; \\ 1, & x \in (\frac{1}{2}, 1]. \end{cases}$$

Then, for $n \in \mathbb{N}$, inductively divide E into 2^n subintervals. We define 2^n functions on E as

$$f_j^{(n)} = \begin{cases} 1, & x \in (\frac{j-1}{2^n}, \frac{j}{2^n}]; \\ 0, & x \notin (\frac{j-1}{2^n}, \frac{j}{2^n}]. \end{cases}$$

Let $f_N = f_j^{(n)}$ for $N = 2^n - 2 + j$ and then $f_N \xrightarrow{m} 0$ on E since for $\epsilon > 0$, $E(|f_N - 0| > \epsilon)$ either is the empty set or $(\frac{j-1}{2^n}, \frac{j}{2^n}]$ and so, $m(E(|f_N - 0| > \epsilon)) \leq \frac{1}{2^n}$.

However, f_N diverges everywhere on $(0, 1]$! In fact, for any $x_0 \in (0, 1]$, no matter how large n is, there is a $j \in \mathbb{N}$ such that $x_0 \in (\frac{j-1}{2^n}, \frac{j}{2^n}]$. Thus, $f_j^{(n)}(x_0) = 1$ and $f_{j+1}^{(n)}(x_0) = 0$ or $f_{j-1}^{(n)}(x_0) = 0$. Therefore, we can always find two subsequences from (f_N) such that one converges to 1 and the other converges to 0.

However, we have the following.

Theorem 2.5.3 [Riesz] Let (f_n) be a sequence of measurable functions that converges in measure to f on E. Then there is a subsequence (f_{n_k}) that converges to f almost everywhere on E.

For $f_n(x) = x^n$, $n = 1, 2, \ldots$, we have that $f_n \to 0$ a.e. on $[0, 1]$. However, f_n is not uniformly convergent to 0 on $[0, 1]$. But for any $E > 0$, we can prove that f_n is uniformly convergent to 0 on $[0, 1 - \epsilon]$. The following theorem provides a relationship between convergence a.e. and uniform convergence.

Theorem 2.5.4 [Egoroff] If (f_n) is a sequence of measurable functions that converges to f a.e. on E with $m(E) < \infty$, then for all $\eta > 0$, there exists $A \subset E$ with $m(A) < \eta$ such that f_n converges to f on $E \setminus A$ uniformly.

A sequence (f_n) of measurable functions on E is called a *Cauchy sequence in measure* if for any $\delta > 0$, we have $\lim_{k, \ell \to \infty} m(E(f_k - f_\ell| \geq \delta)) = 0$. The following gives the Cauchy criterion for convergence in measure.

Theorem 2.5.5 Let (f_k) be a sequence of measurable functions which are finite a.e. on E. Then $f_k \xrightarrow{m} f$ a measurable function f on E if and only if (f_k) is a Cauchy sequence in measure.

Exercises

1. Show that the sequence of functions $(f_n(x)) := (\sin^n x)$ converges to 0 a.e. on $\mathbb{R}$.
2. Let (f_k) be a sequence of measurable functions on E with $m(E) < \infty$. Prove that if $f_k \xrightarrow{m} f$, then for any $p > 0$, $|f_k|^p \xrightarrow{m} |f|$.
3. If f is a measurable function on E with $m(E) < \infty$, then prove there is a sequence (φ_k) of simple functions such that $f_k \xrightarrow{m} f$ and $f_k \to f$ a.e. simultaneously.
4. Suppose $f_k \xrightarrow{m} f$ and $g_k \xrightarrow{m} g$ on domain D. Show that
   (a) $f_k \pm g_k \xrightarrow{m} f \pm g$.
   (b) $|f_k| \xrightarrow{m} |f|$.
   (c) $\min\{f_k, g_k\} \xrightarrow{m} \min\{f, g\}$ and $\max\{f_k, g_k\} \xrightarrow{m} \max\{f, g\}$.
   (d) $f_k g_k \xrightarrow{m} fg$ when $m(D) < \infty$.
5. Let f be a measurable and bounded function on $[a, b]$. Show that there is a sequence (g_k) of continuous functions such that $g_k \to f$ a.e. on $[a, b]$ and $\max_{x \in [a,b]} |g_k(x)| \leq \sup_{x \in [a,b]} |f(x)|$.
6. Show that f is measurable on $[a, b]$ if and only if there is a sequence of polynomials (p_n) such that $p_n \xrightarrow{m} f$ on $[a, b]$.

Chapter 3

The Lebesgue Integral

In the classical theory of integration on $\mathbb{R}$, $\int_a^b f(x)\,dx$ is introduced as the signed area under a continuous curve $y = f(x)$ and is defined as a limit of Riemann sums, which are integrals of piecewise constant functions which approximate f on the interval $[a, b]$. In this chapter we extend the concept of integration to measurable functions which are approximated by simple functions and thus develop the Lebesgue integration theory. We consider only real-valued functions in the following discussion.

1 Riemann Integral and Lebesgue Integral

In this section, we will introduce the Lebesgue integral of bounded measurable functions on E and its basic properties. We begin by recalling the definition of the Riemann integral for a bounded function $f : [a, b] \mapsto \mathbb{R}$. For any *partition*

$$\Delta : a = x_0 < x_1 < \cdots < x_n = b$$

of $[a, b]$, we define the *meshsize* $|\Delta|$ of the partition as $|\Delta| = \max\{\Delta x_k := x_k - x_{k-1} \mid k = 1, \cdots, n\}$. Then the Riemann integral of f on $[a, b]$ is defined by

$$(R) \int_a^b f(x)\,dx = \lim_{|\Delta| \to 0} \sum_{k=1}^n f(\xi_k) \Delta x_k,$$

where $\xi_k \in [x_{k-1}, x_k]$, $k = 1, 2, \cdots, n$.

For a *step function* $\psi \in S_0^{-1}(\Delta)$ on $[a, b]$,

$$\psi(x) = c_k, \quad x_{k-1} < x < x_k, \quad k = 1, 2, \cdots, n,$$

we have

$$(R) \int_a^b \psi(x) \, dx = \sum_{k=1}^n c_k \Delta x_k.$$

For f defined on $[a, b]$ and Δ a partition of $[a, b]$, let $M_k = \sup_{x_{k-1} \leq x \leq x_k} f(x)$ and $m_k = \inf_{x_{k-1} \leq x \leq x_k} f(x)$. Then the upper (Riemann) integral is defined as

$$(R) \overline{\int_a^b} f(x) \, dx = \inf_\Delta \sum_{k=1}^n M_k \Delta x_k.$$

The lower (Riemann) integral is then

$$(R) \int_{\underline{a}}^b f(x) \, dx = \sup_\Delta \sum_{k=1}^n m_k \Delta x_k.$$

Clearly, $(R) \overline{\int_a^b} f(x) \, dx \geq (R) \int_{\underline{a}}^b f(x) \, dx$, and we say that f is Riemann integrable if $(R) \overline{\int_a^b} f(x) \, dx = (R) \int_{\underline{a}}^b f(x) \, dx$. In this case, the Riemann integral of f over $[a, b]$ is

$$(R) \int_a^b f(x) \, dx = (R) \overline{\int_a^b} f(x) \, dx = (R) \int_{\underline{a}}^b f(x) \, dx.$$

Therefore, if f is Riemann integrable over $[a, b]$, then

$$(R) \int_a^b f(x) \, dx = \inf \left\{ \int_a^b \psi(x) \, dx \,\middle|\, \psi \geq f \text{ and } \psi \text{ is a step function} \right\}.$$

The numbers $(R) \overline{\int_a^b} f(x) \, dx$ and $(R) \int_{\underline{a}}^b f(x) \, dx$ will be close if f is continuous. We can show that f is Riemann integrable if f is continuous almost everywhere (see exercises).

EXAMPLE 3.1.1:   Consider the Dirichlet function

$$D(x) = \begin{cases} 0, & \text{if } x \text{ is an irrational in } [0, 1]; \\ 1, & \text{if } x \text{ is a rational in } [0, 1]. \end{cases}$$

We have $(R) \overline{\int_0^1} D(x) \, dx = 1$ and $(R) \int_{\underline{0}}^1 D(x) \, dx = 0$. Therefore, $D(x)$ is not Riemann integrable.

This example shows that f is not Riemann integrable if it has "too many" discontinuities. Notice that though the set of discontinuities of $D(x)$ is $[0, 1]$, $D(x)$ is zero almost everywhere since $\{x \mid D(x) \neq 0\} = \mathbb{Q} \cap [0, 1]$. Lebesgue extended the concept of integration to the space of measurable functions and we would expect that $\int_0^1 D(x)\,dx = 0$.

For a measurable function f defined on a measurable set E with $m(E) < \infty$. Assume that $\ell \leq f(x) \leq L$. Let $\Delta_y : \ell = y_0 < y_1 < \cdots < y_n = L$ and $|\Delta y| = \max_k(y_k - y_{k-1})$. Let

$$E_k = \begin{cases} f^{-1}([y_{k-1}, y_k)), & \text{if } k < n; \\ f^{-1}([y_{k-1}, y_k]), & \text{if } k = n. \end{cases}$$

Then $E_k, k = 1, 2, \cdots, n$ are measurable. For any $\eta_k \in [y_{k-1}, y_k), k = 1, 2, \cdots, n$, if the limit $\lim_{|\Delta y| \to 0} \sum_{k=1}^n \eta_k m(E_k)$ exists, then we say f is *Lebesgue integrable* over E and we denote it by $(L) \int_E f(x)\,dx$ or $\int_E f(x)\,dx$ for short.

For Dirichlet function $D(x)$, if $|\Delta y| < \frac{1}{2}$ it is easy to see that

$$E_k = \begin{cases} [0, 1] \cap Q, & \text{if } 1 \in [y_{k-1}, y_k); \\ [0, 1] \setminus Q, & \text{if } 0 \in [y_{k-1}, y_k); \\ \emptyset, & \text{otherwise.} \end{cases}$$

Therefore, $\int_0^1 D(x)\,dx = 0$.

For a simple function $\varphi(x) = \sum_{k=1}^n c_k \chi_{E_k}$, where $\cup_{k=1}^n E_k = E$ and E_k are mutually disjoint measurable sets, we define $\int_E \varphi(x)\,dx = \sum_{k=1}^n c_k m(E_k)$.

Lemma 3.1.3 If φ and ψ are simple functions on E and $\varphi = \psi$ a.e., then $\int_E \varphi\,dx = \int_E \psi\,dx$.

Proof: Let $\varphi(x) = \sum_{i=1}^n \alpha_i \chi_{E_i}(x)$ and $\psi(x) = \sum_{j=1}^m \beta_j \chi_{F_j}(x)$ for $E = \cup_i E_i = \cup_j F_j$. Then $\alpha_i = \beta_j$ on $E_i \cap F_j$ a.e. Therefore, $\alpha_i m(E_i \cap F_j) = \beta_j m(E_i \cap F_j)$ for all $i = 1, \cdots, n$ and $j = 1, \cdots, n$. Hence,

$$\int_E \varphi\,dx = \sum_{i=1}^n \alpha_i m(E_i) = \sum_{i=1}^n \sum_{j=1}^m \alpha_i m(E_i \cap F_j)$$

$$= \sum_{i=1}^n \sum_{j=1}^m \beta_j m(E_i \cap F_j) = \sum_{j=1}^m \beta_j m(F_j) = \int_E \psi\,dx.$$

∎

The following properties of the integral of simple functions can be proved similarly.

Lemma 3.1.4 Suppose that φ and ψ are simple functions on a measurable set E, then

- (i) If $\varphi \leq \psi$ a.e. on E, then $\int_E \varphi\, dx \leq \int_E \psi\, dx$.
- (ii) $\int_E \varphi dx \leq \max\{\varphi\} \cdot m(E)$.
- (iii) For any real numbers λ and η, $\int_E (\lambda\varphi + \eta\psi)\, dx = \lambda \int_E \varphi\, dx + \eta \int_E \psi\, dx$.
- (iv) If $E = E_1 \cup E_2$ for disjoint measurable sets E_1 and E_2, then $\int_E \varphi\, dx = \int_{E_1} \varphi\, dx + \int_{E_2} \varphi\, dx$.

Similar to upper and lower Riemann integrals, we define the *upper* and *lower Lebesgue integrals* as

$$\overline{\int_E} f(x)\, dx = \inf\left\{ \int_E \psi(x)\, dx, \text{where simple function } \psi(x) \geq f(x) \right\}$$

and

$$\underline{\int_E} f(x)\, dx = \sup\left\{ \int_E \varphi(x)dx, \text{ where simple function } \varphi \leq f(x) \right\},$$

respectively. Therefore we have the following.

Theorem 3.1.1 Let f be a bounded Lebesgue measurable function defined on a measurable set E with $m(E) < \infty$. Then

$$\overline{\int_E} f(x)\, dx = \underline{\int_E} f(x)\, dx.$$

Therefore, f is Lebesgue integrable over E.

Therefore, for a bounded Lebesgue measurable function f on E, we can simply define

$$\int_E f(x)\, dx = \inf_{\psi \geq f, \psi \text{ is simple}} \int_E \psi\, dx.$$

Proof: Let $\alpha \leq f \leq \beta$. For a given $n \in \mathbb{N}$, consider a partition of $[\alpha, \beta]$:

$$\alpha = y_0 < y_1 < \cdots < y_n = \beta$$

satisfying $y_k - y_{k-1} < \frac{1}{n}$. Define the sets

$$E_k = \{x \in E;\ y_{k-1} \leq f(x) < y_k\}$$

for $k = 1, 2, \cdots, n-1$ and $E_n = \{x \in E; f(x) = \beta\}$. Then $E_k, k = 1, 2, \cdots, n$ are measurable, disjoint, and have union E. Thus,

$$\sum_{k=1}^{n} m(E_k) = m(E).$$

The simple functions defined by

$$\psi_n(x) = \sum_{k=1}^{n} y_k \chi_{E_k}(x)$$

and

$$\phi_n(x) = \sum_{k=1}^{n} y_{k-1} \chi_{E_k}(x)$$

satisfy

$$\phi_n(x) \leq f(x) \leq \psi_n(x).$$

Therefore,

$$\inf_{\substack{\psi \geq f \\ \psi \text{ is simple}}} \int_E \psi(x)\, dx \leq \int_E \psi_n(x)\, dx = \sum_{k=1}^{n} y_k m(E_k)$$

and

$$\sup_{\substack{\phi \leq f \\ \phi \text{ is simple}}} \int_E \phi(x)\, dx \geq \int_E \phi_n(x)\, dx = \sum_{k=1}^{n} y_{k-1} m(E_k).$$

Hence

$$0 \leq \inf_{\substack{\psi \geq f \\ \psi \text{ is simple}}} \int_E \psi(x)\, dx - \sup_{\substack{\phi \leq f \\ \phi \text{ is simple}}} \int_E \phi(x)\, dx \leq \frac{\beta - \alpha}{n} \sum_{k=1}^{n} m E_k$$

$$= \frac{\beta - \alpha}{n} m(E).$$

Let $n \to \infty$. We obtain

$$\inf_{\substack{\psi \geq f \\ \psi \text{ is simple}}} \int_E \psi(x)\, dx - \sup_{\substack{\phi \leq f \\ \phi \text{ is simple}}} \int_E \phi(x)\, dx = 0.$$

∎

We sometimes write the integral as $\int_E f$. When E is an interval $[a, b]$, we write $\int_a^b f$ instead of $\int_{[a,b]} f$. From the above theorem, we see that if f is a bounded Lebesgue measurable function on E, then there is a sequence $\langle \psi_n \rangle$ of simple functions such that $\psi_n \to f$ (in fact, the convergence is uniform and the functions ψ_n are uniformly bounded above) and $\int_E \psi_n \to \int_E f$ as $n \to \infty$. On the other hand, if f is bounded on a measurable set E with $m(E) < \infty$, then there are simple functions φ_n and ψ_n such that $\varphi_n(x) \le f(x) \le \psi_n(x)$ and $\int \psi_n - \int \varphi_n \le \frac{1}{n}$. Therefore, except for at most a zero-measure subset of E, f becomes a limit function of simple functions on E. Hence by Theorem 2.4.4, f must be measurable on E. We obtain the following.

Theorem 3.1.2 Let f be a bounded function defined on a measurable set E with $m(E) < \infty$. Then f is Lebesgue integrable on E if and only if f is measurable.

The following result shows that the Lebesgue integral is indeed a generalization of the Riemann integral.

Theorem 3.1.3 If f is bounded and Riemann integrable on $[a, b]$, then f is Lebesgue integrable and

$$\int_a^b f(x)\,dx = (R) \int_a^b f(x)\,dx.$$

Proof: If f is Riemann integrable then

$$(R) \int_a^b f(x)\,dx = \sup_{s \le f} \left\{ \int_a^b s(x)\,dx \,\middle|\, s \text{ is a step function} \right\}$$

$$= \inf_{S \ge f} \left\{ \int_a^b S(x)\,dx \,\middle|\, S \text{ is a step function} \right\}.$$

Noticing that the step functions are simple functions, we have

$$\sup_{s \le f} \left\{ \int_a^b s(x)\,dx \right\} \le \sup_{\varphi \le f} \left\{ \int_a^b \varphi\,dx \right\} \le \inf_{\psi \ge f} \left\{ \int_a^b \psi\,dx \right\} \le \inf_{S \ge f} \left\{ \int_a^b S(x)\,dx \right\}$$

for the step functions s and S and the simple functions φ and ψ. Therefore

$$(R) \int_a^b f(x)dx \le \sup_{\varphi \le f} \left\{ \int_a^b \varphi\,dx \right\} \le \inf_{\psi \ge f} \left\{ \int_a^b \psi\,dx \right\} \le (R) \int_a^b f(x)\,dx.$$

Thus, $\int_a^b f(x)\,dx = (R) \int_a^b f(x)\,dx$ if f is Riemann integrable on $[a, b]$. ∎

Using Lemma 3.1.3 and Lemma 3.1.4, we can prove the following properties of $\int_E f$ for a bounded measurable function f on E.

Theorem 3.1.4 Suppose f and g are bounded and measurable on E with $m(E) < \infty$. Then we have the following properties:

(i) $\int_E (af \pm bg)\,dx = a \int_E f\,dx \pm b \int_E g\,dx$.

(ii) If $f = g$ a.e., then $\int_E f\,dx = \int_E g\,dx$.

(iii) If $f \le g$ a.e., then $\int_E f\,dx \le \int_E g\,dx$. Thus, $|\int_E f(x)\,dx| \le \int_E |f(x)|\,dx$.

(iv) If $l \le f \le L$, then $l\,m(E) \le \int_E f\,dx \le L\,m(E)$.

(v) If $E = E_1 \cup E_2$, $E_1 \cap E_2 = \emptyset$, and E_1, E_2 are measurable, then $\int_E f(x)\,dx = \int_{E_1} f(x)\,dx + \int_{E_2} f(x)\,dx$.

For a convergent sequence of measurable functions, we have the following.

Theorem 3.1.5 [Bounded Convergence Theorem] Let $\langle f_n \rangle$ be a sequence of measurable functions on E with $m(E) < \infty$ and $f_n \to f$ for $x \in E$ as $n \to \infty$. If $|f_n(x)| \le M$ for $x \in E$ and all $n \in \mathbb{N}$, then

$$\lim_{n \to \infty} \int_E f_n(x)\,dx = \int_E f(x)\,dx.$$

That is, $\lim_{n \to \infty} \int_E f_n(x)\,dx = \int_E \lim_{n \to \infty} f_n(x)\,dx$.

Proof: It is clear that the theorem holds if $f_n \to f$ uniformly on E. By Egoroff's Theorem (Theorem 2.5.4) for $\delta = \frac{\epsilon}{4M}$ ($\epsilon > 0$), we have $E_\delta \subset E$ such that $m(E \backslash E_\delta) < \delta$ and $f_n \to f$ on E_δ uniformly. That is, on E_δ for $\epsilon > 0$ there exists $N \in \mathbb{N}$ such that $|f_n(x) - f(x)| < \frac{\epsilon}{2m(E_\delta)}$. For the given $\epsilon > 0$,

$$\left| \int_E f_n(x)\,dx - \int_E f(x)\,dx) \right| = \left| \int_{E_\delta} (f_n - f)\,dx + \int_{E \backslash E_\delta} (f_n - f)\,dx \right|$$

$$\le \int_{E_\delta} |f_n - f|\,dx + \int_{E \backslash E_\delta} |f_n - f|\,dx$$

$$\le \frac{\epsilon}{2m(E_\delta)} m(E_\delta) + 2Mm(E \backslash E_\delta) < \frac{\epsilon}{2} + 2M\delta$$

$$= \frac{\epsilon}{2} + 2M\frac{\epsilon}{4M} = \frac{\epsilon}{2} + \frac{\epsilon}{2} = \epsilon.$$

Thus, $\lim_{n \to \infty} \int_E f_n(x)\,dx = \int_E f(x)\,dx$. ∎

Using a similar proof, we can see that the conclusion of the bounded convergence theorem is still true if the condition $f_n \to f$ is relaxed to $f_n \to f$ a.e.

Exercises

1. For f defined on $[a, b]$ and Δ is a partition of $[a, b]$, let $\omega_k = M_k - m_k$. Prove that f is Riemann integrable if and only if $\lim_{|\Delta| \to 0} \sum_{k=1}^{n} \omega_k \Delta x_k = 0$.

2. If f and g are simple functions on a measurable set E and $f \leq g$ a.e. on E, prove that $\int_E f \, dx \leq \int_E g \, dx$.

3. Prove Theorem 3.1.4.

4. Let f be a bounded nonnegative measurable function on $[a, b]$ and E, F measurable subsets of $[a, b]$. Show that if $E \subset F$, then $\int_E f(x) \, dx \leq \int_F f(x) \, dx$.

5. If f is a bounded measurable function on $[a, b]$ and $\int_a^b [f(x)]^2 \, dx = 0$, prove that $f = 0$ a.e. on $[a, b]$.

6. Let $E_k, k = 1, \cdots, n$ be measurable subsets of $[0, 1]$. If each point of $[0, 1]$ belongs to at least three of these sets, show that at least one of the sets has measure $\geq 3/n$. [Hint: Consider the characteristic functions of E_k and first show that $\sum_k \chi_{E_k} \geq 3$.]

7. Suppose $m(E) < \infty$ and $\langle f_n \rangle$ is a sequence of measurable functions which are bounded a.e. on E. Show that $f_n \xrightarrow{m} 0$ if and only if $\int_E \frac{|f_n|}{1 + |f_n|} \, dx \to 0$ as $n \to \infty$.

2 The General Lebesgue Integral

For a μ-measurable set E, if E has a decomposition $E = \bigcup_{k=1}^{\infty} E_k$ with $\mu(E_k) < \infty$ for all $k \in \mathbb{N}$, then we say E has a σ-finite measure. For any measurable set with σ-finite measure, there is a nondecreasing sequence $\langle F_k \rangle$ of measurable subsets of E with $\mu(F_k) < \infty$, for example $F_k = \bigcup_{j=1}^{k} E_j$, such that $E = \bigcup_k F_k$. Any Lebesgue measurable set E is σ-finite since E can be expressed as $E = \sup_{n \in \mathbb{N}} E \cap (n, n + 1]$ and $m(E \cap (n, n + 1]) \leq 1$ for all $n \in \mathbb{N}$. In this section, we will extend the definition of Lebesgue integral to a general measurable function over arbitrary measurable sets. Let f be a real-valued function defined on E. We define its positive part $f^+ = \max\{f, 0\}$ and its negative part $f^- = \max\{-f, 0\}$. Then both f^+ and f^- are nonnegative and $f = f^+ - f^-$ and $|f| = f^+ + f^-$. For any $N \in \mathbb{N}$, let $[f]_N = \max\{\min\{f(x), N\}, -N\}$ be an N-truncation function of f. On the points that $|f(x)| \leq N$, we have $[f(x)]_N = f(x)$, if $f(x) > N$, then $[f(x)]_N = N$, and if $f(x) < -N$, $[f(x)]_N = -N$.

For a measurable set E with a σ-finite measure, if $\langle F_k \rangle$ is a sequence of nondecreasing measurable subsets of E with $m(F_k) < \infty$ and $E = \bigcup_k F_k$, then $\langle F_k \rangle$ is called a *finite-measure monotone covering* of E.

Definition 3.2.1 Let f be a measurable function defined on E with a σ-finite measure. If f is nonnegative and $\langle F_k \rangle$ is a finite-measure monotone covering of E, then f is said to be Lebesgue integrable if the limit $\lim_{N \to \infty} \int_{F_N} [f(x)]_N \, dx$ is finite. We write $\int_E f(x) \, dx = \lim_{N \to \infty} \int_{F_N} [f(x)]_N \, dx$.

It seems that the integral $\int_E f(x) \, dx$ depends on the choices of finite-measure monotone coverings of E. The following lemma shows that $\int_E f(x) \, dx$ is well-defined on E. When $m(E) < \infty$, this definition concise with the Lebesgue integral definition given in the previous section.

Lemma 3.2.1 Let f be a measurable function defined on E with a σ-finite measure. If f is nonnegative and $\langle F_k \rangle$, $\langle E_k \rangle$ are two finite-measure monotone coverings of E, then

$$\lim_{k \to \infty} \int_{F_k} [f(x)]_k \, dx = \lim_{n \to \infty} \int_{E_n} [f(x)]_n \, dx$$

whenever one of the two limits is finite.

Proof: Let's write $\ell_k = \int_{F_k} [f(x)]_k \, dx$ and $r_n = \int_{E_n} [f(x)]_n \, dx$. Then both $\langle \ell_k \rangle$ and $\langle r_n \rangle$ are nondecreasing sequences of real numbers. If $\ell = \lim_{k \to \infty} \ell_k$ exists, then we have $\ell_k \leq \ell$ for all $k \in \mathbb{N}$.

Now for any fixed N, $F_N \subset E$ with $m(F_N) < \infty$. Then $F_N \setminus E_k$ is a decreasing sequence of measurable sets. By Theorem 2.3.7, we have $\lim_{k \to \infty} m(F_N \setminus E_k) = 0$. Therefore, taking limit as $k \to \infty$ from the inequality

$$\int_{F_N} [f]_N \, dx = \int_{F_N \cap E_k} [f]_N \, dx + \int_{F_N \setminus E_k} [f]_N \, dx$$
$$\leq \int_{E_k} [f]_N \, dx + Nm(F_N \setminus E_k)$$
$$\leq \ell + Nm(F_N \setminus E_k),$$

we obtain that

$$r_N = \int_{F_N} [f]_N \, dx \leq \ell.$$

Therefore, $\langle r_n \rangle$ converges and $\lim_n r_n \leq \ell$. Now, we interchange the positions between r_n and ℓ_k, we have $\lim_k \ell_k \leq \lim_n r_n$ and thus, $\lim_k \ell_k = \lim_n r_n$. ∎

EXAMPLE 3.2.1: If

$$f(x) = \begin{cases} \frac{1}{\sqrt{x}}, & x \in (0,1], \\ \frac{1}{x^2}, & x \in (1,\infty), \end{cases}$$

then, for $F_k = [\frac{1}{k^2}, k]$,

$$\int_{F_N} [f(x)]_N \, dx = (R) \int_{F_N} [f(x)]_N \, dx = \int_{1/N^2}^{1} \frac{1}{\sqrt{x}} \, dx + \int_{1}^{N} \frac{1}{x^2} dx = 3 - \frac{3}{N}.$$

Therefore, $\int_0^\infty f \, dx = 3$.

For a general measurable function on E with a σ-finite measure, its positive part and negative part are both Lebesgue integrable.

Definition 3.2.2 If f is a measurable function on E with a σ-finite measure and both f^+ and f^- are Lebesgue integrable, then f is said to be *Lebesgue integrable* $\int_E f \, dx$ is defined as $\int_E f \, dx = \int_E f^+ \, dx - \int_E f^- \, dx$.

Clearly, f is Lebesgue integrable on E if and only if $\int_E |f| \, dx < \infty$. This cannot be true for Riemann integrals and thus, it shows a difference between the Riemann integral and the Lebesgue integral in the general setting.

We denote by $L(E)$ the class of all Lebesgue integrable functions on E and write $L[a,b]$ if $E = [a,b]$.

EXAMPLE 3.2.2:

(1) Consider a function f defined on $(0,\infty)$ by $f(x) = \frac{\sin x}{x}$. Then it is well known that its Riemann improper integral $(R) \int_0^\infty \frac{\sin x}{x} \, dx = \frac{\pi}{2}$. However, $(R) \int_0^\infty |\frac{\sin x}{x}| \, dx$ diverges and thus, $f(x) \notin L(0,\infty)$.

(2) Define f on $[0,1)$ by

$$f(x) = \begin{cases} 0, & x = 0 \\ (-1)^{n+1}n, & \frac{1}{n+1} < x \leq \frac{1}{n}, n \in \mathbb{N}. \end{cases}$$

Then, its Riemann improper integral $(R) \int_0^1 f(x) \, dx = \sum_{n=1}^{\infty} (-1)^{n+1} \frac{1}{n+1} = 1 - \ln 2$. However, $(R) \int_0^1 |f(x)| \, dx = \infty$ and thus, $f(x) \notin L[0,1]$.

The following result is often used to check integrability of measurable functions.

Lemma 3.2.2 Suppose $f \in L(E)$ and E has a σ-finite measure. Let g be a measurable function on E. If $|g| \le f$, then g is also in $L(E)$.

Proof: Since $|g| \le f$, we have $g^+ \le f$ and $g^- \le f$. For a finite-measure monotone covering $\langle F_k \rangle$ of E and positive integers k, we have

$$0 \le \int_{F_k} [g^+]_k \, dx \le \int_{F_k} [f]_k \, dx \le \int_E f \, dx < \infty.$$

Therefore, g^+ is integrable. Similarly, g^- is integrable and so is g. ∎

Next, we would like to verify the integral properties listed in Theorem 3.1.4 for the general integrable functions defined on E with a σ-finite measure. First we prove the linearity property.

Theorem 3.2.1 Let $f, g \in L(E)$ and E a σ-finite measure set. Then for any real numbers α and β, $\alpha f + \beta g$ is also integrable and

$$\int_E (\alpha f + \beta g) \, dx = \alpha \int_E f \, dx + \beta \int_E g \, dx.$$

Proof: First, we verify that for $\alpha \ge 0$, αf is integrable. In this case, $(\alpha f)^+ = \alpha f^+$ and we'd like to show $\int_E (\alpha f)^+ dx = \alpha \int_E f^+ dx$. It is true when $\alpha = 0$. For $\alpha > 0$, $[\alpha f]_N = \alpha [f]_{N/\alpha}$ and thus,

$$\int_E (\alpha f)^+ \, dx = \lim_{n \to \infty} \int_{F_k} [(\alpha f)^+]_k \, dx = \alpha \lim_{n \to \infty} \int_{F_k} [f^+]_{k/\alpha} \, dx = \alpha \int_E f^+ \, dx.$$

From $|\alpha f| = |\alpha||f|$ and Lemma 3.2.2, we see that αf is measurable on E, since f integrable implies that $|f|$ is also integrable.

Now it is sufficient to prove the theorem only for $\alpha = \beta = 1$. Since $|f+g| \le |f| + |g|$, we have that $f + g$ is integrable. From the definition, we see that the values $\int_E (f+g) \, dx = \int_E (f+g)^+ \, dx - \int_E (f+g)^- \, dx$ and $\int_E f^+ \, dx - \int_E f^- \, dx + \int_E g^+ \, dx - \int_E g^- \, dx$ should be equal.

One notes that if $f = h - k$, $h \ge 0$, $k \ge 0$, and f is integrable on E, then $\int_E f = \int_E h - \int_E k$. In fact, $f = f^+ - f^- = h - k$, so $f^+ + k = h + f^-$. Thus, $\int f^+ \, dx + \int k \, dx = \int h \, dx + \int f^- \, dx$. That is, $\int f = \int f^+ - \int f^- = \int h - \int k$. Therefore $\int (f + g) \, dx = \int f^+ - \int f^- + \int g^+ - \int g^- = \int f + \int g$. Since $f + g = f^+ - f^- + g^+ - g^-$. ∎

We can also prove the following properties.

Theorem 3.2.2 Let $f, g \in L(E)$ and E a σ-finite measure set. Then
  (1) $\int_E f \, dx \le \int_E g \, dx$ if $f \le g$;
  (2) If $E = E_1 \cup E_2$ and $E_1 \cap E_2 = \emptyset$ then $\int_E f \, dx = \int_{E_1} f \, dx + \int_{E_2} f \, dx$.

The following is called the *absolute continuity* property of Lebesgue integrals.

Theorem 3.2.3 If $f \in L(E)$ and E has a σ-finite measure, then for any $\epsilon > 0$, there exists $\delta > 0$ such that for any measurable subset A of E with $m(A) < \delta$, we have $|\int_A f \, dx| < \epsilon$.

Proof: Since $|f|$ is measurable, for any finite-measure monotone covering $\langle F_k \rangle$ of E, there is a positive integer N such that

$$\int_E |f| \, dx - \int_{F_N} [|f|]_N \, dx < \frac{\epsilon}{2}.$$

On the other hand, $\int_{F_N} [|f|]_N \, dx \le \int_E [|f|]_N \, dx$ and thus,

$$\int_E (|f| - [|f|]_N) dx < \frac{\epsilon}{2}.$$

Choosing $\delta = \frac{\epsilon}{2(N+1)}$, we have

$$\left| \int_A f \, dx \right| \le \int_A |f| \, dx = \int_A (|f| - [|f|]_N) \, dx + \int_A [|f|]_N \, dx$$

$$\le \frac{\epsilon}{2} + Nm(A) < \frac{\epsilon}{2} + N\delta < \epsilon.$$

∎

Exercises

1. For a measurable function f on E with a σ-finite measure, $f \in L(E)$ if and only if $|f| \in L(E)$ and $|\int_E f dx| \le \int_E |f| dx$. Show by an example that if the condition that f is measurable on E is removed, then there is a function f such that $|f| \in L(E)$ but $f \notin L(E)$.
2. Let $f \in L(E)$ and nonnegative such that $\int_E f dx = 0$. Then prove $f = 0$ a.e. on E.
3. Let $f \in L(E)$ and nonnegative such that $m(\{x \in E \mid f(x) > 0\}) > 0$. Then $\int_E f(x) \, dx > 0$.
4. Directly prove (without using Theorem 3.2.3) that if $f \ge 0$ is integrable on E, then for all $\epsilon > 0$, there exists $\delta > 0$, such that $\int_A f \, dx < \epsilon$ whenever $A \subset E$ and $m(A) < \delta$.

5. If $f \in L(E)$, then prove $|f| < \infty$ a.e. on E.
6. For every $\epsilon > 0$ and $f \in L(E)$, show that

$$m(E(|f| \geq \epsilon)) \leq \frac{1}{\epsilon} \int_E |f|\, dx.$$

This is called *Chebyshev's inequality*.
7. Let f and g be Lebesgue integrable functions defined on E with a σ-finite measure. Then
   (1) $\int_E f\, dx \leq \int_E g\, dx$ if $f \leq g$;
   (2) If $E = E_1 \cup E_2$ and $E_1 \cap E_2 = \emptyset$ then $\int_E f\, dx = \int_{E_1} f\, dx + \int_{E_2} f\, dx$.
8. If $f \in L[a, b]$ and for any $c \in [a, b]$, $\int_a^c f(x)\, dx = 0$, then $f = 0$ a.e. on $[a, b]$.
9. If $f \in L[a, b]$, then $F(x) := \int_a^x f(t)\, dt$ is continuous.
10. Suppose E_k, $k \in \mathbb{N}$ are mutually disjoint measurable sets and $E = \cup_k E_k$ has a σ-finite measure. If $f \in L(E)$, then prove

$$\sum_{k=1}^{\infty} \int_{E_k} f(x)\, dx = \int_E f(x)\, dx.$$

3 Convergence and Approximation of Lebesgue Integrals

We discuss next the convergence properties of sequences of Lebesgue integrals and also the approximation properties of Lebesgue integrable functions. We assume that E has a σ-finite measure.

Generalizing Theorem 3.1.5, the following is the most frequently used theorem which allows us to interchange the operations of integration and limits.

Theorem 3.3.1 [Lebesgue Dominated Convergence Theorem] Suppose g is Lebesgue integrable on E. The sequence $\langle f_n \rangle$ of measurable functions satisfies:
   (i) $|f_n| \leq g$ a.e. on E for $n \in \mathbb{N}$;
   (ii) $f_n \xrightarrow{m} f$ a.e. on E.

Then, $f \in L(E)$ and

$$\lim_{n \to \infty} \int_E f_n \, dx = \int_E f \, dx. \tag{3.1}$$

Proof: Since $f_n \overset{m}{\to} f$, f is measurable on E and also there is a subsequence $\langle f_{n_k} \rangle$ of $\langle f_n \rangle$ such that $f_{n_k} \to f$ a.e. on E. Therefore, $|f| \leq g$ a.e. on E and thus, f is integrable from Theorem 3.2.2.

Next, we first prove (3.1) for $m(E) < \infty$. For $\epsilon > 0$, let $E_n = E\left(|f_n - f| \geq \dfrac{\epsilon}{2(m(E) + 1)}\right)$. Then,

$$\left| \int_E (f_n - f) \, dx \right| \leq \left| \int_{E \setminus E_n} (f_n - f) \right| + \left| \int_{E_n} (f_n - f) \right|$$

$$\leq \int_{E \setminus E_n} |f_n - f| + \left| \int_{E_n} (f_n - f) \right|. \tag{3.2}$$

The first term in (3.2) is bounded from above by $\dfrac{\epsilon}{2(m(E) + 1)} \cdot m(E \setminus E_n) < \dfrac{\epsilon}{2}$. For such an $\epsilon > 0$, applying Theorem 3.2.3 to g, there is $\delta > 0$ such that $\int_A g \, dx < \frac{\epsilon}{4}$. Since $f_n \overset{m}{\to} f$ a.e. on E, there is $N \in \mathbb{N}$ such that $m(E_N) < \delta$. Hence, the second term in (3.2) is bounded from above by $2 \int_{E_n} g \, dx < \frac{\epsilon}{2}$ for $n \geq N$. Therefore, (3.1) holds for $m(E) < \infty$.

If E has a σ-finite measure and $\langle F_k \rangle$ is a finite-measure monotone covering of E, then for $\epsilon > 0$, there is $F_k \subset E$ with $m(F_k) < \infty$ such that

$$\int_E g \, dx < \int_{F_k} [g]_k \, dx + \frac{\epsilon}{4}$$

and thus,

$$\int_{E \setminus F_K} g = \int_E g - \int_{F_k} g \leq \int_E g - \int_{F_k} [g]_k \leq \frac{\epsilon}{4}.$$

Therefore,

$$\left| \int_E (f_n - f) \right| \leq \left| \int_{F_k} (f_n - f) \right| + \left| \int_{E \setminus F_k} (f_n - f) \right|$$

$$\leq \left| \int_{F_k} (f_n - f) \right| + \int_{E \setminus F_k} 2g$$

$$< \left| \int_{F_k} (f_n - f) \right| + \frac{\epsilon}{2}. \tag{3.3}$$

Noticing that $m(F_k) < \infty$ and thus (3.1) holds, there is $N \in \mathbb{N}$ such that for $n \geq N$, we have $| \int_{F_k} (f_n - f)| \leq \frac{\epsilon}{2}$. Therefore,

$$\left| \int_E (f_n - f) \right| < \epsilon$$

for $n \geq N$. ∎

Using a similar proof, we can obtain the Lebesgue Dominated Convergence Theorem for $f_n \to f$ a.e. on E.

Theorem 3.3.2 Suppose g is Lebesgue integrable on E. The sequence $\langle f_n \rangle$ of measurable functions satisfies:
 (i) $|f_n| \leq g$ a.e. on E for $n \in \mathbb{N}$;
 (ii) $f_n \to f$ a.e. on E.
Then, $f \in L(E)$ and

$$\lim_{n \to \infty} \int_E f_n \, dx = \int_E f \, dx. \tag{3.4}$$

The following two theorems are equivalent to the Lebesgue Dominated Convergence Theorem. We omit the proofs.

Theorem 3.3.3 [Monotone Convergence Theorem] If $\langle f_n \rangle$ is an increasing sequence of integrable functions on E, and $\lim_n \int_E f_n < \infty$, then $\lim_{n \to \infty} f_n$ exists a.e. on E and $\lim_{n \to \infty} \int_E f_n \, dx = \int_E \lim_{n \to \infty} f_n \, dx$.

Theorem 3.3.4 [Fatou's Lemma] Suppose that $\langle f_n \rangle$ is a sequence of integrable functions on E. If $f_n \geq h$ a.e. on E for an integrable function h and $\underline{\lim}_{n \to \infty} \int_E f_n \, dx < \infty$. Then

$$\int_E \underline{\lim}_{n \to \infty} f_n \, dx \leq \underline{\lim}_{n \to \infty} \int_E f_n \, dx.$$

We present the following application of the Lebesgue Dominated Convergence Theorem.

Theorem 3.3.5 Let $f(x, y)$ be a function defined on $E \times (a,b)$. Suppose f is integrable with respect to x on E and f is differentiable with respect to y on (a, b), and there exists an integrable function $F(x)$ on E such that $|\frac{\partial}{\partial y} f(x, y)| \leq F(x)$. Then

$$\frac{d}{dy} \int_E f(x, y)\, dx = \int_E \frac{\partial}{\partial y} f(x, y)\, dx.$$

Proof: Fix $y \in (a, b)$. Choose $h_k \to 0$ as $k \to \infty$. Then

$$\lim_{k \to \infty} \frac{f(x, y + h_k) - f(x, y)}{h_k} = \frac{d}{dy} f(x, y).$$

On the other hand, $|\frac{\partial}{\partial y} f(x, y)| \leq F(x)$ for $(x, y) \in E \times (a, b)$ and $\frac{f(x, y + h_k) - f(x, y)}{h_k} = \frac{\partial}{\partial y} f(x, \xi_k)$ by the Mean Value Theorem from calculus — here ξ is between y and $y + h_k$. Therefore, $|\frac{f(x, y + h_k) - f(x, y)}{h_k}| \leq F(x)$. Applying the Lebesgue Dominated Convergence Theorem, we have

$$\frac{d}{dy} \int f(x, y)\, dx = \lim_{k \to \infty} \int_E \frac{f(x, y + h_k) - f(x, y)}{h_k}\, dx$$

$$= \int_E \lim_{k \to \infty} \frac{f(x, y + h_k) - f(x, y)}{h_k}\, dx = \int_E \frac{\partial}{\partial y} f(x, y)\, dx.$$

∎

Let f and f_n be integrable on E. If $\lim_n \int_E |f_n - f|\, dx = 0$, then $\langle f_n \rangle$ is said to be *convergent in mean* (of order 1) to f or $\langle f_n \rangle$ is said to be L^1-*convergent* to f. Later in Chapter 5, we shall consider convergence in mean of order p (or L^p-*convergence*). The proof of the following result is left as an exercise.

Theorem 3.3.6 If $\langle f_n \rangle$ converges to f in mean on E, then $f_n \xrightarrow{m} f$ on E.

Recall that Lusin's Theorem (Theorem 2.4.10) provides an approximation to a measurable function using continuous functions. We next prove an approximation theorem for integrable functions.

Theorem 3.3.7 If $f \in L(E)$, then for every $\epsilon > 0$ there exists a continuous function g on $\mathbb{R}$ such that $\int_a^b |f(x) - g(x)|\, dx < \epsilon$.

Proof: For $\epsilon > 0$, by the definition of integrals of f^+ and f^-, we have $N \in \mathbb{N}$ such that

$$\int_E |f - [f]_N|\, dx = \int_E (f - [f^+]_N)\, dx + \int_E (f - [f^-]_N)\, dx < \frac{\epsilon}{3}.$$

Applying Lusin's Theorem to $[f]_N$, there is a continuous function g on $\mathbb{R}$ with $|g| \leq N$, and a measurable subset $E\delta$ such that $m(E\delta) < \delta$ and $g = [f]_N$ for $x \in E \setminus E\delta$. Choose $\delta = \frac{\epsilon}{3N+1}$. We have

$$\int_E |[f]_N - g|\, dx = \int_{E\delta} |[f]_N - g|\, dx \leq 2N \cdot m(E\delta) < \frac{2\epsilon}{3}.$$

Therefore,

$$\int_E |f - g|\, dx \leq \int_E |f - [f]_N|\, dx + \int_E |[f]_N - g|\, dx < \epsilon.$$

∎

The above theorem shows that if $f \in L(E)$, then for any $\epsilon > 0$, f has the decomposition $f = f_1 + f_2$, where f_1 is continuous on $\mathbb{R}$ and $\int_E |f_2|\, dx < \epsilon$. As an application, we have the following average continuity property of Lebesgue integrals.

Theorem 3.3.8 If f is Lebesgue integrable on $\mathbb{R}$, then

$$\lim_{h \to 0} \int_{\mathbb{R}} |f(x + h) - f(x)|\, dx = 0.$$

Proof: For every $\epsilon > 0$, there exists a continuous function g on $\mathbb{R}$ such that $\{x \mid g(x) \neq 0\} \subset [a, b]$ for $-\infty < a \leq b < \infty$ and $\int_{\mathbb{R}} |f - g| dx < \epsilon$. Therefore,

$$\int_{\mathbb{R}} |f(x + h) - f(x)|\, dx \leq \int_{\mathbb{R}} |f(x + h) - g(x + h)|\, dx + \int_{\mathbb{R}} |g(x + h) - g(x)|\, dx$$

$$+ \int_{\mathbb{R}} |g(x) - f(x)|\, dx < \epsilon + \int_{\mathbb{R}} |g(x + h) - g(x)|\, dx + \epsilon$$

$$= 2\epsilon + \int_{\mathbb{R}} |g(x + h) - g(x)|\, dx.$$

Noticing that g can be chosen as a continuous function on $\mathbb{R}$ and $g(x) = 0$ for $x \in \mathbb{R} \setminus [a, b]$ for some $a, b \in \mathbb{R}$, we have that g is uniformly continuous on $\mathbb{R}$. So, for $\epsilon > 0$, there exists $\delta > 0$ such that for $|h| < \delta$, $|g(x + h) - g(x)| dx < \frac{\epsilon}{b-a}$. So,

$$\int_{\mathbb{R}} |g(x + h) - g(x)|\, dx = \int_a^b |g(x + h) - g(x)|\, dx < \frac{\epsilon}{b - a} \cdot (b - a) = \epsilon.$$

Therefore, as long as $|h| < \delta$, we have $\int_{\mathbb{R}} |f(x+h) - f(x)| \, dx < 3\epsilon$. ϵ is arbitrary, so $\lim_{h \to 0} \int_{\mathbb{R}} |f(x+h) - f(x)| \, dx = 0$. ∎

Exercises

1. Let $\langle f_n \rangle$ be a sequence of measurable functions on E. If (i) there exists an integrable function F on E such that $|f_n| \leq F$ a.e. on E for $n \in \mathbb{N}$, and (ii) $f_n \to f$ a.e. on E, then prove $f_n \overset{m}{\to} f$ as $n \to \infty$.

2. Suppose $\langle g_n \rangle$ is a sequence of integrable functions on E and $g_n \to g$ a.e. with g integrable. If the sequence $\langle f_n \rangle$ of measurable functions satisfies:
   (i) $|f_n| \leq g_n$ for $n \in \mathbb{N}$;
   (ii) $f_n \to f$ a.e. on E,
   then prove $\lim_{n \to \infty} \int_E f_n \, dx = \int_E f \, dx$.

3. Let $L[a, b]$ denote the class of all Lebesgue integrable functions on $[a, b]$. Show that for $f \in L[a, b]$ and for all $\epsilon > 0$, there is
   (i) a bounded measurable function g such that $\int_a^b |f(x) - g(x)| \, dx < \epsilon$;
   (ii) a continuous function h such that $\int_a^b |f(x) - h(x)| \, dx < \epsilon$;
   (iii) a polynomial function p such that $\int_a^b |f(x) - p(x)| \, dx < \epsilon$;
   (iv) a step function s such that $\int_a^b |f(x) - s(x)| \, dx < \epsilon$.

4. If $f \in L[a, b]$, then prove $\lim_{n \to \infty} \int_a^b f(x) \cos nx \, dx = 0$ and $\lim_{n \to \infty} \int_a^b f(x) \sin nx \, dx = 0$. [Hint: Consider a step function f first.]

5. Show by example that the inequality in Fatou's Lemma is not in general an equality even if the sequence of functions $\langle f_n \rangle$ converges everywhere.

6. If $f \in L[a, b]$, then prove $\lim_{n \to \infty} \int_a^b f(x) |\cos nx| \, dx = \frac{2}{\pi} \int_a^b f(x) \, dx$.

7. If $f \in L[a, b]$ and for $k \in \mathbb{N}$, $\int_a^b x^k f(x) \, dx = 0$, then prove $f = 0$ a.e. on $[a, b]$.

8. If $f \in L(\mathbb{R})$ and for any compact supported continuous function g, $\int_{\mathbb{R}} f(x) g(x) \, dx = 0$, then prove $f = 0$ a.e.

9. If $\langle f_n \rangle$ is a sequence of integrable functions on E satisfying $f_n \overset{m}{\to} f$, then prove $\int_E f(x) \, dx \leq \underline{\lim}_{n \to \infty} \int_E f_n(x) \, dx$.

10. Let $f \in L[0, 1]$. Then $x^n f(x) \in L[0, 1]$ for any $n \in \mathbb{N}$ and $\lim_{n \to \infty} \int_0^1 x^n f(x) \, dx = 0$.

11. If $f_n, f \in L(E)$ and $\int_E |f_n - f| \, dx \to 0$, then $f_n \overset{m}{\to} f$.

4 Lebesgue Integrals in the Plane

In this section, we outline the theory of measure and integration in the plane $\mathbb{R}^2$. The procedure is very much analogous to that for the real line $\mathbb{R}$. Rectangles in $\mathbb{R}^2$ play the role of intervals in $\mathbb{R}$.

Let C_0 be the collection of rectangles in the form $(a, b] \times (c, d] = \{(x, y) \mid a < x \leq b, c < y \leq d\}$ and define a set function m on C_0 by

$$m((a, b] \times (c, d]) = (b - a)(d - c).$$

We can show that m can be uniquely extended to be a measure on $\mathcal{R}(C_0)$. Using a method similar to the construction of m on $\mathbb{R}$, we can define a class of Lebesgue measurable sets in $\mathbb{R}^2$.

A function $f(x, y)$ defined on a measurable set E in the plane is said to be *measurable* if for any $\alpha \in \mathbb{R}$, the set $\{(x, y) \in E \mid f(x, y) \leq \alpha\}$ is measurable. We may then define the Lebesgue integral on E. We let $L(E)$ denote the class of Lebesgue integrable functions on E. If $f \in L(E)$, the integral of f over E is denoted by $\int \int_E f(x, y) \, dx \, dy$. Such integrals are often called *double integrals*.

For a measurable function f on E, the following theorem provides the connection between measurable functions on the plane and the measurable functions on the real line. For simplicity, we may assume E is a rectangle domain.

Theorem 3.4.1 If f is a measurable function $f(x, y)$ on $T = [a, b] \times [c, d]$, then $f(x, y)$ is a measurable function of y for any fixed $x \in [a, b]$ and $f(x, y)$ is a measurable function of x for any fixed $y \in [c, d]$.

In practice, the integral $\int \int_T f(x, y) \, dx \, dy$ is computed by integrating first with respect to x and then with respect to y or vice versa. This method is usually called *iterated integrals* or *repeated integrals*. This requires some caution to go beyond continuous functions.

EXAMPLE 3.4.1: Let

$$f(x, y) = \begin{cases} \frac{x^2 - y^2}{(x^2 + y^2)^2}, & (x, y) \in (0, 1] \times (0, 1]; \\ 0, & (x, y) = (0, 0). \end{cases}$$

Then

$$\int_0^1 \left\{ \int_0^1 f(x, y) dx \right\} dy = -\frac{\pi}{4}$$

but that

$$\int_0^1 \left\{ \int_0^1 f(x,y)dy \right\} dx = \frac{\pi}{4}.$$

The fact that a double integral can be evaluated by iterated integration does not follow immediately from the definition of $\int \int_T f(x,y)\,dx\,dy$, but rather is a famous and difficult theorem called Fubini's Theorem.

Theorem 3.4.2 [Fubini's Theorem] Let f be an integrable function on $\mathbb{R}^2$. Then the double integral and iterated integrals are equal. In particular, $f(x,y)$ is integrable with respect to x for almost every y and $\int f(x,y)\,dx$ is integrable as a function of y and the same with x and y reversed.

Proof: First, we assume that f is nonnegative. Then similar to the one variable setting, there is a sequence $\langle \varphi_n \rangle$ of nonnegative simple functions converging to f and thus, $\int f dx = \int \varphi_n\,dx$. Since the double and iterated integrals are equal for simple functions, we obtain

$$\int \int \varphi_n(x,y)\,dx\,dy = \int \left\{ \int \varphi_n(x,y)\,dx \right\} dy.$$

Applying the Monotone Convergence Theorem to $\varphi_n(x,y)$ as a function of x for a fixed y, we have $\lim_n \int \varphi_n dx = \int f(x,y)\,dx$. Noticing that $\int \varphi_n\,dx$ is measurable and so is $\int f(x,y)$, we apply the Monotone Convergence Theorem one more time and obtain

$$\lim_n \int \left\{ \int \varphi_n(x,y)\,dx \right\} dy = \int \left\{ \int f(x,y)\,dx \right\} dy.$$

Similarly, we have

$$\lim_n \int \left\{ \int \varphi_n(x,y)\,dy \right\} dx = \int \left\{ \int f(x,y)\,dy \right\} dx.$$

Next, for a general integrable function f, we apply the proof in the first step to f^+ and f^-, respectively. Using the linearity of the double integral and iterated integrals we obtain

$$\int \left\{ \int f(x,y)\,dy \right\} dx = \int \left\{ \int f(x,y)\,dx \right\} dy.$$

■

If $f \in L(T)$ with $T = [a, b] \times [c, d]$, then Fubini's Theorem gives that

$$\int\int_T f(x, y)\, dx\, dy = \int_a^b \left\{ \int_c^d f(x, y)\, dy \right\} dx = \int_c^d \left\{ \int_a^b f(x, y)\, dx \right\} dy.$$

That is, $f(x, y) \in L[c, d]$ for almost every $x \in [a, b]$ and $f(x, y) \in L[a, b]$ for almost every $y \in [c, d]$. Moreover, $\int_c^d f(x, y)\, dy \in L[a, b]$ and $\int_a^b f(x, y)\, dx \in L[c, d]$ and the iterated integrals are equal to the double integral.

In practice, it is much easier to check the existence of iterated integrals instead of double integrals. The following version of Fubini's Theorem is the most useful.

Theorem 3.4.3 Let f be a measurable function on $\mathbb{R}^2$. Suppose one of the iterated integrals for $|f|$ exists, say $\int f(x, y)\, dx$ exists a.e. on y and $\int \{ \int |f(x, y)|\, dx \}\, dy$ exists. Then the double integral and iterated integrals on f exist and are equal.

Proof: From the first part of the proof of Theorem 3.4.2, we see that the double integral of $|f|$ is equal to its iterated integrals and so, $|f|$ is integrable. This means f is integrable. Then applying Theorem 3.4.2 to f completes the proof. ∎

If $f \in L(0, \infty)$, then the function $e^{-xt}f(t)$ is also in $L(0, \infty)$ for $x \in (0, \infty)$ since $|e^{-xt}f(t)| \le |f(t)|$. The function of x defined by $\int_0^\infty e^{-xt}f(t)\, dt$, denoted by $\mathcal{L}(f)$, is called the *Laplace transform* of f.

Definition 3.4.1 For $f, g \in L(\mathbb{R})$, if the integral $\int_{-\infty}^\infty f(x - y)g(y)\, dy$ exists, then is it called the *convolution* of f and g, denoted $f * g$.

Theorem 3.4.4 If $f, g \in L(\mathbb{R})$, then $(f * g)(x)$ exists a.e. for x and

$$\int_{-\infty}^\infty |(f * g)(x)|\, dx \le \left(\int_{-\infty}^\infty |f(x)|\, dx \right) \left(\int_{-\infty}^\infty |g(x)|\, dx \right).$$

Therefore, $f * g \in L(\mathbb{R})$.

Proof: Noticing that

$$\int_{-\infty}^\infty |f(x - y)|\, dx = \int_{-\infty}^\infty |f(u)|\, du$$

by the change of variable $u = x - y$, we obtain that the iterated integral

$$\int_{-\infty}^{\infty} \left\{ \int_{-\infty}^{\infty} |f(x - y)g(y)| \, dx \right\} dy = \int_{-\infty}^{\infty} |f(u)| \, du \int_{-\infty}^{\infty} |g(y)| \, dy.$$

Thus, by Theorem 3.4.3, $f(x - y)g(y) \in L(\mathbb{R}^2)$. By Fubini's Theorem, $f * g$ exists for all most every x and is integrable. ∎

We next state a famous result involving Laplace transforms and convolutions.

Theorem 3.4.5 Let $f, g \in L(0, \infty)$. Then the Laplace transform of the convolution $f * g$ is the product of the Laplace transforms of f and g. That is,

$$\mathcal{L}(f * g) = \mathcal{L}(f)\mathcal{L}(g).$$

Proof: Exercise. ∎

Exercises

1. Let $f(x, y) \in L([0, 1] \times [0, 1])$. Then prove

$$\int_0^1 \left\{ \int_0^x f(x, y) \, dy \right\} dx = \int_0^1 \left\{ \int_y^1 f(x, y) \, dx \right\} dy.$$

2. Let

$$f(x, y) = \begin{cases} \frac{4xy - x^2 - y^2}{(x+y)^4}, & x > 0, y > 0); \\ 0, & (x, y) = (0, 0). \end{cases}$$

   Show that

$$\int_0^{\infty} \left\{ \int_0^{\infty} f(x, y) \, dx \right\} dy = \int_0^{\infty} \left\{ \int_0^{\infty} f(x, y) \, dy \right\} dx = 0$$

   but that $\int \int_{[0,\infty) \times [0,\infty)} f(x, y) \, dx \, dy$ does not exist. Which hypotheses of the Fubini's theorem have been violated?

3. Let

$$f(x, y) = \begin{cases} \frac{xy}{(x^2+y^2)}, & x^2 + y^2 > 0); \\ 0, & (x, y) = (0, 0). \end{cases}$$

Show that

$$\int_0^1 \left\{ \int_0^1 f(x,y)\, dx \right\} dy = \int_0^1 \left\{ \int_0^1 f(x,y)\, dy \right\} dx = 0$$

but that $\int \int_{[0,\infty)\times[0,\infty)} f(x,y)\, dx\, dy$ does not exist.

4. Prove Theorem 3.4.5.

5. If $f \in L[a,b]$ and $g \in L[c,d]$, then prove $h(x,y) = f(x)g(y) \in L([a,b] \times [c,d])$ and

$$\int \int_{[a,b]\times[c,d]} h(x,y)\, dx\, dy = \int_a^b f(x)\, dx \int_c^d g(y)\, dy.$$

6. Evaluate $\mathcal{L}(x^k)$, the Laplace transform of the power functions x^k for $k = 0, 1, \cdots$.

7. Let $B_1(x) = \chi_{[-1/2,1/2]}(x)$ and $B_n(x)$ be defined using the recursion formula

$$B_n(x) = (B_{n-1} * B_1)(x)$$

for $n = 2, 3, \cdots$. The functions $B_n(x)$ are called *B-splines*. Find explicit expressions for B_2, B_3, and B_4.

8. Verify the following properties of B-splines:
   (1) B_n is in spline space S_{n-1}^{n-2}, the space of splines with total degree $n-1$ and smoothness order of $n-2$;
   (2) $B_n \geq 0$ and $B_n > 0$ for $x \in (-n/2, n/2)$;
   (3) $\sum_j B_n(x-j) = 1$ for all $x \in \mathbb{R}$;
   (4) $\int_{-\infty}^{\infty} B_n(x)\, dx = 1$;
   (5) $B_n'(x) = B_{n-1}(x + \frac{1}{2}) - B_{n-1}(x - \frac{1}{2})$.

9. Suppose that $g \in L(\mathbb{R})$ and f is bounded and continuous. Then $f * g$ is also bounded and continuous.

10. If $g \in L(\mathbb{R})$ and f' is bounded and continuous, then $f * g$ is differentiable and

$$\frac{d}{dx}(f * g)(x) = \int_{-\infty}^{\infty} f'(x-y)g(y)\, dy.$$

Chapter 4

Special Topics of Lebesgue Integral and Applications

This chapter contains first a brief discussion of the relationship between differentiation and integration on $\mathbb{R}$. We explore the conditions under which the fundamental theorem of calculus $\int_a^b F(x)\,dx = F(b) - F(a)$ is valid. This exploration leads to interesting and perhaps unexpected measure-theoretic ideas. The main result in this direction is the Lebesgue–Radon–Nikodym theorem, which we cannot include in this book and refer reader to other texts, [23] for example, for the material. The remainder of the chapter presents some of the main theorems in probability which are closely related to measure theory and Lebesgue integration.

I Differentiation and Integration

In order to discuss derivatives of a Lebesgue integrable function we need the following definitions.

Definition 4.1.1 Let $x \in \mathbb{R}$ and f a real-valued function and $\delta > 0$. If f is defined on $[x, x + \delta)$, then define

$$D^+ f(x) = \overline{\lim}_{h \to 0^+} \frac{f(x+h) - f(x)}{h},$$

$$D_+f(x) = \underline{\lim}_{h\to 0^+} \frac{f(x+h) - f(x)}{h}.$$

If f is defined on $(x - \delta, x]$, then define

$$D^-f(x) = \overline{\lim}_{h\to 0^-} \frac{f(x+h) - f(x)}{h},$$

$$D_-f(x) = \underline{\lim}_{h\to 0^-} \frac{f(x+h) - f(x)}{h}.$$

These four extended real numbers are usually *Dini derivatives* of f. We say f is *differentiable* at x if $D^+f(x) = D^-f(x) = D_+f(x) = D_+f(x)$ and write $f'(x)$ or $\frac{d}{dx}f(x)$.

EXAMPLE 4.1.1: Let $f(x) = x \sin \frac{1}{x}$ for $x \neq 0$ and $f(0) = 0$. Then $D^+f(0) = 1, D_+f(0) = -1, D^-f(0) = 1$, and $D_-f(0) = -1$.

A real-valued function f on $\mathbb{R}$ is said to be *nondecreasing (strictly increasing)* if $f(x_1) \leq f(x_2)$ ($f(x_1) < f(x_2)$) whenever $x_1 < x_2$ and *nonincreasing (strictly decreasing)* if $f(x_1) \geq f(x_2)$ ($f(x_1) > f(x_2)$) whenever $x_1 < x_2$; in either case f is called a *monotone function*. For a monotone function f and an interior point a of the domain E, the *jump* of f at a, denoted by $J(f, a)$ is the absolute difference between the right limit value $f(a+) = \lim_{x\to a+} f(x)$ and the left limit value $f(a-) = \lim_{x\to a-} f(x)$. That is $J(f, a) = |f(a+) - f(a-)|$.

Theorem 4.1.1 If f is continuous on $[a, b]$ and $D^+f(x) \geq 0$ for all $x \in (a, b)$ then f is monotone nondecreasing on $[a, b]$.

Proof: We leave the proof as an exercise. ■

Our next goal is to show Lebesgue's famous theorem that a monotone function has a finite derivative almost everywhere. The main tool used in the proof is the so-called Vitali's Covering Theorem. Vitali's theorem has many applications in classical analysis, particularly in the theory of differentiation.

Definition 4.1.2 Let $\mathcal{V}$ be a collection of intervals of $\mathbb{R}$, each having positive length. We say that $\mathcal{V}$ is a *Vitali cover* of $E \subset \mathbb{R}$ if $\mathcal{V}$ covers E finely. This means

$E \subset \cup_{I \in \mathcal{V}} I$ and for every $x \in E$ and $\epsilon > 0$, there exists $I \in \mathcal{V}$ such that $x \in I$ and its length $\lambda(I) = m(I) < \epsilon$.

EXAMPLE 4.1.2: Let $E = [a, b]$ and $\langle r_n \rangle$ be the sequence of all rational numbers in E. Then $\mathcal{V} = \{[r_n - \frac{1}{m}, r_n + \frac{1}{m}] \mid m, n \in \mathbb{N}\}$ is a Vitali's cover of E.

Theorem 4.1.2 [Vitali's Covering Theorem] Let E be a set with $m^*(E) < \infty$. Let $\mathcal{V}$ be a Vitali cover for E. Then for every $\epsilon > 0$ there exists disjoint intervals (finitely many) $I_1, I_2, \cdots, I_n$ in $\mathcal{V}$ such that $m^*(E \setminus \cup_{k=1}^n I_k) < \epsilon$.

Proof: We can assume that $\mathcal{V}$ is a set of closed intervals, for otherwise we replace each interval by its closure which does not change the length of the intervals.

Because $m^*(E) < \infty$, we can have an open set G such that $G \supset E$ with $m(G) < \infty$. Since $\mathcal{V}$ is a Vitali cover of E, we may assume that each $I \in \mathcal{V}$ is contained in G, for otherwise, $\mathcal{W} = \{I \in \mathcal{I} \mid I \subset \mathcal{W}\}$ is a Vitali cover of E. We choose a sequence $\langle I_n \rangle$ of disjoint intervals of $\mathcal{V}$ by induction as follows: Let I_1 be an arbitrary interval in $\mathcal{V}$. If $E \subset I_1$, then we are done. If not, suppose $I_1, \cdots, I_k$ were chosen such that they are mutually disjoint. If $E \subset \cup_{j=1}^k I_j$, then again, we are done. Otherwise, let $\delta_k = \sup\{\lambda(I) \mid I \in \mathcal{V}, I \cap I_j = \emptyset, j = 1, \cdots, k\}$, the supremum of the lengths of the intervals of $\mathcal{V}$ that do not meet any of the intervals $I_1, \cdots, I_k$. Then $\delta_k \leq m(G) < \infty$. Therefore, we can find $I_{k+1} \in \mathcal{V}$ such that its length $\lambda(I_{k+1}) > \frac{\delta_k}{2}$ and $I_j, j = 1, \cdots, k+1$ are mutually disjoint. In this way, we find a sequence $\langle I_n \rangle$ of disjoint intervals of $\mathcal{V}$ such that $\cup I_k \subset G$ and thus, $\sum_k \lambda(I_k) \leq m(G) < \infty$. Therefore, for $\epsilon > 0$ there is $N \in \mathbb{N}$ such that $\sum_{j=n}^\infty \lambda(I_j) < \frac{\epsilon}{5}$ for $n \geq N$. Next, let $S = E \setminus \cup_{j=1}^N I_j$. Then it is sufficient to show that $m^*(S) < \epsilon$. Since $\cup_{j=1}^N I_j$ is closed and thus, for any $x \in S$, the distance between x and $\cup_{j=1}^N I_j$ is positive, we can find an interval I of $\mathcal{V}$ which contains x and whose length is so small that I does not meet any of the intervals $I_j, j = 1, \cdots, N$. If $I \cap I_j = \emptyset$ for $j \leq n$, from the definition of δ_k, we must have $\lambda(I) \leq \delta_n \leq 2\lambda(I_{n+1})$. Since $\lambda(I_n) \to 0$, the interval I must meet at least one of the intervals I_n. Let n be the smallest integer such that I meets I_n. We have $n > N$, and $\lambda(I) \leq \delta_{n-1} \leq 2\lambda(I_n)$. Since $x \in I$ and I has a common point with I_n, it follows that the distance from x to the midpoint c_n of I_n is at most $\lambda(I) + \frac{1}{2}\lambda(I_n) \leq \frac{5}{2}\lambda(I_n)$. Hence, x is in the interval $J_n = [c_n - \frac{5\lambda(I_n)}{2}, c_n + \frac{5\lambda(I_n)}{2}]$ having the same midpoint as I_n and five times the length. Thus, $S \subset \cup_{n=N+1}^\infty J_n$. Therefore,

$$m^*(S) \leq \sum_{n=N+1}^{\infty} |J_n| = 5 \sum_{n=N+1}^{\infty} |I_n| < \epsilon. \qquad \blacksquare$$

Theorem 4.1.3 [Lebesgue] If f is increasing on $[a, b]$ then f is differentiable a.e. on $[a, b]$. Furthermore, f' is integrable on $[a, b]$ and

$$\int_{[a,b]} f' \, dx \leq f(b) - f(a).$$

Proof: Using the Vitali Covering Theorem, we can show that f' exists a.e. To this end, let us first show that the sets where any two derivatives are unequal have measure zero. In this case, we only show the set $E := \{x \in (a, b) \mid D^+f(x) > D_-f(x)\}$ has measure 0; the sets arising from other combinations of derivatives can be proved similarly. For rational numbers r and s, we write

$$E_{r,s} = \{x \mid D^+f(x) > r > s > D_-f(x)\}$$

and then, $E = \cup_{r,s \in \mathbb{Q}} E_{r,s}$. Therefore, it suffices to prove that $m^*E_{r,s} = 0$.

Choose an open set $G \supset E_{r,s}$ such that $m(G) < m^*(E_{r,s}) + \epsilon$. For any $x \in E_{r,s}$, since $D_-f(x) < s$, there is $h > 0$ such that

$$\frac{f(x - h) - f(x)}{-h} < s$$

for sufficiently small h, and so we assume $[x - h, x] \subset G$. All such intervals form a Vitali cover of $E_{r,s}$. By Theorem 4.1.2, for any $\epsilon > 0$, there are disjoint intervals $[x_1 - h_1, x_1], \cdots, [x_N - h_N, x_N]$ such that

$$m^*(E_{r,s} \setminus \cup_{i=j}^{N} [x_j - h_j, x_j]) < \epsilon.$$

Set $U = \cup_{j=1}^{N} (x_j - h_j, x_j)$. Then $m^*(E_{r,s} \setminus U) < \epsilon$. Also, $U \subset G$ implies that

$$\sum_{j=1}^{N} h_j = m(U) \leq m(G) < m^*(E_{r,s}) + \epsilon.$$

Summing the inequalities

$$f(x_j) - f(x_j - h_j) < sh_j, j = 1, 2, \cdots N,$$

we obtain

$$\sum_{j=1}^{N} [f(x_j) - f(x_j - h_j)] < s \sum_{j=1}^{N} h_j < sm(G) < s(m^*(E_{r,s}) + \epsilon).$$

Next, for each point $y \in E_{r,s} \cap U$, $D^+f(y) > r$ implies that $f(y + k) - f(y) > rk$ for sufficiently small k, and so, we assume $[y, y + k] \subset U \cap E_{r,s}$. All such intervals form a Vitali cover of $U \cap E_{r,s}$. Using Theorem 4.1.2 again, we can pick out disjoint intervals $J_1, \cdots, J_M$ with $J_i = [y_i, y_i + k_i]$ such that

$$m^*\left((E_{r,s} \cap U) \setminus \cup_{i=1}^M J_i\right) < \epsilon.$$

Thus, $m^*(E_{r,s}) \leq m^*(E_{r,s} \cap U) + m^*(E_{r,s} \setminus U) \leq \epsilon + \sum_{i=1}^M \lambda(J_i) + \epsilon$. Hence,

$$\sum_{i=1}^M k_i > m^*\left(E_{r,s} \cap (\cup_{j=1}^N (x_j - h_j, x_j))\right) - \epsilon$$

and thus, summing

$$f(y_i + k_i) - f(y_i) > rk_i,$$

over i, we obtain

$$\sum_{i=1}^M [f(y_i + k_i) - f(y_i)] > r(m^*(E_{r,s}) - 2\epsilon).$$

Noticing that f is increasing, we have

$$\sum_{j=1}^N [f(x_j) - f(x_j - h_j)] \geq \sum_{i=1}^M [f(x_i + k_i) - f(x_i)]$$

and so

$$r(m^*(E_{r,s}) - 2\epsilon) < s(m^*(E_{r,s}) + \epsilon).$$

Since ϵ is arbitrary, we see that $rm^*(E_{r,s}) \leq sm^*(E_{r,s})$. However, $r > s$, so we must have $m^*(E_{r,s}) = 0$.

Next, let $A = \{x \in (a, b) \mid f'(x) = \infty\}$. Then we prove $m^*(A) = 0$. For any $N \in \mathbb{N}$ and $x \in A$, $D^+f(x) = \infty$ implies that there is $h > 0$ small enough such that

$$f(x + h) - f(x) > Nh.$$

We can assume $[x, x + h] \subset [a, b]$ and thus, all such intervals form a Vitali cover of A. Using Theorem 4.1.2 again, we can pick out disjoint intervals $I_1, \cdots, I_n$ with $I_i = [x_i, x_i + k_i]$ such that

$$m^*(A \setminus \cup_{i=1}^n I_i) < \frac{1}{N}.$$

Using a similar argument as before, we have

$$Nm^*(A) < 1 + f(b) - f(a)$$

for any $N \in \mathbb{N}$ and thus, $m^*(A) = 0$. Therefore, $f'(x)$ exists a.e. on $[a, b]$.

Finally, consider

$$f_n(x) = \frac{f(x + 1/n) - f(x)}{1/n},$$

where $f(x) = f(b)$ if $x > b$. We have $f_n(x) \to f'(x)$ a.e. and $f_n(x) \geq 0$ since f is increasing. Then by Fatou's Lemma (Theorem 3.3.4), we obtain

$$\int_a^b f'(x)\,dx \leq \underline{\lim}_{n \to \infty} \int_a^b f_n(x)\,dx$$

$$= \underline{\lim}_{n \to \infty} \int_a^b n[f(x + 1/n) - f(x)]\,dx$$

$$= n[\underline{\lim}_{n \to \infty} \int_b^{b+1/n} f(x)\,dx - \int_a^{a+1/n} f(x)\,dx]$$

$$= f(b) - \overline{\lim}_{n \to \infty} n \int_a^{a+1/n} f(x)\,dx \leq f(b) - f(a).$$

The last inequality holds because f is increasing.   ∎

Corollary 4.1.1 If f is the difference of two monotone functions, then f is differentiable almost everywhere.

In the study of integrals, often the function $\int_a^x f(t)\,dt$ for $f \geq 0$ is an increasing function on $[a, b]$. For a general integrable function on $[a, b]$, we have

$$F(x) := \int_a^x f(t)\,dt = \int_a^x f^+(t)\,dt - \int_a^x f^-(t)\,dt,$$

the difference of two increasing functions. Let us investigate for a moment which functions can express the difference of two increasing functions.

Let f be a real-valued function on $[a, b]$. For any partition

$$\triangle : a = x_0 < x_1 < x_2 < \cdots < x_n = b,$$

we define

$$p(f, \triangle) = \sum_{k=1}^{n} [f(x_k) - f(x_{k-1})]^+$$

and

$$n(f, \triangle) = \sum_{k=1}^{n} [f(x_k) - f(x_{k-1})]^-$$

where, $y^+ = \max\{y, 0\}$, $y^- = |y| - y^+$.

Clearly,

$$p(f, \triangle) - n(f, \triangle) = \sum_{k=1}^{\infty} [f(x_k) - f(x_{k-1})] = f(b) - f(a)$$

and

$$p(f, \triangle) + n(f, \triangle) = \sum_{k=1}^{\infty} |f(x_k) - f(x_{k-1})|.$$

Define

$$P_a^b(f) = \sup_{\triangle} p(f, \triangle)$$

$$N_a^b(f) = \sup_{\triangle} n(f, \triangle)$$

and

$$T_a^b(f) = \sup_{\triangle} \{p(f, \triangle) + n(f, \triangle)\}.$$

Definition 4.1.3 $T_a^b(f)$ is called the *total variation* of f over $[a, b]$. If $T_a^b(f) < \infty$, then f is said to be of *bounded variation* over $[a, b]$, denoted by $f \in BV[a, b]$.

Theorem 4.1.4 $T_a^b(f) = P_a^b(f) + N_a^b(f)$ and $P_a^b(f) - N_a^b(f) = f(b) - f(a)$. If $f \in BV[a, b]$, then f can be written as a difference of two monotone increasing functions.

Proof: We have $p(f, \triangle) = n(f, \triangle) + f(b) - f(a)$. Thus, $P_a^b(f) = N_a^b(f) + f(b) - f(a)$. So, $P_a^b(f) - N_a^b(f) = f(b) - f(a)$. Therefore $p(f, \triangle) + n(f, \triangle) = 2n(f, \triangle) + f(b) - f(a)$, and $T_a^b(f) = 2N_a^b(f) + f(b) - f(a) = P_a^b(f) + N_a^b(f)$.

If $f \in BV[a, b]$, then $f(x) - f(a) = P_a^x(f) - N_a^x(f)$. So, $f(x) = P_a^x(f) - (N_a^x(f) - f(a))$ and $f(x) = \frac{1}{2}(T_a + f(x)) - \frac{1}{2}(T_a - f(x))$ are two decompositions of the difference of two increasing functions. Noticing $P_a^x(f)$ and $N_a^x(f)$ are increasing, the conclusion follows. ∎

Corollary 4.1.2 If $f \in BV[a, b]$, then f' exists a.e. on $[a, b]$.

We leave the proof of the following properties of bounded variation functions to the reader.

Theorem 4.1.5 For bounded variation functions, the following properties hold.

(1) If $f \in BV[a, b]$, then f is bounded.
(2) If $f, g \in BV[a, b]$, then $\alpha f + \beta g \in BV[a, b]$ for $\alpha, \beta \in \mathbb{R}$ and

$$T_a^b(\alpha f + \beta g) \leq |\alpha| T_a^b(f) + |\beta| T_a^b(g).$$

(3) If $f, g \in BV[a, b]$, then $f * g \in BV[a, b]$. The convolution is given by $f * g(x) = \int_a^b f(x - y) g(y) \, dy$.
(4) If $f \in BV[a, b]$, and $T_a^b(f) = 0$, then f is a constant.
(5) If $f \in BV[a, b]$ and $c \in [a, b]$, then

$$T_a^b(f) = T_a^c(f) + T_c^b(f).$$

(6) If $f \in \text{Lip M}$, then $f \in BV[a, b]$.
(7) If f is monotone increasing, then $T_a^b(f) = f(b) - f(a)$.
(8) If f is integrable on $[a, b]$, then $F(x) = \int_a^x f(t) \, dt \in BV[a, b]$.

Recall that differentiation and integration are inverse operations to each other on the space of smooth functions. That is, if f is continuous on $\mathbb{R}$, we have

$$\frac{d}{dx} \int_a^x f(t) \, dt = f(x).$$

Conversely, if the derivative function f' is continuous, then the integration of the derivative coincides the original function:

$$\int_a^x f'(t) \, dt = f(x) - f(a).$$

Those are fundamental theorems of the calculus for continuous functions. We will investigate next the fundamental theorems for the calculus for Lebesgue integrable functions. First, we consider whether

$$\frac{d}{dx} \int_a^x f(t) \, dt = f(x)$$

holds for an integrable function f on $[a, b]$. Of course, we may expect to have the following:

$$\frac{d}{dx} \int_a^x f(t) \, dt = f(x) \text{ a.e. on } [a, b].$$

Lemma 4.1.1 If f is integrable on $[a, b]$ and $\int_0^x f(t) \, dt = 0$ for any $x \in [a, b]$, then $f = 0$ a.e. on $[a, b]$.

Proof: If $f > 0$ on a set E with $m(E) > 0$, then there exists a closed set $F \subset E$ such that $m(F) > 0$. It is easy to see that $\int_E f(x) \, dx > 0$ (see exercise 2.3 in

Chapter 3). Let $O = [a, b] \setminus E$. Then either $\int_a^b f(x)dx \neq 0$, or else

$$0 = \int_a^b f(x)\,dx = \int_F f\,dx + \int_O f(x)\,dx,$$

and

$$\int_O f(x)\,dx = -\int_F f(x)\,dx \neq 0.$$

From the open set structure, we know that O is a union of countably many disjoint open intervals: $O = \cup_n(a_n, b_n)$ and thus

$$\int_O f(x)\,dx = \sum_n \int_{a_n}^{b_n} f(x)\,dx.$$

Hence, for some n we have $\int_{a_n}^{b_n} f \neq 0$. Then, either $\int_a^{a_n} f(x)dx \neq 0$ or $\int_a^{b_n} f(x)\,dx \neq 0$. In any case we see there is an x such that $\int_a^x f(t)\,dt \neq 0$ if $f > 0$ on a set of positive measure, and similarly for $f < 0$ on a set of positive measure. Therefore, $f = 0$ a.e. on $[a, b]$. ∎

Let $F(x) = \int_a^x f(t)\,dt$ for an integrable function f. If $f \geq 0$, then $F(x)$ is absolutely continuous (by Theorem 3.2.3) and increasing, and, therefore, F' exists a.e. (by Theorem 4.1.3).

Suppose f is bounded, that is, $0 \leq f \leq K$ for a constant K. Define

$$g_n(x) = \frac{F(x + 1/n) - F(x)}{1/n}.$$

Then $g_n(x) \geq 0$ and $g_n(x) = n \int_x^{x+1/n} f(t)\,dt \leq K$ and $g_n(x) \to F'(x)$ as $n \to \infty$ (since $F \nearrow$, F' exists a.e.).

Applying the Bounded Convergence Theorem (Theorem 3.1.5), we have

$$\lim_{n \to \infty} \int_a^c g_n(x)\,dx = \int_a^c F'(x)\,dx$$

for any $c \in [a, b]$.

On the other hand,

$$\int_a^c g_n(x) = \int_a^c \frac{F(x + 1/n) - F(x)}{1/n}\,dx$$

$$= -n \int_a^{a+1/n} F(x)\,dx + n \int_c^{c+1/n} F(x)\,dx$$

$$= -F(\xi_n) + F(\eta_n),$$

where $\xi_n \in (a, a + 1/n), \eta_n \in (c, c + 1/n)$ according to the mean value theorem for continuous functions. Therefore, as $n \to \infty$ we have

$$\lim_{n\to\infty} \int_a^c g_n(x)\, dx = F(c) - F(a) = \int_a^c f(x)\, dx.$$

So $\int_a^c F'(x)\, dx = \int_a^c f(x)\, dx$ for any $c \in [a, b]$. That is, $\int_a^c (F'(x) - f(x))\, dx = 0$. By Lemma 4.1.1, we have $F'(x) = f(x)$ a.e.

If f is assumed only to be nonnegative, then let $f_n(x) = [f(x)]_n$ where $n \in \mathbb{N}$. $\langle f_n(x) \rangle$ is an increasing sequence converges to f and thus, $\langle F_n \rangle$ also increasingly converges to F, where

$$F_n(x) = \int_a^x f_n(t)\, dt$$

and $F(x) = \int_a^x f(t)dt$ by the Monotone Convergence Theorem (Theorem 3.3.3). Let $G_n(x) = F(x) - F_n(x)$. Then $G_n(x)$ is a monotone increasing function of x and $G_n'(x) \geq 0$ a.e. We have

$$F(x) = G_n(x) + F_n(x), \text{ and } F'(x) = G_n'(x) + F_n'(x).$$

Clearly, $F'(x) \geq F_n'(x)$ a.e. on $[a, b]$.

Since $f_n(x)$ is increasing on x, again from Theorem 4.1.3 we have $f_n(x) = F_n'(x)$ a.e. and thus,

$$f(x) = \lim_{n\to\infty} f_n(x) = \lim_{n\to\infty} F_n'(x) \leq F'(x)$$

a.e. on $[a, b]$. Therefore, for any $c \in [a, b]$

$$F(c) - F(a) = \int_a^c f(x)\, dx \leq \int_a^c F'(x)\, dx.$$

However, F is increasing on x. From the last part of Theorem 4.1.3, we know that $\int_a^c F'(x)\, dx \leq F(c) - F(a)$. Therefore, $\int_a^c F'(x)\, dx = F(c) - F(a) = \int_a^c f(x)\, dx$ and so, $F' = f$ a.e. on $[a, b]$.

For a general function f, we notice that $f = f^+ - f^-$. The reasoning is still valid because Lemma 4.1.1 is true for any integrable function on $[a, b]$. Therefore, we have proved the following.

Theorem 4.1.6 If f is integrable on $[a, b]$ and $F(x) = F(a) + \int_a^x f(t)\, dt$, then $F' = f$ a.e. on $[a, b]$.

So far, we have proved the first part of the fundamental theorem of calculus: for a function $f \in L[a, b]$, its *indefinite (Lebesgue) integral* $F(x) := \int_a^x f(t)\, dt$ is differentiable a.e. on $[a, b]$ and at these points $F'(x) = f(x)$. Next we would like to look at the converse question: what function $F : [a, b] \mapsto \mathbb{R}$ is the indefinite integral of its derivative? To answer this question, we introduce a new class of functions which are "stronger" continuous than continuous functions, called absolutely continuous functions.

Definition 4.1.4 f is said to be *absolutely continuous* on $[a, b]$, if for all $\epsilon > 0$, there exists $\delta > 0$, such that if $\{(x_i, y_i)\}_{i=1}^{k}$ is a disjoint collection of intervals with $\sum_{i=1}^{k}(y_i - x_i) < \delta$, then

$$\sum_{i=1}^{k} |f(y_i) - f(x_i)| < \epsilon.$$

The class of absolutely continuous functions on $[a, b]$ is denoted by $AC[a, b]$.

It is easy to see directly from the definition that if $f \in AC[a, b]$, then $f \in C[a, b]$.

Theorem 4.1.7 If $f \in AC[a, b]$, then $f \in BV[a, b]$.

Proof: $f \in AC[a, b]$, take $\epsilon = 1$, there exists $\delta > 0$, such that $\sum_{i=1}^{k} |f(b_i) - f(a_i)| < 1$ as long as $\sum_{i=1}^{k} |b_i - a_i| < \delta$ for the disjoint intervals (a_i, b_i), $i = 1, \cdots, k$.

Now, consider any partition $\triangle$ of $[a, b]$: $a = x_0 < x_1 < \cdots x_m = b$ with its meshsize $|\triangle| = \max_j\{|x_{j+1} - x_j|\} < \delta$. It is clear from the absolutely continuity that $T_{x_j}^{x_{j+1}}(f) < 1$ and thus,

$$T_a^b(f) = \sum_{j=0}^{n} T_{x_j}^{x_{j+1}}(f) < m < \infty.$$

That is, $f \in BV[a, b]$. ∎

Lemma 4.1.2 If $f \in AC[a, b]$ and $f' = 0$ a.e. then $f = $ constant.

Proof: We leave the proof as an exercise. ∎

Theorem 4.1.8 $F \in AC[a, b]$, if and only if F is an indefinite integral of a Lebesgue integrable function f, i.e.,

$$F(x) = F(a) + \int_a^x f(t)\, dt.$$

Proof: If F is an indefinite integral, then F is absolutely continuous from Theorem 3.2.3. Suppose on the other hand that F is absolutely continuous on

$[a, b]$. Then F is in $BV[a, b]$ and thus, it can be expressed as the difference of two increasing functions:

$$F(x) = F_1(x) - F_2(x)$$

and

$$|F'(x)| \leq F_1'(x) + F_2'(x).$$

Thus,

$$\int_a^b |F'(x)|\, dx \leq F_1(b) + F_2(b) - F_1(a) - F_2(a).$$

Therefore, F' is integrable on $[a, b]$. Let $G(x) = \int_a^x F'(t)dt$. Then $G \in AC[a, b]$. Applying Theorem 4.1.3 to the function $f' = F' - G'$, we obtain that $f' = 0$ a.e. and thus f is a constant by Lemma 4.1.1. Therefore,

$$F(x) = \int_a^x F'(t)\, dt + F(a). \qquad \blacksquare$$

Corollary 4.1.3 Every AC function F is the indefinite integral of its derivative. In fact, $F(x) = F(a) + \int_a^x F'(t)\, dt$.

Exercises

1. Show that (a) $D^+[-f(x)] = -D_+f(x)$. (b) $D^+[f(-x)] = -D_-f(-x)$.
2. Suppose f is defined on an open interval containing a and f attains its minimum at a point a. Show that $D_+f(a) \geq 0 \geq D^-f(x)$.
3. If f is monotone on $[a, b]$, then the set $\{x \in [a, b] \mid J(f, x) > 0\}$ is countable.
4. Prove that if f is continuous on $[a, b]$ and $D^+f(x) \geq 0$ for all $x \in (a, b)$ then f is monotone nondecreasing on $[a, b]$. [Hint: consider g with $D^+g \geq \epsilon > 0$ first and then apply it to $g(x) = f(x) + \epsilon x$.]
5. Let f be differentiable on (a, b) such that f' is bounded, then $f \in BV[a, b]$.
6. Show that the function

$$f(x) = \begin{cases} x \sin \frac{\pi}{x}, & x \neq 0; \\ 0, & x = 0 \end{cases}$$

is uniformly continuous on $[0, 1]$ but is not in $BV[0, 1]$.

7. Let f be bounded variation on $[a, b]$. Show that

$$\int_a^b |f'(x)|dx \leq T_a^b(f).$$

8. Prove that if f is an absolutely continuous function on $[a, b]$, then $T_a^b(f) = \int_a^b |f'|dx$.

9. Prove that a function f satisfying a Lipschitz condition is of bounded variation and absolutely continuous.

10. Let f be an integrable function on E with $\int_E f(x)dx = r > 0$. Then there is a subset $e \subset E$ such that $\int_e f(x)dx = \frac{r}{3}$. [Hint: Consider the function $\int_{E \cap [a,x]} f(t)dt$.]

11. A function f on $[a, b]$ is said to be *convex* if for any $x, y \in [a, b]$ and $r \in [0, 1]$, we have

$$f(rx + (1 - r)y) \leq rf(x) + (1 - r)f(y).$$

Show that for convex function f:
  (i) For any points x, x', y, y' satisfying $x \leq x' \leq y \leq y'$, we have

$$\frac{f(y) - f(x)}{y - x} \leq \frac{f(y') - f(x')}{y' - x'}.$$

  (ii) f' exists a.e. on $[a, b]$.
  (iii) If f has second derivative then f is convex, if and only if $f''(x) \geq 0$ for all x.

12. (Jensen's Inequality) Let f be a convex function on $(-\infty, \infty)$ and g an integrable function on $[0,1]$. Then

$$\int_0^1 f(g(t)) \, dt \geq f\left(\int_0^1 g(t)dt\right).$$

Let g be an integrable function on $[0,1]$. Then

$$\int_0^1 e^{g(t)}dt \geq e^{\int_0^1 g(t) \, dt}.$$

13. If $f \in \text{Lip } \alpha$, then $f \in AC[a, b]$.

14. If $f \in AC[a, b]$, then $f \in BV[a, b]$ and $T_a^b(f) = \int_a^b |f'| \, dx$.

15. Prove Lemma 4.1.2.

2 Mathematical Models for Probability

In this section, we briefly describe probability models which use measure theory and Lebesgue integration. A measure theoretic foundation for probability theory was set up by a Russian mathematician, Andrei Kolmogorov (1903–1987). It has been widely accepted that probabilities should be studied as special sorts of measures.

Imagine that some experiment involving random chance takes place, and use the symbol Ω, called the *sample space*, to denote the set of all possible outcomes. An *event* is a collection of outcomes. We say that an event occurs when any of the outcomes that belong to it occur. The probability of an event A gives a scale that measures opinions on the likehoods of events happening. If $\mathcal{E}$ denotes the collection of events, then (i) $\Omega \in \mathcal{E}$, (ii) whenever $A \in \mathcal{E}$, then $A^c \in \mathcal{E}$, and (iii) if $\langle A_k \rangle$ are events in $\mathcal{E}$, then $\cup_k A_k \in \mathcal{E}$.

There are three fundamental laws of probability. Probabilities are real numbers and whatever the event A, its probability $P(A)$ satisfies (1) $0 \leq P(A) \leq 1$, (2) $P(\Omega) = 1$, and (3) for pairwise disjoint events $A_k, k = 1, \cdots, P(\cup A_k) = \sum_k P(A_k)$. This property reminds us of the countable additivity of a measure. In general, in probability theory, for the collection $\mathcal{E}$ of μ-measurable sets on the sample space Ω, the triple $(\Omega, \mathcal{E}, \mu)$ is usually called a *measurable space*. If $\mu(\Omega) < \infty$, then μ is called a *finite measure* and $(\Omega, \mathcal{E}, \mu)$ is called a *finite measurable space*. If $\mu(\Omega) = 1$, then μ is called a *probability measure* and $(\Omega, \mathcal{E}, \mu)$ is called a *probability space*. As mentioned above, the class $\mathcal{E}$ of events usually is a σ-algebra.

> **EXAMPLE 4.2.1:** Let $X = \{\omega_k \mid k \in \mathbb{N}\}$ be a countable set and take $\mathcal{E}$ to be the collection of all subsets of X. Let $\sum_{k=1}^{\infty} p_k$ be a series of nonnegative terms, and assume that the series converges to 1. Define a set function μ on $\mathcal{E}$ by $\mu(\emptyset) = 0$ and $\mu(E) = \sum_{\omega_{k_i} \in E} p_{k_i}$, where the term p_{k_i} is included in the sum if the corresponding ω_{k_i} is in E. The function μ is a probability measure. The probability space $(X, \mathcal{E}, \mu)$ serves to describe any probability experiment of a discrete nature.

Many experiments with finitely many outcomes have such a degree of symmetry that it is reasonable to assume that all these outcomes are equally likely.

> **EXAMPLE 4.2.2:** Some typical examples of discrete probability models are:
>
> (1) The binomial distribution:

$$\omega = 0, 1, \cdots, n, \ P(\omega = k) = \binom{n}{k} p^k (1-p)^{n-k}, \ p \in [0,1].$$

(2) The Poisson distribution:

$$\omega = 0, 1, \cdots, \ P(\omega = k) = \frac{\lambda^k}{k!} e^{-\lambda}, \ \lambda > 0.$$

Because a probability space is a finite measure space, we can immediately infer for a probability space any properties of a finite measure space. For future reference, we list some important properties of probability in the following.

Theorem 4.2.1 Suppose that $(\Omega, \mathcal{E}, P)$ is a probability space. If $A, B \in \mathcal{E}$, then the following hold.

(i) If $A \subset B$, then $P(A) \leq P(B)$ and $P(B \setminus A) = P(B) - P(A)$.

(ii) $P(A^c) = 1 - P(A)$.

(iii) $P(A \cup B) = P(A) + P(B) - P(A \cap B)$.

(iv) If $E_k \in \mathcal{E}$, then $P(\cup E_k) \leq \sum_k P(E_k)$.

A real-valued measurable function defined on Ω is called a *random variable* and is usually denoted by symbols such as $X, Y \cdots$ rather than the symbols $f, g, \cdots$. Subsets of Ω such as $\{\omega \mid X(\omega) \leq x\}$ are written as $\{X \leq x\}$ and the probability $P(\{X \leq x\})$ is written simply as $P(X \leq x)$.

A *distribution function* corresponding to a random variable X is the function defined by

$$F_X(x) = P(X \leq x).$$

It is easy to see that the distribution function is nondecreasing and continuous from the right; moreover, it is bounded since $F_X(-\infty) = 0$ and $F_X(\infty) = 1$.

A random variable X is said to be *(absolutely) continuous* if and only if its distribution function F_X is (absolutely) continuous on $\mathbb{R}$. From Theorem 4.1.3 we see that F_X is differentiable a.e. on $x \in \mathbb{R}$. Applying Theorem 4.1.8, we can show the following.

Theorem 4.2.2 If X is a random variable and if B is a Borel set of $\mathbb{R}$, then

$$P(X \in B) = \int_B F_X'(x) \, dx. \tag{2.1}$$

Proof: Let $\mathcal{E}$ be the collection of sets for which (2.1) is valid. Then it is easy to verify that $\mathcal{E}$ is a σ-field. From the definition of F_X, we see that all intervals of the form $(-\infty, x]$ are members of $\mathcal{E}$, and thus by considering complements and unions it follows that all intervals are in $\mathcal{E}$. So E is a σ-algebra, and it contains the Borel sets. This, together with Theorem 4.1.8, yields the conclusion.   ∎

The derivative function $f(x) := F'_X(x)$ is called the *density function* of random variable X. Since F_X is increasing and $F_X(\infty) = 1$, we obtain the basic properties of a density function f: (i) $f \geq 0$ and (ii) $\int_{-\infty}^{\infty} f(x)\,dx = 1$.

EXAMPLE 4.2.3: Some typical examples of density functions are:
(1) The uniform density on $[a, b]$:

$$f(x) = \begin{cases} \frac{1}{b-a}, & x \in [a, b], \\ 0, & \text{elsewhere.} \end{cases}$$

(2) The exponential density ($\lambda > 0$):

$$f(x) = \begin{cases} \lambda e^{-\lambda x}, & x \geq 0, \\ 0, & \text{elsewhere.} \end{cases}$$

(3) The normal density function:

$$f(x) = \frac{1}{\sqrt{2\pi\sigma}} \exp\left[-\frac{(x-\mu)^2}{2\sigma^2} \right], \ \sigma > 0, \ \mu \in \mathbb{R}.$$

Events A and B are said to be *independent* if $P(A \cap B) = P(A)P(B)$. For more than two events, we should be very careful to distinguish between two types of independence, *pairwise independence* and *mutual independence*. Events $A_k, k = 1, \cdots, A_n$, are said to be *pairwise independent* if for $i \neq j$, A_i and A_j are independent. However, the concept of mutual independence means that for any subset of $\{1, 2, \cdots, n\}$, say $\{i_1, \cdots i_k\}$, we have

$$P(A_{i_1} \cap \cdots \cap A_{i_k}) = P(A_{i_1}) \cdots P(A_{i_k}).$$

Clearly, mutual independence implies pairwise independence, but the converse is not true.

Using properties of a measure, we can prove the following results on limits of events.

Theorem 4.2.3 [Borel–Canteli Lemma] Suppose that $(\Omega, \mathcal{E}, P)$ is a probability space and that $E_n \in \mathcal{E}$ for $n = 1, 2, \cdots$.
(1) If $\sum_n P(E_n) < \infty$, then $P(\overline{\lim}_{n\to\infty} E_n) = 0$.
(2) If $E_1, \cdots, E_n, \cdots$ are mutually independent and $\sum_{n=1}^{\infty} P(E_n) = \infty$, then $P(\overline{\lim}_{n\to\infty} E_n) = 1$.

Proof: We leave the proof as an exercise. ∎

Definition 4.2.1 Let $(\Omega, \mathcal{E}, P)$ be a probability space and $E \in \mathcal{E}$ with $P(E) > 0$. Then for $F \in \mathcal{E}$, the *conditional probability* of F given E is defined to be

$$P(F|E) = \frac{P(F \cap E)}{P(E)}.$$

Corollary 4.2.1 If events E and F are independent and $P(E) > 0$, then $P(F|E) = P(F)$.

Let $X_k, k = 1, \cdots, n$ be n random variables defined on the same probability space $(\Omega, \mathcal{E}, P)$. Then the *joint probability distribution* of $X := (X_1, \cdots, X_n)$ is

$$F(x) = F(x_1, x_2, \cdots, x_n) = P(x; X_1 \leq x_1, \cdots, X_n \leq x_n),$$

where $x = (x_1, \cdots, x_n) \in \mathbb{R}^n$.

Theorem 4.2.4 The random variables $X_k, k = 1, \cdots, n$, are independent if and only if

$$F(x_1, x_2, \cdots, x_n) = F_1(x_1)F_2(x_2) \cdots F_n(x_n), \text{ for all } x_1, \cdots, x_n.$$

We may also characterize independence in terms of densities.

Theorem 4.2.5 If $X = (X_1, \cdots, X_n)$ has a density f and each X_i has a density f_i, then $X_1, \cdots, X_n$ are independent if and only if

$$f(x_1, \cdots, x_n) = f_1(x_1) \cdots f_n(x_n)$$

for all $(x_1, \cdots, x_n) \in \mathbb{R}^n$ except possibly for a Borel set of measure zero.

In the study of probability distributions, we would like to examine the center point and the overall spread of information of the distribution. These usually can be obtained by computing the so-called *mean value* and *standard deviation*.

Consider a random variable taking only finitely many values: $x_1, \cdots, x_n$ with probabilities $p_1, \cdots, p_n$, respectively. The "average" value

$$\mu := \frac{1}{n} \sum_{k=1}^{n} p_k x_k$$

is called the *expected value* of X. The *variance* of X is

$$\sigma^2 := (x_1 - \mu)^2 p_1 + (x_2 - \mu)^2 p_2 + \cdots + (x_k - \mu)^2 p_k = \sum_{i=1}^{n}(x_i - \mu)^2 p_i.$$

The *standard deviation* σ is the square root of the variance.

Suppose that X is a continuous random variable with probability density $f(x)$. The *expected value* of X is

$$\mu = E(X) = \int xf(x)\,dx$$

and the *variance* of X is

$$\text{Var}(X) = \int (x - \mu)^2 f(x)\,dx,$$

where the integrals are taken over all possible values of X. Again, the *standard deviation* σ is the square root of the variance.

The following properties of expected value and variance (standard deviation) are easy to verify.

Theorem 4.2.6 Assume that X and Y are random variables and $c \in \mathbb{R}$. Then
   (1) $E(X + Y) = E(X) + E(Y)$.
   (2) $E(cX) = cE(X)$.
   (3) $E(c) = c$.
   (4) $\text{Var}(X) = E(X^2) - [E(X)]^2$.

   EXAMPLE 4.2.4:

   (i) If X is a binomial distribution, then $E(X) = np$ and $\text{Var}(X) = np(1 - p)$.
   (ii) If X is a normal distribution, then $E(X) = \mu$ and $\text{Var}(X) = \sigma^2$.

Definition 4.2.2 Let X be a random variable on a probability space $(\Omega, \mathcal{E}, P)$. If $k > 0$, then the number $E(X^k)$ is called the kth *moment* of X.

Note that $E(X^k)$ is finite if and only if $E(|X|^k)$ is finite. The following shows that the finiteness of the kth moments implies finiteness of the r moments.

Lemma 4.2.1 If $k > 0$ and $E(X^k)$ is finite, then $E(X^j)$ is finite for $0 < j < k$.

Proof:

$$E(|X|^j) = \int |x|^j f(x)\, dx$$

$$= \int_{\{x|^j < 1\}} + \int_{\{x|^j \geq 1\}} (|x|^j f(x))\, dx$$

$$\leq P(|X|^j < 1) + \int_\Omega |x|^k f(x)\, dx < \infty. \qquad \blacksquare$$

The following result is related to measure theory.

Theorem 4.2.7 If X is a random variable with $E(|X|^p) < \infty$, then for any $\epsilon > 0$, we have *Markov's inequality*

$$P(|X| \geq \epsilon) \leq \frac{E(|X|^p)}{\epsilon^p}.$$

In particular, if X has a finite mean value and variance σ^2, we have *Chebyshev's inequality*

$$P(|X - E(X)| \geq \epsilon) \leq \frac{\sigma^2}{\epsilon^2}.$$

Proof:

$$P(|X| \geq \epsilon) = \int_{|x| \geq \epsilon} f(x)\, dx$$

$$\leq \int_{|x| \geq \epsilon} \frac{|x|^p}{\epsilon^p}\, dx \leq \frac{1}{\epsilon^p} \int_{-\infty}^{\infty} |x|^p f(x)\, dx = \frac{E(|X|^p)}{\epsilon^p}.$$

Choosing $p = 2$ and replacing X by $X - E(X)$, we obtain the Chebyshev inequality. $\qquad \blacksquare$

The following result is a direct consequence of Fubini's Theorem (Theorem 3.4.2).

Theorem 4.2.8 Let $X_1, \cdots, X_n$ be independent random variables on $(\Omega, \mathcal{E}, P)$. If all X_i are nonnegative or if $E(X_i)$ is finite for each i, then $E(X_1 \cdots X_n)$ exists and is equal to $E(X_1) \cdots E(X_n)$.

Exercises

1. Set up a probability space for tossing a fair coin three times. What is the chance all three tosses give the same result?
2. Prove Theorem 4.2.1.
3. Let f be a nonnegative integrable function and assume that $\int_{-\infty}^{\infty} f(x)\, dx = 1$. Prove that there exists a random variable having f as its density function.
4. For the normal distribution density

$$f(x) = \frac{1}{\sqrt{2\pi}\sigma} \exp\left[-\frac{(x-\mu)^2}{2\sigma^2}\right], \ \sigma > 0, \ \mu \in \mathbb{R}.$$

   Verify that $\int_{-\infty}^{\infty} f(x) = 1$. [Hint: Use $[\int f(x)\,dx]^2 = \int f(x)\,dx \int f(y)\,dy$ and a polar coordinate substitution.]
5. Let X be an exponential random variable with parameter λ. Find the probability distribution function $F(x)$.
6. Show that if A and B are independent events, then so are the events A^c and B^c.
7. Let S and T be independent with $P(S) = P(T)$ and $P(S \cup T) = \frac{1}{2}$. What is $P(S)$?
8. Assume that the proportion of commercial vehicles among users of the Humber Bridge varies randomly from day to day, with density $f(x) = cx(1-x)^2$ for $0 < x < 1$, where c is a constant. Show that $c = 12$, then find the distribution function. On what fraction of days is the proportion of commercial vehicles between 20% and 50%?
9. Prove the Borel–Canteli Lemma. [Hint: In the second part, consider the complements and use the fact that for $x > 0, e^{-x} \geq 1 - x$.]
10. Show that $E(X) = \lambda$ and $\mathrm{Var}(X) = \lambda^2$ if X is a Poisson distribution.
11. Evaluate $E(X)$ and $\mathrm{Var}(X)$ for the exponential distribution.
12. If $X_1, \cdots, X_n$ are random variables with finite $E(X_i X_j)$ for all i, j with $i \neq j$, then prove

$$\mathrm{Var}(X_1 + \cdots + X_n) = \sum_{i-1}^{n} \mathrm{Var}(X_i) + 2 \sum_{1 \leq i < j \leq n}^{n} \mathrm{Cov}(X_i, X_j),$$

where $\mathrm{Cov}(X, Y) = E[(X - E(X))(Y - E(Y))] = E(XY) - E(X)E(Y)$ is the *covariance* of X and Y.

13. If $X_1, \cdots, X_n$ are random variables with finite $E(X_i X_j)$ for all i, j with $i \neq j$, then prove for $a_i \in \mathbb{R}, i = 1, \cdots, n,$

$$\mathrm{Var}(a_1 X_1 + \cdots + a_n X_n) =$$
$$\sum_{i-1}^{n} a_i^2 \mathrm{Var}(X_i) + 2 \sum_{1 \leq i < j \leq n} a_i a_j \mathrm{Cov}(X_i, X_j).$$

14. If $X_1, \cdots, X_n$ are pairwise independent random variables with finite variances, then for $a_i \in \mathbb{R}, i = 1, \cdots, n,$

$$\mathrm{Var}(a_1 X_1 + \cdots + a_n X_n) = \sum_{i-1}^{n} a_i^2 \mathrm{Var}(X_i).$$

3 Convergence and Limit Theorems

In this section, we will discuss convergence of random variables and distributions.

Definition 4.3.1 Let $\langle X_n(\omega) \rangle$ be a sequence of random variables on the probability space $(\Omega, \mathcal{E}, P)$.
   (i) If $X_n \to X$ a.e. on Ω, that is,

$$P\left(\omega; \lim_{n \to \infty} X_n(\omega) = X(\omega)\right) = 1$$

   a.e. on Ω, then X_n is said to be convergent to X *almost surely*.
   (ii) If $X_n \to X$ in probability measure P ($X_n \overset{P}{\to} X$), that is,

$$\lim_{n \to \infty} P(\omega; |X_n(\omega) - X(\omega)| \geq \epsilon) = 0$$

   for every $\epsilon > 0$, then $\langle X_n \rangle$ is said to X *in probability*.

The following are corollaries of the corresponding theorems in Chapter 3.

Theorem 4.3.1

(1) Random variable sequence $X_n \to X$ almost surely if and only if for $\epsilon > 0$,

$$\lim_{n \to \infty} P\left(\cup_{k=n}^{\infty}(|X_k - X| \geq \epsilon)\right) = 0.$$

(2) If $X_n \to X$ almost surely, then $X_n \xrightarrow{P} X$.

From the fact that convergence in measure cannot imply convergence a.e., we know that convergence in probability does not imply convergence almost surely.

Definition 4.3.2 Let $\langle X_n(\omega) \rangle$ be a sequence of random variables on the probability space $(\Omega, \mathcal{E}, P)$. $F_n(x)$ are the corresponding distribution functions of X_n, $n \in \mathbb{N}$.

(i) If there is a nondecreasing function $F(x)$ such that

$$\lim_{n \to \infty} F_n(x) = F(x)$$

holds for every point of continuity of F, then $\langle F_n \rangle$ is said to *weakly converge* to F, denoted by $F_n \to F$ (w).

If F becomes a probability distribution function of a random variable X on $(\Omega, \mathcal{E}, P)$, then we say that $\langle X_n \rangle$ *weakly converges* to X, denoted by $X_n \to X$ (w).

(ii) Let the random variables X_n and X satisfy $E(|X_n|^r) < \infty$ and $E(|X|^r) < \infty$ for $r > 0$. If

$$\lim_{n \to \infty} E(|X_n - X|^r) = 0,$$

then $\langle X_n \rangle$ is said to converges to X *of order r* (or *in the rth mean*), denoted by $X_n \to X$ (order r).

EXAMPLE 4.3.1:

(1) The limit of distribution function F_n may not be a probability distribution at all. Define

$$F_n(x) = \begin{cases} 0, & x < n, \\ 1, & x \geq n \end{cases}$$

and $F(x) = 0$. Then we have $F_n \to F$ (w), but F is not a distribution function.

(2) A random variable sequence may converge weakly, but may not converge in probability. For example, let X and X_n be independent discrete random variables with identical probability distribution: $p(0) = p(1) = \frac{1}{2}$. Then, of course $X_n \to X$ (w) since they have identical distributions. But for $\epsilon > 0$,

$$P(|X_n - X| > \epsilon) = P(X_n = 1, X = 0) + P(X_n = 0, X = 1)$$
$$= P(X_n = 1)P(X = 0) + P(X_n = 0)P(X = 1)$$
$$= \frac{1}{2} \cdot \frac{1}{2} + \frac{1}{2} \cdot \frac{1}{2} = \frac{1}{2}.$$

Therefore, $X_n \not\to X$ in probability.

Theorem 4.3.2 If the random variables $X_n \to X$ in probability, then $X_n \to X$ weakly.

Proof: For $x, y \in \mathbb{R}$, we have

$$F(y) \leq F_n(x) + P(X_n > x, X \leq y)$$

since

$$(X \leq y) = (X_n \leq x, X \leq y) \cup (X_n > x, X \leq y)$$
$$\subset (X_n \leq x) \cup (X_n > x, X \leq y).$$

From $X_n \to X$ in probability, we see that for $y < x$,

$$P(X_n > x, X \leq y) \leq P(|X_n - X| \geq x - y) \to 0$$

as $n \to \infty$. Therefore,

$$F(y) \leq \underline{\lim}_{n \to \infty} F_n(x).$$

Similarly, we can have for $x < z$,

$$\overline{\lim}_{n \to \infty} F_n(x) \leq F(z).$$

Hence for $y < x < z$,

$$F(y) \leq \underline{\lim}_{n \to \infty} F_n(x) \leq \overline{\lim}_{n \to \infty} F_n(x) \leq F(z).$$

Therefore, for any point of continuity x of F, we have

$$\lim_{n \to \infty} F_n(x) = F(x). \qquad \blacksquare$$

We leave the proof of the following result to the reader.

Theorem 4.3.3 If distribution functions $F_n \to F$ (w) and F is a continuous probability distribution function, then $\lim_n F_n(x) = F(x)$ uniformly on $\mathbb{R}$.

For the relationship between convergence of order r and convergence in probability, we have the following.

Theorem 4.3.4 If $X_n \to X$ (order r), then $X_n \xrightarrow{P} X$.

Proof: The conclusion follows from the Markov inequality

$$P(|X_n - X| \geq \epsilon) \leq \frac{E(|X_n - X|^r}{\epsilon^r}. \qquad \blacksquare$$

The converse of Theorem 4.3.4 is not true (exercise).

In summary, we have the following implications of convergences:

Corollary 4.3.1 On $(\Omega, \mathcal{E}, P)$, convergence almost surely (or convergence of order r) implies convergence in probability, and convergence in probability implies convergence weakly.

Next, let $\langle X_n \rangle$ be random variables on the same probability space $(\Omega, \mathcal{E}, P)$ which are independent with identical distribution (iid). Write $S_n = \sum_{k=1}^{n} X_k$, $\bar{X} = \frac{S_n}{n}$ and $S^2 = \sum_{k=1}^{n} \frac{(X_i - \bar{X})^2}{(n-1)}$. We would like to investigate the probability distribution of $\bar{X}$ when n is large.

Lemma 4.3.1 Suppose $\langle X_n \rangle$ is a sequence of iid random variables, with $E(X_i) = \mu$ and $\text{Var}(X_i) = \sigma^2$ both finite. Then $E(\bar{X}) = \mu$ and

$$\text{Var}(\bar{X}) = \frac{\sigma^2}{N}, \text{ and } E(S^2) = \sigma^2.$$

Proof: We leave the proof as an exercise. $\qquad \blacksquare$

Theorem 4.3.5 [Weak Law of Large Numbers] Suppose $\langle X_n \rangle$ are iid with mean μ and finite variance σ^2. Then $\bar{X} \xrightarrow{P} \mu$.

Proof: Using Chebyshev's inequality and Lemma 4.3.1, we have

$$P(|\bar{X} - \mu| > \epsilon) \leq \frac{\sigma^2}{n\epsilon^2}.$$

The conclusion follows by letting $n \to \infty$. ■

Using Markov's inequality and the Borel–Canteli Lemma, we can prove the Strong Law of Large Numbers:

Theorem 4.3.6 [Strong Law of Large Numbers] Suppose $\langle X_n \rangle$ are iid with mean μ and finite fourth moment $E(|X_i|^4)$. Then $\bar{X} \xrightarrow{P} \mu$ almost surely.

Proof: Notice that

$$E\left((S_n - n\mu)^4\right) = E\left(\sum_{i=1}^{n}(X_i - \mu)^4\right) + E\left(\sum_{i<j}(X_i - \mu)^2(X_j - \mu)^2\right),$$

as all other terms in the expansion are zero, based on their independence. Thus,

$$E((S_n - n\mu)^4) = nE((X_1 - \mu)^4) + 3n(n-1)\sigma^4 \leq Cn^2$$

for some constant C.

From Markov's inequality, we have

$$P\left(|\frac{S_n}{n} - \mu| > \epsilon\right) = P((S_n - n\mu)^4 > n^4\epsilon^4) \leq \frac{Cn^2}{n^4\epsilon^4} = \frac{C_1}{n^2},$$

where $C_1 = C/\epsilon^4$. Since $\sum_n \frac{1}{n^2}$ converges, the first part of the Borel–Cantelli Lemma shows that, with probability one, only finitely many of the events $|S_n/n - \mu| > \epsilon$ occur for any $\epsilon > 0$. Therefore, $S_n/n \to \mu$ almost surely. ■

Exercises

1. Prove that if $X_n \to X$ and $Y_n \to Y$, then $X_n + Y_n \to X + Y$, provided the mode of convergence is either in probability, almost surely, or of order r throughout.
2. Prove that if $f \in C(0, \infty)$ and f is bounded and strictly increasing with $f(0) = 0$, then $X_n \xrightarrow{P} 0$ if and only if $\lim_{n \to \infty} E(f|X_n|) = 0$.

3. Prove that if X_n is a Poisson distribution with parameter n, then $\frac{X_n - n}{\sqrt{n}}$ converges weakly to X having the normal distribution.

4. Show by an example that the converse of Theorem 4.3.4 is not true.

Chapter 5

Vector Spaces, Hilbert Spaces, and the L^2 Space

1 Groups, Fields, and Vector Spaces

The theory of vector spaces has a tremendous number of applications, as is seen in a standard sophomore level linear algebra class. In this section, we review the definitions and properties of finite dimensional vector spaces. We do so by taking an algebraic approach and start from scratch.

Definition 5.1.1 A *group* is a set of elements G along with a mapping (called a *binary operation*) $\star : G \times G \mapsto G$ such that

(1) There exists an element $e \in G$ such that for all $g \in G$, $e \star g = g \star e = g$. This element e is called the *identity element* of group G.

(2) For any element $g \in G$ there exists a unique element $h \in G$ such that $g \star h = h \star g = e$. Element h is the *inverse* of g and is denoted $h = g^{-1}$.

(3) For all $g, h, j \in G$, $g \star (h \star j) = (g \star h) \star j$. That is, $\star$ is *associative*.

We denote the group as $\langle G, \star \rangle$. If, in addition, for all $g, h \in G$, $g \star h = h \star g$ then G is an *Abelian* (or *commutative*) group. A *subgroup* of group $\langle G, \star \rangle$ is a subset S of G such that S is closed under $\star$.

EXAMPLE 5.1.1: Examples of *additive* groups include:

(a) The integers under addition modulo n: $\langle \mathbb{Z}_n, + \rangle$.

(b) The integers under addition: $\langle \mathbb{Z}, + \rangle$.

(c) The rational numbers under addition: $\langle \mathbb{Q}, + \rangle$.
(d) The real numbers under addition: $\langle \mathbb{R}, + \rangle$.
(e) The complex numbers under addition: $\langle \mathbb{C}, + \rangle$.

Notice that each of these groups is Abelian. Notice that $\langle \mathbb{Z}, + \rangle$ is a subgroup of $\langle \mathbb{Q}, + \rangle$, $\langle \mathbb{Q}, + \rangle$ is a subgroup of $\langle \mathbb{R}, + \rangle$, and $\langle \mathbb{R}, + \rangle$ is a subgroup of $\langle \mathbb{C}, + \rangle$.

Definition 5.1.2 A *field* is a set of elements $\mathbb{F}$ along with two mappings, called *addition*, denoted $+$, and *multiplication*, denoted $\cdot$, where $+ : \mathbb{F} \times \mathbb{F} \mapsto \mathbb{F}$ and $\cdot : \mathbb{F} \times \mathbb{F} \mapsto \mathbb{F}$, such that $\langle \mathbb{F}, + \rangle$ is an Abelian group with identity element 0 and $\langle \mathbb{F} \setminus \{0\}, \cdot \rangle$ is an Abelian group. The identity element of $\langle \mathbb{F} \setminus \{0\}, \cdot \rangle$ is denoted by 1 and called *unity*. We denote the field as $\langle \mathbb{F}, +, \cdot \rangle$. A *subfield* of $\langle \mathbb{F}, +, \cdot \rangle$ is a subset $\mathbb{S}$ of $\mathbb{F}$ such that $\langle \mathbb{S}, +, \cdot \rangle$ is a field.

EXAMPLE 5.1.2:   Examples of fields are:

(a) The integers modulo p, where p is prime, under addition and multiplication: $\langle \mathbb{Z}_p, +, \cdot \rangle$.
(b) The rational numbers under addition and multiplication: $\langle \mathbb{Q}, +, \cdot \rangle$.
(c) The real numbers under addition and multiplication: $\langle \mathbb{R}, +, \cdot \rangle$.
(d) The complex numbers under addition and multiplication: $\langle \mathbb{C}, +, \cdot \rangle$.

$\langle \mathbb{Z}_p, +, \cdot \rangle$ is an example of a *finite field* and the remaining examples are infinite fields. Notice that $\langle \mathbb{Q}, +, \cdot \rangle$ is a subfield of $\langle \mathbb{R}, +, \cdot \rangle$, and $\langle \mathbb{R}, +, \cdot \rangle$ is a subfield of $\langle \mathbb{C}, +, \cdot \rangle$.

We are now in the position to define a vector space. Since we have established a bit of algebraic background, we can give a definition slightly more concise than that encountered in a sophomore-level linear algebra class.

Definition 5.1.3 A *vector space* over field $\mathbb{F}$ (the elements of which are called *scalars*) is a set V of elements called *vectors* such that

(a) A mapping called *addition*, denoted $+$, is defined such that $+ : V \times V \mapsto V$ and $\langle V, + \rangle$ is an Abelian group. The identity element of this group is denoted **0**.
   We also define a mapping $\mathbb{F} \times V \mapsto V$ called *scalar multiplication*. For $f \in \mathbb{F}$ and $\mathbf{v} \in V$ we denote the scalar product of f and $\mathbf{v}$ as $f\mathbf{v}$. Scalar multiplication satisfies the following properties: for all $a, b \in \mathbb{F}$ and for all $\mathbf{u}, \mathbf{v} \in V$:

(b) $a(\mathbf{u} + \mathbf{v}) = a\mathbf{u} + a\mathbf{v}$ (distribution of scalar multiplication over vector addition),
(c) $(a + b)\mathbf{v} = a\mathbf{v} + b\mathbf{v}$ (distribution of scalar multiplication over scalar addition),

(d) $a(b\mathbf{v}) = (a \cdot b)\mathbf{v}$ (associativity of scalar multiplication),

(e) $1\mathbf{v} = \mathbf{v}$, and

(f) $0\mathbf{v} = \mathbf{0}$.

We denote this vector space as $\langle V, \mathbb{F} \rangle$.

EXAMPLE 5.1.3: Examples of vector spaces include:

(a) $\mathbb{Q}^n = \langle V, \mathbb{Q} \rangle$ where $V = \{(q_1, q_2, \ldots, q_n) \mid q_i \in \mathbb{Q} \text{ for } 1 \leq i \leq n\}$, and scalar multiplication and vector addition are defined componentwise.

(b) $\mathbb{R}^n = \langle V, \mathbb{R} \rangle$ where $V = \{(r_1, r_2, \ldots, r_n) \mid r_i \in \mathbb{R} \text{ for } 1 \leq i \leq n\}$, and scalar multiplication and vector addition are defined componentwise.

(c) $\mathbb{C}^n = \langle V, \mathbb{C} \rangle$ where $V = \{(c_1, c_2, \ldots, c_n) \mid c_i \in \mathbb{C} \text{ for } 1 \leq i \leq n\}$, and scalar multiplication and vector addition are defined componentwise.

(d) $\mathbb{F}^n = \langle V, \mathbb{F} \rangle$ where $V = \{(f_1, f_2, \ldots, f_n) \mid f_i \in \mathbb{F} \text{ for } 1 \leq i \leq n\}$, and scalar multiplication and vector addition are defined componentwise.

(e) $l^2(\mathbb{R}) = \langle V, \mathbb{R} \rangle$ where

$$V = \left\{ (r_1, r_2, r_3, \ldots) \mid r_i \in \mathbb{R} \text{ for } i \geq 1 \text{ and } \sum_{i=1}^{\infty} r_i^2 < \infty \right\},$$

and scalar multiplication and vector addition are defined componentwise.

(f) $l^2(\mathbb{C}) = \langle V, \mathbb{C} \rangle$ where

$$V = \left\{ (c_1, c_2, c_3, \ldots) \mid c_i \in \mathbb{C} \text{ for } i \geq 1 \text{ and } \sum_{i=1}^{\infty} |c_i|^2 < \infty \right\},$$

and scalar multiplication and vector addition are defined componentwise.

The reader is probably familiar with the vector spaces of Example 5.1.3 (b) and (c). However, the vector spaces of (e) and (f) may be new to you. It may not even be clear that the sets V in these examples are closed under vector addition. We will explore these examples in much more detail in the following sections, and find that they play a role as fundamental as the other examples.

We would like to classify vector spaces and see what they "look like." In that direction, we introduce several definitions.

Definition 5.1.4 Suppose $\langle V, \mathbb{F} \rangle$ is a vector space. A *linear combination* of vectors $\mathbf{v}_1, \mathbf{v}_2, \ldots, \mathbf{v}_n \in V$ is a sum of the form $f_1\mathbf{v}_1 + f_2\mathbf{v}_2 + \cdots + f_n\mathbf{v}_n$

where $f_1, f_2, \ldots, f_n \in \mathbb{F}$ are scalars. A set of vectors $\{v_1, v_2, \ldots, v_n\}$ is *linearly independent* if $f_1 v_1 + f_2 v_2 + \cdots + f_n v_n = 0$ only when $f_1 = f_2 = \cdots = f_n = 0$. The *span* of a set of vectors $\{v_1, v_2, \ldots, v_n\} \subset V$ is the set of all linear combinations of the vectors: $\mathrm{span}\{v_1, v_2, \ldots, v_n\} = \{f_1 v_1 + f_2 v_2 + \cdots + f_n v_n \mid f_1, f_2, \ldots, f_n \in \mathbb{F}\}$. A *basis* for a vector space is a linearly independent spanning set of the vector space. A vector space is *finite dimensional* if it has a basis of finite cardinality.

We follow the method of Lang [17] in our classification of finite-dimensional vector spaces. First, we need a preliminary result concerning systems of equations. We will use this result to show that all bases of a given finite-dimensional vector space are of the same cardinality.

Lemma 5.1.1 Consider the homogeneous system of equations

$$
\begin{array}{ccccccc}
a_{11}x_1 & + & a_{12}x_2 & + & \cdots & + & a_{1n}x_n & = & 0 \\
a_{21}x_1 & + & a_{22}x_2 & + & \cdots & + & a_{2n}x_n & = & 0 \\
\vdots & & & & \ddots & & \vdots \\
a_{m1}x_1 & + & a_{m2}x_2 & + & \cdots & + & a_{mn}x_n & = & 0
\end{array}
$$

with coefficients a_{ij} $(1 \le i \le m, 1 \le j \le n)$ and unknowns x_k $(1 \le k \le n)$ from field $\mathbb{F}$. If $n > m$ then the system has a nontrivial solution (that is, a solution $x_1, x_2, \ldots, x_n$ where $x_k \ne 0$ for some $1 \le k \le n$).

Proof: We prove the result by induction on the number of equations m. First, suppose we have $m = 1$ equation in $n > 1$ unknowns: $a_{11}x_1 + a_{12}x_2 + \cdots + a_{1n}x_n = 0$. If $a_{1j} = 0$ for $1 \le j \le n$, then we have the nontrivial solution $x_1 = x_2 = \cdots = x_n = 1$. If some coefficient $a_{1j^*} \ne 0$, then we have the nontrivial solution

$$
x_k = \begin{cases} 1 & \text{if} \quad k \ne j^*; \\ -(a_{1j^*})^{-1}(a_{11} + a_{12} + \cdots + a_{1n} - a_{1j^*}) & \text{if} \quad k = j^*. \end{cases}
$$

This proves the result for $m = 1$ and $n > m$.

Next suppose the result holds for a system of $m - 1$ equations in $n - 1 > m - 1$ unknowns. If all coefficients $a_{ij} = 0$, then $x_1 = x_2 = \cdots = x_n = 1$ is a nontrivial solution. If some $a_{i^*j^*} \ne 0$, then consider the system of equations

$$
\begin{aligned}
(a_{11} - (a_{i^*j^*})^{-1}a_{1j^*}a_{i^*1})x_1 &+ (a_{12} - (a_{i^*j^*})^{-1}a_{1j^*}a_{i^*2})x_2 + \cdots \\
&+ (a_{1j^*} - (a_{i^*j^*})^{-1}a_{1j^*}a_{i^*j^*})x_{j^*} + \cdots \\
&+ (a_{1n} - (a_{i^*j^*})^{-1}a_{1j^*}a_{i^*n})x_n = 0;
\end{aligned}
$$

$$(a_{21} - (a_{i^*j^*})^{-1}a_{2j^*}a_{i^*1})x_1 \; + \; (a_{22} - (a_{i^*j^*})^{-1}a_{2j^*}a_{i^*2})x_2 + \cdots$$
$$+ \; (a_{2j^*} - (a_{i^*j^*})^{-1}a_{2j^*}a_{i^*j^*})x_{j^*} + \cdots$$
$$+ \; (a_{2n} - (a_{i^*j^*})^{-1}a_{2j^*}a_{i^*n})x_n = 0;$$

$$\vdots \quad \vdots \quad \vdots$$

$$(a_{m1} - (a_{i^*j^*})^{-1}a_{mj^*}a_{i^*1})x_1 \; + \; (a_{m2} - (a_{i^*j^*})^{-1}a_{mj^*}a_{i^*2})x_2 + \cdots$$
$$+ \; (a_{mj^*} - (a_{i^*j^*})^{-1}a_{mj^*}a_{i^*j^*})x_{j^*} + \cdots$$
$$+ \; (a_{mn} - (a_{i^*j^*})^{-1}a_{mj^*}a_{i^*n})x_n = 0.$$

(This system is obtained from the original one by eliminating the variable x_{j^*} from all equations.) Notice that the coefficient of x_{j^*} is 0 in each equation and that the j^* equation is $0 = 0$. Therefore, this is a system of $m-1$ equations in the $n-1$ variables $x_1, x_2, \ldots, x_{j^*-1}, x_{j^*+1}, x_{j^*+2}, \ldots, x_n$. By the induction hypothesis, this system has a nontrivial solution, and this solution along with

$$x_{j^*} \; = \; -(a_{i^*j^*})^{-1}(a_{j^*1}x_1 + a_{j^*2}x_2 + \cdots + a_{j^*(j^*-1)}x_{j^*-1} + a_{j^*(j^*+1)}x_{j^*+1}$$

$$+ \; a_{j^*(j^*+2)}x_{j^*+2} + \cdots + a_{j^*n}x_n)$$

forms a nontrivial solution to the original system of equations. Hence, by induction, the result holds for all $m \geq 1$ and all $n > m$. ∎

Theorem 5.1.1 Let $\langle V, \mathbb{F} \rangle$ be a vector space with bases $\{v_1, v_2, \ldots, v_m\}$ and $\{w_1, w_2, \ldots, w_n\}$. Then $n = m$.

Proof: Suppose $n > m$. Since $\{v_1, v_2, \ldots, v_m\}$ is a basis, then for some a_{ij} where $1 \leq i \leq m, 1 \leq j \leq n$ we have

$$w_1 \; = \; a_{11}v_1 + a_{21}v_2 + \cdots + a_{m1}v_m$$
$$w_2 \; = \; a_{12}v_1 + a_{22}v_2 + \cdots + a_{m2}v_m$$

$$\vdots \quad \vdots \quad \vdots$$

$$w_n \; = \; a_{1n}v_1 + a_{2n}v_2 + \cdots + a_{mn}v_m.$$

Let $x_1, x_2, \ldots, x_n$ be ("unknown") elements of $\mathbb{F}$. Then

$$x_1w_1 + x_2w_2 + \cdots + x_nw_n = (x_1a_{11} + x_2a_{12} + \cdots + x_na_{1n})v_1 + (x_1a_{21} + x_2a_{22} +$$

$$\cdots + x_na_{2n})v_2 + \cdots + (x_1a_{m1} + x_2a_{m2} + \cdots + x_na_{mn})v_m.$$

The system of equations

$$x_1 a_{11} + x_2 a_{12} + \cdots + x_n a_{1n} = 0$$
$$x_1 a_{21} + x_2 a_{22} + \cdots + x_n a_{2n} = 0$$
$$\vdots \quad \vdots \quad \vdots$$
$$x_1 a_{m1} + x_2 a_{m2} + \cdots + x_n a_{mn} = 0$$

has a nontrivial solution $x_1, x_2, \ldots, x_n$ by Lemma 5.1.1, since $n > m$. Therefore $x_1 \mathbf{w}_1 + x_2 \mathbf{w}_2 + \cdots + x_n \mathbf{w}_n = 0$ for $x_1, x_2, \ldots x_n$ where $x_k \neq 0$ for some $1 \leq k \leq n$. That is, the set of vectors $\{\mathbf{w}_1, \mathbf{w}_2, \ldots, \mathbf{w}_m\}$ is linearly dependent. But this is a contradiction since $\{\mathbf{w}_1, \mathbf{w}_2, \ldots, \mathbf{w}_n\}$ is a basis for $\langle V, \mathbb{F} \rangle$, and hence is a linearly independent set. Therefore $n \leq m$. Similarly, $m \leq n$ and we conclude that $n = m$. ∎

We will make extensive use of Theorem 5.1.1. It will be the whole basis (!) of our classification of finite-dimensional vector spaces.

Definition 5.1.5 If vector space $\langle V, \mathbb{F} \rangle$ is a finite dimensional vector space, then the *dimension* of the vector space is the cardinality of a basis.

Definition 5.1.6 Two vector spaces over the same field $\mathbb{F}$, $\langle V, \mathbb{F} \rangle$ and $\langle W, \mathbb{F} \rangle$, are *isomorphic* if there is a one-to-one and onto mapping $\phi : V \mapsto W$ such that for all $f, f' \in \mathbb{F}$ and $\mathbf{v}, \mathbf{v}' \in V$, we have: $\phi(f\mathbf{v} + f'\mathbf{v}') = f\phi(\mathbf{v}) + f'\phi(\mathbf{v}')$.

Informally, an isomorphism is a one-to-one and onto mapping between two mathematical entities which preserves the structure of those entities (whether the structure is connectivity in a graph, the binary operation in a group, and so forth). The structure in a vector space consists of scalar multiplication and vector addition, and that is why we define isomorphism in this setting as we do.

We are now prepared to completely classify finite-dimensional vector spaces. The following result gives us the answer to the question "What does a finite-dimensional vector space *look like*?" More precisely, this result tells us, up to isomorphism, what an n-dimensional vector space is. We raise this result to the status of a "fundamental theorem" and declare it the *Fundamental Theorem of Finite-Dimensional Vector Spaces*.

Theorem 5.1.2 [The Fundamental Theorem of Finite-Dimensional Vector Spaces] If $\langle V, \mathbb{F} \rangle$ is an n-dimensional vector space, then $\langle V, \mathbb{F} \rangle$ is isomorphic to $\mathbb{F}^n = \langle V^*, \mathbb{F} \rangle$ where $V^* = \{(f_1, f_2, \ldots, f_n) \mid f_1, f_2, \ldots, f_n \in \mathbb{F}\}$, and scalar multiplication and vector addition are defined componentwise.

Proof: Let $\{v_1, v_2, \ldots, v_n\}$ be a basis of $\langle V, \mathbb{F} \rangle$. Define $\phi : V \mapsto V^*$ as

$$\phi((f_1 v_1 + f_2 v_2 + \cdots + f_n v_n)) = (f_1, f_2, \ldots, f_n).$$

Since $\{v_1, v_2, \ldots, v_n\}$ is a linearly independent set, then ϕ is one-to-one. Since $\{v_1, v_2, \ldots, v_n\}$ is a spanning set of $\langle V, \mathbb{F} \rangle$ then ϕ is onto. Finally, for any $f, f' \in \mathbb{F}$ and $v, v' \in V$ we have

$$\phi(f v + f' v') = \phi(f(f_1 v_1 + f_2 v_2 + \cdots + f_n v_n) + f'(f_1' v_1 + f_2' v_2 + \cdots$$
$$+ f_n' v_n)) \text{ where } v = f_1 v_1 + f_2 v_2 + \cdots + f_n v_n$$

and

$$\begin{aligned}
v' &= f_1' v_1 + f_2' v_2 + \cdots + f_n' v_n \\
&= \phi((ff_1 + f'f_1') v_1 + (ff_2 + f'f_2') v_2 + \cdots (ff_n + f'f_n') v_n) \\
&= (ff_1 + f'f_1', ff_2 + f'f_2', \ldots, ff_n + f'f_n') \\
&= (ff_1, ff_2, \ldots, ff_n) + (f'f_1', f'f_2', \ldots, f'f_n') \\
&= f(f_1, f_2, \ldots, f_n) + f'(f_1', f_2', \ldots, f_n') \\
&= f\phi(f_1 v_1 + f_2 v_2 + \cdots + f_n v_n) + f'\phi(f_1' v_1 + f_2' v_2 + \cdots + f_n' v_n) \\
&= f\phi(v) + f'\phi(v').
\end{aligned}$$

Therefore ϕ is an isomorphism. ∎

The Fundamental Theorem of Finite-Dimensional Vector Spaces tells us, in particular, that an n-dimensional vector space over scalar field $\mathbb{R}$ is isomorphic to $\mathbb{R}^n$. Similarly, an n-dimensional vector space over scalar field $\mathbb{C}$ is isomorphic to $\mathbb{C}^n$. We therefore can completely determine a finite-dimensional vector space simply by knowing its dimension and scalar field. We will see in Section 5.4 that this idea can be extended to infinite-dimensional spaces as well. However, this extension will require that we modify some of the definitions encountered in this section.

Now that we know what an n-dimensional vector space "looks like," we use the Fundamental Theorem of Finite-Dimensional Vector Spaces to classify certain transformations of these vector spaces.

Definition 5.1.7 A transformation T mapping one vector space $\langle V, \mathbb{F} \rangle$ into another $\langle W, \mathbb{F} \rangle$ is a *linear transformation* if for all $v, v' \in V$ and for all $f, f' \in \mathbb{F}$, we have $T(f v + f' v') = fT(v) + f'T(v')$.

Notice that an isomorphism between two vector spaces over the same field is just a one-to-one and onto mapping which is linear. The following result allows us to classify all linear transformations between two finite-dimensional vector spaces. We do so by considering the behavior of linear transformations on the *standard basis* of $\mathbb{F}^n$: $\{(1,0,0,\dots,0), (0,1,0,\dots,0), \dots, (0,0,0,\dots,0,1, 0,\dots,0), \dots, (0,0,\dots,0,1)\}$. We commonly represent the set of standard basis vectors as $\{\mathbf{e}_1, \mathbf{e}_2, \dots, \mathbf{e}_n\}$. If $\mathbf{v} \in \mathbb{F}^n$ and $\mathbf{v} = v_1\mathbf{e}_1 + v_2\mathbf{e}_2 + \cdots + v_n\mathbf{e}_n$, then we represent $\mathbf{v}$ as $(v_1, v_2, \dots, v_n)$.

Theorem 5.1.3 If T is a linear transformation from n-dimensional vector space $\langle V, \mathbb{F} \rangle$ to m-dimensional vector space $\langle W, \mathbb{F} \rangle$ then T is equivalent to the action of an $m \times n$ matrix $A_T : \mathbb{F}^n \mapsto \mathbb{F}^m$.

Proof: Let $\mathbf{v} \in V$ and consider the representation of $\mathbf{v}$ with respect to the standard basis of $\langle V, \mathbb{F} \rangle$, $\mathbf{v} = v_1\mathbf{e}_1 + v_2\mathbf{e}_2 + \cdots + v_n\mathbf{e}_n := (v_1, v_2, \dots, v_n)$. Then applying T to $\mathbf{v}$ yields

$$
\begin{aligned}
T(\mathbf{v}) &= T(v_1\mathbf{e}_1 + v_2\mathbf{e}_2 + \cdots + v_n\mathbf{e}_n) \\
&= v_1 T(\mathbf{e}_1) + v_2 T(\mathbf{e}_2) + \cdots + v_n T(\mathbf{e}_n).
\end{aligned}
$$

The vectors $T(\mathbf{e}_i)$, $1 \leq i \leq n$ are elements of W. Suppose that, with respect to the standard basis for $\langle W, \mathbb{F} \rangle$, we have the representation $T(\mathbf{e}_i) = (a_{1i}, a_{2i}, \dots, a_{mi})$ for $1 \leq i \leq n$. Then defining

$$
A_T = \begin{bmatrix} a_{11} & a_{12} & \cdots & a_{1n} \\ a_{21} & a_{22} & \cdots & a_{2n} \\ \vdots & & \ddots & \vdots \\ a_{m1} & a_{m2} & \cdots & a_{mn} \end{bmatrix},
$$

we see that vector $\mathbf{v}$ is mapped equivalently under T and A_T. ∎

Surprisingly, Theorem 5.1.3 can be extended to linear transformations between infinite-dimensional vector spaces as well. We will see this in Section 5.4.

The Fundamental Theorem of Finite-Dimensional Vector Spaces (Theorem 5.1.2) gives a complete classification of finite-dimensional vector spaces simply in terms of the dimension and the scalar field. We now turn our attention to infinite-dimensional vector spaces. We will offer some alternative definitions of *basis*, *span*, and *linear combination* in the infinite-dimensional setting. If we follow Definition 5.1.4 in the infinite-dimensional setting (where *linear combi-*

nation, linearly independent, and *span* are all defined in terms of *finite* sums), then the type of basis that results is called a *Hamel basis.* Every vector space has a Hamel basis.

Theorem 5.1.4 Let $\langle V, \mathbb{F} \rangle$ be a vector space. Then there exists a set of vectors $B \subset V$ such that (1) B is linearly independent and (2) for any $v \in V$ there exists finite sets $\{b_1, b_2, \dots, b_n\} \subset B$ and $\{f_1, f_2, \dots, f_n\}$ such that $v = f_1 b_1 + f_2 b_2 + \cdots + f_n b_n$. That is, B is a Hamel basis for $\langle V, \mathbb{F} \rangle$.

Proof: Let P be the class whose members are the linearly independent sub-sets of V. Then define the partial order $\prec$ on P as $A \prec B$ for $A, B \in P$ if $A \subset B$. Now for $v \neq 0$, $\{v\} \in P$ and so P is nonempty. Next, suppose Q is a totally ordered subset of P. Define M to be the union of all the sets in Q. Then $M \in P$ is an upper bound of Q. Hence by Zorn's Lemma (see Theorem 1.2.2), P has a maximal element, call it B. Since B is in P, B is linearly independent. Also, any vector v must be a linear combination of ele-ments of B, for if not, then the set $B \bigcup \{v\}$ would be in P and B would not be maximal. ∎

Unfortunately, the proof of Theorem 5.1.4 requires Zorn's Lemma (or equiv-alently the Axiom of Choice). This makes the construction of a Hamel basis quite a delicate affair! A much more useful idea was introduced by J. Schauder. As opposed to writing any element of the vector space as a *finite* linear com-bination of basis vectors, it is desired to write any element as a *series* of basis vectors. Therefore we are required to deal with vector spaces which have a metric.

Definition 5.1.8 Let $\langle V, \mathbb{F} \rangle$ be a vector space with metric m. Then a set of vectors $B \subset V$ is a *Schauder basis* for $\langle V, \mathbb{F} \rangle$ if for each $v \in V$ there is a unique set of scalars $\{f_1, f_2, \dots\} \subset \mathbb{F}$ such that $v = \sum_{n=1}^{\infty} f_n b_n$. That is, $\lim_{n \to \infty} m\left(v, \sum_{i=1}^{n} f_i b_i\right) = 0$.

The requirement that the set of scalars $\{f_1, f_2, \dots\}$ be unique for any given vector v guarantees that $0 = \sum_{n=1}^{\infty} f_n b_n$ only when $f_n = 0$ for all $n \in \mathbb{N}$ (this can be interpreted as a requirement that set B be linearly independent). Therefore, a Schauder basis of a vector space with a metric is a linearly independent spanning set where *linear combination* and *span* are defined in terms of *series* as opposed to *finite* linear combinations.

> **EXAMPLE 5.1.4:** Define $\mathbb{R}^\infty = \langle V, \mathbb{R} \rangle$ where $V = \{(r_1, r_2, \ldots) \mid r_i \in \mathbb{R} \text{ for } 1 \le n < \infty\}$, and scalar multiplication and vector addition are defined componentwise. Then $\mathbb{R}^\infty$ is a vector space and each vector can be written as an infinite linear combination of the vectors $(1, 0, 0, 0, \ldots), (0, 1, 0, 0, \ldots), (0, 0, 1, 0, \ldots), \ldots$ (the presence of so many zeros in this collection of vectors allows us to avoid the need for a metric — in general, though, we should avoid infinite sums in this space). However, we will see that we cannot define an *inner product* (or *dot* product) on $\mathbb{R}^\infty$.

If we are taking infinite sums of elements of a vector space, we must be concerned with convergence. We'll explore this more in the next section.

Exercises

1. Verify that the examples of groups, fields, and vector spaces given in this section are what they are claimed to be.
2. Prove that the matrix A_T of Theorem 5.1.3 is unique.
3. Show that if B_1 and B_2 are Hamel bases for a given infinite-dimensional vector space, then B_1 and B_2 are of the same cardinality.
4. Consider the vector space $\mathcal{P}_n$ of all polynomials of degree at most n. Find a basis for this space and find the matrix which represents the differentiation operator (with respect to the basis you choose).
5. Consider the nth order linear homogeneous differential equation

$$f_n(x)y^{(n)} + f_{n-1}(x)y^{(n-1)} + \cdots + f_1(x)y' + f_0(x)y = 0.$$

   Show that the set of solutions of this differential equation forms a vector space.
6. Let $\langle V_1, \mathbb{F} \rangle$ and $\langle V_2, \mathbb{F} \rangle$ be vector spaces, and let $V = V_1 \cap V_2$. Show that $\langle V, \mathbb{F} \rangle$ is a vector space (called the *intersection* of the two given vector spaces). Give an example of V_1 and V_2 where $V_1 \cup V_2$ is *not* a vector space.
7. Consider the vector space of functions continuous on the interval $[a, b]$, denoted $C([a, b])$. Show that $f_1(x) = \cos x$ and $f_2(x) = \sin x$ are linearly independent in this space. Show that the set $\{1, x, x^2, \ldots, x^n\}$ is a linearly independent set in $C([a, b])$.

8. Prove that if $\langle V, \mathbb{F} \rangle$ is an n-dimensional vector space and $\{\mathbf{v}_1, \mathbf{v}_2, \ldots, \mathbf{v}_n\}$ is a linearly independent set, then $\{\mathbf{v}_1, \mathbf{v}_2, \ldots, \mathbf{v}_n\}$ is a spanning set of $\langle V, \mathbb{F} \rangle$.

9. Let T be a linear transformation from one vector space $\langle V, \mathbb{F} \rangle$ into another $\langle W, \mathbb{F} \rangle$. Show that $\langle T(V), \mathbb{F} \rangle$ is a subspace of $\langle W, \mathbb{F} \rangle$.

10. If T is a linear transformation from one vector space to another, then the set of vectors mapped to $\mathbf{0}$ under T is called the *kernel* of T. Prove that the kernel of T is a vector space.

11. Let T be a linear transformation from $\langle V, \mathbb{F} \rangle$ into $\langle V, \mathbb{F} \rangle$. Vector $\mathbf{v}$ is an *eigenvector* of T if there exists $\lambda \in \mathbb{F}$ such that $T(\mathbf{v}) = \lambda\mathbf{v}$. Suppose $\{\mathbf{v}_1, \mathbf{v}_2, \ldots, \mathbf{v}_n\}$ is a basis for $\langle V, \mathbb{F} \rangle$ where each $\mathbf{v}_i$ is an eigenvector and $T(\mathbf{v}_i) = \lambda_i\mathbf{v}_i$. Find the $n \times n$ matrix A_T which represents T.

2 Inner Product Spaces

As seen in a linear algebra class, the idea of dot product allows one to discuss certain geometric properties such as orthogonality and distance. In this section, we take these properties as motivation in defining the concept of an inner product space.

Definition 5.2.1   A vector space with complex scalars $\langle V, \mathbb{C} \rangle$ is an *inner product space* (also called a *Pre-Hilbert space*) if there is a function $\langle \cdot, \cdot \rangle : V \times V \mapsto \mathbb{C}$ such that for all $\mathbf{u}, \mathbf{v}, \mathbf{w} \in V$ and $a \in \mathbb{C}$ we have:

(a) $\langle \mathbf{v}, \mathbf{v} \rangle \in \mathbb{R}$ and $\langle \mathbf{v}, \mathbf{v} \rangle \geq 0$ with $\langle \mathbf{v}, \mathbf{v} \rangle = 0$ if and only if $\mathbf{v} = \mathbf{0}$,

(b) $\langle \mathbf{u}, \mathbf{v} + \mathbf{w} \rangle = \langle \mathbf{u}, \mathbf{v} \rangle + \langle \mathbf{u}, \mathbf{w} \rangle$,

(c) $\langle a\mathbf{u}, \mathbf{v} \rangle = a\langle \mathbf{u}, \mathbf{v} \rangle$, and

(d) $\langle \mathbf{u}, \mathbf{v} \rangle = \overline{\langle \mathbf{v}, \mathbf{u} \rangle}$ where the over line represents the operation of complex conjugation.

The function $\langle \cdot, \cdot \rangle$ is called an *inner product*. (Some texts, such as Reed and Simon [24], replace property (c) with the requirement that $\langle \mathbf{u}, a\mathbf{v} \rangle = a\langle \mathbf{u}, \mathbf{v} \rangle$.)

Notice that properties (b), (c), and (d) of Definition 5.2.1 combine to imply that

$$\langle a\mathbf{u} + b\mathbf{v}, \mathbf{w} \rangle = a\langle \mathbf{u}, \mathbf{w} \rangle + b\langle \mathbf{u}, \mathbf{w} \rangle$$

and

$$\langle \mathbf{u}, a\mathbf{v} + b\mathbf{w} \rangle = \overline{a}\langle \mathbf{u}, \mathbf{v} \rangle + \overline{b}\langle \mathbf{u}, \mathbf{w} \rangle$$

for all relevant vectors and scalars. That is, $\langle \cdot, \cdot \rangle$ is linear in the first position and "conjugate-linear" in the second position.

> **EXAMPLE 5.2.1:** The vector space $\mathbb{C}^n$ is an inner product space with the inner product defined for $\mathbf{u} = (u_1, u_2, \ldots, u_n)$ and $\mathbf{v} = (v_1, v_2, \ldots, v_n)$ as $\langle \mathbf{u}, \mathbf{v} \rangle = \sum_{j=1}^{n} u_j \overline{v}_j$.

We can also define an inner product on a vector space with real scalars by requiring that $\langle \cdot, \cdot \rangle : V \times V \mapsto \mathbb{R}$ and by replacing property (d) in Definition 5.2.1 with the requirement that the inner product be symmetric: $\langle \mathbf{u}, \mathbf{v} \rangle = \langle \mathbf{v}, \mathbf{u} \rangle$. Then $\mathbb{R}^n$ with the usual dot product is an example of a real inner product space.

Definition 5.2.2 For inner product space $\langle V, \mathbb{C} \rangle$ with inner product $\langle \cdot, \cdot \rangle$, define the *norm* induced by the inner product as $\|\mathbf{v}\| = \langle \mathbf{v}, \mathbf{v} \rangle^{1/2}$ for all $\mathbf{v} \in V$.

From Definition 5.2.1 we see that $\|\mathbf{v}\|$ is real, $\|\mathbf{v}\| \geq 0$ and $\|\mathbf{v}\| = 0$ if and only if $\mathbf{v} = \mathbf{0}$. Also,

$$\|a\mathbf{v}\| = \langle a\mathbf{v}, a\mathbf{v} \rangle^{1/2} = (a\bar{a}\langle \mathbf{v}, \mathbf{v} \rangle)^{1/2} = |a|\langle \mathbf{v}, \mathbf{v} \rangle^{1/2} = |a|\|\mathbf{v}\|.$$

Of course, if we use the term "norm," then we must verify that $\| \cdot \| : V \mapsto \mathbb{R}$ satisfies another property, the "triangle inequality." We now take steps in that direction.

Theorem 5.2.1 [The Schwarz inequality] For all $\mathbf{u}, \mathbf{v}$ in inner product space $\langle V, \mathbb{C} \rangle$, we have

$$|\langle \mathbf{u}, \mathbf{v} \rangle| \leq \|\mathbf{u}\|\|\mathbf{v}\|.$$

Proof: We know that for all $a \in \mathbb{C}$

$$\|\mathbf{u} + a\mathbf{v}\| = \langle \mathbf{u} + a\mathbf{v}, \mathbf{u} + a\mathbf{v} \rangle \geq 0.$$

In particular, this inequality holds for $a = b\dfrac{\overline{\langle \mathbf{u}, \mathbf{v} \rangle}}{|\langle \mathbf{u}, \mathbf{v} \rangle|}$ where b is real. Therefore

$$\begin{aligned} \langle \mathbf{u} + a\mathbf{v}, \mathbf{u} + a\mathbf{v} \rangle &= \|\mathbf{u}\|^2 + a\langle \mathbf{u}, \mathbf{v} \rangle + \overline{a\langle \mathbf{u}, \mathbf{v} \rangle} + |a|^2\|\mathbf{v}\|^2 \\ &= \|\mathbf{u}\|^2 + 2b|\langle \mathbf{u}, \mathbf{v} \rangle| + b^2\|\mathbf{v}\|^2 \geq 0. \end{aligned} \qquad (2.1)$$

Solving the equality

$$\|u\|^2 + 2b|\langle u, v \rangle| + b^2\|v\|^2 = 0$$

for b, we get

$$b = \frac{-2|\langle u, v \rangle| \pm \sqrt{(2|\langle u, v \rangle|)^2 - 4\|v\|^2\|u\|^2}}{2\|v\|^2}. \qquad (2.2)$$

Therefore, inequality (2.1) holds for all u and v if and only if the discriminant of (2.2) is nonpositive:

$$(2|\langle u, v \rangle|)^2 - 4\|v\|^2\|u\|^2 \leq 0.$$

That is, $|\langle u, v \rangle| \leq \|u\|\|v\|$. ■

The Schwarz Inequality puts us in the position to show that $\| \cdot \|$ satisfies the triangle inequality.

Theorem 5.2.2 [The Triangle inequality] For all u, v in an inner product space $\langle V, \mathbb{C} \rangle$ we have $\|u + v\| \leq \|u\| + \|v\|$.

Proof: We have by the Schwarz Inequality (Theorem 5.2.1):

$$\begin{aligned}
\|u + v\|^2 &= \langle u + v, u + v \rangle = \langle u, u \rangle + \langle u, v \rangle + \langle v, u \rangle + \langle v, v \rangle \\
&= \|u\|^2 + \langle u, v \rangle + \overline{\langle u, v \rangle} + \|v\|^2 \\
&= \|u\|^2 + 2\mathrm{Re}\langle u, v \rangle + \|v\|^2 \\
&\leq \|u\|^2 + 2|\langle u, v \rangle| + \|v\|^2 \\
&= (\|u\| + \|v\|)^2.
\end{aligned}$$

Taking square roots yields the result. ■

We have now seen that $\| \cdot \|$ satisfies for all $u, v \in V$ and for all $a \in \mathbb{C}$:
  (1) $\|v\| \geq 0$ and $\|v\| = 0$ if and only if $v = 0$,
  (2) $\|av\| = |a|\|v\|$, and
  (3) $\|u + v\| \leq \|u\| + \|v\|$.
Therefore we are justified in calling $\| \cdot \|$ a norm. So for any vector space with an inner product, there is also a norm. The converse does not hold, however (the classical Banach spaces are normed but the norm is not induced by an inner

product in most cases — see Exercise 5 of Section 5.5). Therefore we have the following hierarchy:

$$\left(\begin{array}{c} \text{vector} \\ \text{spaces} \end{array} \right) \supset \left(\begin{array}{c} \text{normed} \\ \text{vector} \\ \text{spaces} \end{array} \right) \supset \left(\begin{array}{c} \text{inner} \\ \text{product} \\ \text{spaces} \end{array} \right).$$

Analogous to the use of dot product in $\mathbb{R}^n$ to define the angle between two vectors, we have the following definition.

Definition 5.2.3 Two vectors u, v in an inner product space are *orthogonal* if $\langle u, v \rangle = 0$. A set of vectors $\{v_1, v_2, \dots\}$ is orthogonal if $\langle v_i, v_j \rangle = 0$ for $i \neq j$. This orthogonal set of vectors is *orthonormal* if in addition $\langle v_i, v_i \rangle = \|v_i\|^2 = 1$ for all i and, in this case, the vectors are said to be *normalized*.

An important property of an orthonormal system of vectors is the following.

Theorem 5.2.3 [The Pythagorean Theorem] Let $\{v_1, v_2, \dots, v_n\}$ be an orthonormal set of vectors in an inner product space $\langle V, \mathbb{C} \rangle$. Then for all $u \in V$

$$\|u\|^2 = \sum_{j=1}^{n} |\langle u, v_j \rangle|^2 + \left\| u - \sum_{j=1}^{n} \langle v_j, u \rangle v_j \right\|^2.$$

Proof: Trivially

$$u = \sum_{j=1}^{n} \langle v_j, u \rangle v_j + \left(u - \sum_{j=1}^{n} \langle v_j, u \rangle v_j \right).$$

(We will see later that this is a rather fundamental decomposition of u.) Since

$$\left\langle \sum_{j=1}^{n} \langle v_j, u \rangle v_j, u - \sum_{j=1}^{n} \langle v_j, u \rangle v_j \right\rangle$$

$$= \left\langle \sum_{j=1}^{n} \langle v_j, u \rangle v_j, u \right\rangle - \left\langle \sum_{j=1}^{n} \langle v_j, u \rangle v_j, \sum_{j=1}^{n} \langle v_j, u \rangle v_j \right\rangle$$

$$= \sum_{j=1}^{n} \overline{\langle v_j, u \rangle} \langle v_j, u \rangle - \sum_{j=1}^{n} \left\langle \langle v_j, u \rangle v_j, \sum_{k=1}^{n} \langle v_k, u \rangle v_k \right\rangle$$

$$= \sum_{j=1}^{n} \overline{\langle v_j, u \rangle} \langle v_j, u \rangle - \sum_{j=1}^{n} \sum_{k=1}^{n} \langle \langle v_j, u \rangle v_j, \langle v_k, u \rangle v_k \rangle$$

$$= \sum_{j=1}^{n} \overline{\langle v_j, u \rangle} \langle v_j, u \rangle - \sum_{j=1}^{n} \sum_{k=1}^{n} \langle v_j, u \rangle \overline{\langle v_k, u \rangle} \langle v_j, v_k \rangle$$

$$= \sum_{j=1}^{n} |\langle v_j, u \rangle|^2 - \sum_{j=1}^{n} |\langle v_j, u \rangle|^2 = 0,$$

then these two vectors are orthogonal. Therefore

$$\|u\|^2 = \langle u, u \rangle = \left\langle \sum_{j=1}^{n} \langle v_j, u \rangle v_j + \left(u - \sum_{j=1}^{n} \langle v_j, u \rangle v_j \right), \right.$$

$$\left. \sum_{j=1}^{n} \langle v_j, u \rangle v_j + \left(u - \sum_{j=1}^{n} \langle v_j, u \rangle v_j \right) \right\rangle$$

$$= \left\langle \sum_{j=1}^{n} \langle v_j, u \rangle v_j, \sum_{j=1}^{n} \langle v_j, u \rangle v_j \right\rangle$$

$$+ \left\langle u - \sum_{j=1}^{n} \langle v_j, u \rangle v_j, u - \sum_{j=1}^{n} \langle v_j, u \rangle v_j \right\rangle$$

$$= \sum_{j=1}^{n} |\langle v_j, u \rangle|^2 + \left\| u - \sum_{j=1}^{n} \langle v_j, u \rangle v_j \right\|^2.$$

Notice that if we have v and w orthogonal and set $u = v + w$ then the Pythagorean Theorem implies the familiar result that $\|u\|^2 = \|v\|^2 + \|w\|^2$.

We immediately have from the Pythagorean Theorem the following:

Corollary 5.2.1 [Bessel's inequality] Let $\{v_1, v_2, \dots, v_n\}$ be an orthonormal set in an inner product space $\langle V, \mathbb{C} \rangle$. Then for all $u \in V$ we have

$$\|u\|^2 \geq \sum_{j=1}^{n} |(u, v_j)|^2.$$

Since inner product spaces have the induced norm $\| \cdot \| = \langle \cdot, \cdot \rangle^{1/2}$, we can define the metric $d : V \times V \mapsto \mathbb{R}$ as $d(u, v) = \|u - v\| = \langle u - v, u - v \rangle^{1/2}$. Therefore in an inner product space there is the concept of "distance" and "closeness." As a consequence, we can address several topics that are familiar in the setting of $\mathbb{R}^n$: limits and convergence of sequences and series, Cauchy sequences, and completeness. An axiomatic development of the real numbers requires (as seen in Chapter 1) an Axiom of Completeness. It is this completeness of $\mathbb{R}$ which leads us to several desired properties, such as the convergence of Cauchy sequences. We parallel this development by focusing our study on complete inner product spaces. However, our approach is necessarily a bit different from the setting of $\mathbb{R}$.

Definition 5.2.4 In inner product space $\langle V, \mathbb{C} \rangle$ a sequence $(\mathbf{v}_1, \mathbf{v}_2, \dots)$, where $\mathbf{v}_i \in V$ for all i, is *Cauchy* if for any $\epsilon > 0$ there exists $N \in \mathbb{N}$ such that for all $m, n \geq N$ we have $d(\mathbf{v}_m, \mathbf{v}_n) < \epsilon$. The sequence is *convergent* if there exists $\mathbf{v} \in V$ such that, for any $\epsilon > 0$ there exists $N \in \mathbb{N}$ satisfying the property that for all $n > N$ we have $d(\mathbf{v}, \mathbf{v}_n) < \epsilon$. Element $\mathbf{v}$ is called the *limit* of the sequence.

In real n-dimensional space, $\mathbb{R}^n$, Definition 5.2.4 reduces to the familiar concepts. So, with Theorem 1.4.6 as motivation, we define completeness in an inner product space in terms of Cauchy sequences:

Definition 5.2.5 An inner product space is *complete* if Cauchy sequences converge.

Intuitively (and *very informally*), a complete space has no holes in it! For example, $\mathbb{Q}$ is not complete, since we can take the sequence $(1, 1.4, 1.41, 1.414, \dots)$ of progressively better decimal approximations of $\sqrt{2}$, which we know to be Cauchy (for it is a convergent sequence in $\mathbb{R}$ and a sequence of real numbers is Cauchy if and only if it is convergent to a real number). However, the sequence does not converge in $\mathbb{Q}$! This is because, again informally, $\mathbb{Q}$ has a hole in it at $\sqrt{2}$. In inner product spaces we will be interested in finding a basis in the sense of Schauder, and therefore we must address the issue of convergence of series (which, of course, is defined in terms of the convergence of sequences of partial sums). Hence, from this point on, we focus our attention on complete inner product spaces.

Definition 5.2.6 A complete inner product space is a *Hilbert space*.

We have reached a climax in terms of defining mathematical structures! Much of the previous material in this book will be aimed at exploring the structure of Hilbert spaces, and all of the remaining material in this book is a direct application of Hilbert space theory!

EXAMPLE 5.2.2: Since $\mathbb{R}$ and $\mathbb{C}$ are complete, then $\mathbb{R}^n$ and $\mathbb{C}^n$ are examples of Hilbert spaces (with the familiar dot product as the inner product on $\mathbb{R}^n$ and the inner product on $\mathbb{C}^n$ as defined in Example 5.2.1).

We are quite familiar with the structure of $\mathbb{R}^n$ and $\mathbb{C}^n$ from our studies of linear algebra. In fact, every real vector space of dimension n is isomorphic to $\mathbb{R}^n$ and every complex vector space of dimension n is isomorphic to $\mathbb{C}^n$ (see Theorem 5.1.2). So we will now focus our attention on the infinite-dimensional vector spaces which are also Hilbert spaces.

Exercises

1. Use Bessel's inequality to give an alternative proof of the Schwarz inequality.
2. Prove that in an inner product space, every convergent sequence is Cauchy.
3. Let $\mathbf{u}$ and $\mathbf{v}$ be vectors in an inner product space. Prove that $\|\mathbf{u} + \mathbf{v}\| + \|\mathbf{u} - \mathbf{v}\| = 2(\|\mathbf{u}\|^2 + \|\mathbf{v}\|^2)$. This is called the *Parallelogram Law*. We can interpret this geometrically in $\mathbb{R}^n$ in the sense that $\mathbf{u}$ and $\mathbf{v}$ determine a parallelogram with diagonals of length $\|\mathbf{u} + \mathbf{v}\|$ and $\|\mathbf{u} - \mathbf{v}\|$.
4. Let $\mathbf{u}$ and $\mathbf{v}$ be vectors in an inner product space. Prove that

$$\langle \mathbf{u}, \mathbf{v} \rangle = \frac{1}{4} \left\{ \|\mathbf{u} + \mathbf{v}\|^2 - \|\mathbf{u} - \mathbf{v}\|^2 + i \left(\|\mathbf{u} + i\mathbf{v}\|^2 - \|\mathbf{u} - i\mathbf{v}\|^2 \right) \right\}.$$

This is called the *Polarization Identity* and, in the event that a norm is induced by an inner product, allows us to express the inner product directly in terms of the norm.
5. Let $\mathbf{u}$, $\mathbf{v}$, and $\mathbf{w}$ be vectors in an inner product space. Prove that $\|\mathbf{u} - \mathbf{v}\| + \|\mathbf{v} - \mathbf{w}\| = \|\mathbf{u} - \mathbf{w}\|$ if and only if $\mathbf{v} = t\mathbf{u} + (1 - t)\mathbf{w}$ for some $t \in [0, 1]$.
6. Prove that the inner product is continuous. That is, if $(\mathbf{u}_n) \to \mathbf{u}$ and $(\mathbf{v}_n) \to \mathbf{v}$, then $\langle \mathbf{u}_n, \mathbf{v}_n \rangle \to \langle \mathbf{u}, \mathbf{v} \rangle$.

3 The Space L^2

We now study infinite-dimensional Hilbert spaces. We will see in the Fundamental Theorem of Infinite-Dimensional Vector Spaces (Theorem 5.4.9) of the next section that all infinite-dimensional Hilbert spaces (with a particular type of basis) are isomorphic. Therefore there is, up to isomorphism, only one such space. In this section, we give a few different "incarnations" of this space.

First, for the interval $[a, b]$, we define the space $L^2([a, b])$ as

$$L^2([a, b]) = \left\{ f \,\middle|\, \int_{[a,b]} |f|^2 < \infty \right\},$$

where the inner product is $\langle f, g \rangle = \int_a^b fg$. Since sets of measure zero will not affect inner products or norms, we draw no distinctions between functions which are equal almost everywhere. To be precise, we partition $L^2([a, b])$ into equivalence classes with $f \equiv g$ if and only if $f = g$ a.e. on $[a, b]$ (we will encounter this concept again in Section 5.5). With this convention, it is clear that $L^2([a, b])$ satisfies Definition 5.2.1 and hence is an inner product space. We show that $L^2([a, b])$ is complete in the following theorem, and hence it is a Hilbert space.

Theorem 5.3.1 $L^2([a, b])$ is complete.

Proof: Let (f_N) be a Cauchy sequence in $L^2([a, b])$. Consider a subsequence (f_n) where

$$\|f_n - f_{n+1}\|_2 = \sqrt{\int |f_n - f_{n+1}|^2} < 2^{-n}.$$

Let

$$g_m(x) = \sqrt{\sum_{n=1}^m |f_n(x) - f_{n+1}(x)|^2}.$$

Let $g_\infty = \lim_{m \to \infty} g_m$. Then (g_m) monotonically approaches g_∞ and

$$\int |g_m|^2 = \int \sum_{n=1}^m |f_n - f_{n+1}|^2 = \sum_{n=1}^m \int |f_n - f_{n+1}|^2 = \sum_{n=1}^m \|f_n - f_{n+1}\|_2^2 < 1.$$

So we have

$$\int |g_\infty|^2 = \int \left| \lim_{m \to \infty} g_m \right|^2 = \int \lim_{m \to \infty} |g_m|^2$$

$$= \lim_{m \to \infty} \int |g_m|^2$$

(by the Monotone Convergence Theorem [Theorem 3.3.3])

$$\leq 1.$$

So $g_\infty \in L^2([a,b])$. Therefore $|g_\infty(x)| < \infty$ almost everywhere on $[a,b]$, and

$$f_m(x) = f_n(x) - \sum_{n=1}^{m-1} (f_n(x) - f_{n+1}(x))$$

converges pointwise a.e. to a function f. Now we show that f is the limit of (f_n) in the L^2 sense and that $f \in L^2$. For each x, we have

$$
\begin{aligned}
|f_m(x)| &\leq |f_1(x)| + \sum_{n=1}^{m-1} |f_n(x) - f_{n+1}| \\
&= |f_1(x)| + |g_m(x)| \leq |f_1(x)| + |g_\infty(x)|
\end{aligned}
$$

and

$$|f_m(x)|^2 \leq \left(|f_1(x)| + |g_\infty(x)|\right)^2 \leq 4\max\left\{|f_1(x)|^2, |g_\infty(x)|^2\right\}.$$

Now $f_1, g_\infty \in L^2$, so by the Lebesgue Dominated Convergence Theorem (Theorem 3.3.1)

$$\lim_{m\to\infty} \int |f_m|^2 = \int \lim_{m\to\infty} |f_m|^2 = \int |f|^2$$

and

$$\int |f|^2 = \lim_{m\to\infty} \int |f_m|^2 < \infty$$

and hence $f \in L^2$. Next,

$$|f(x) - f_n(x)| \leq |f(x)| + |f_1(x)| + |g_\infty(x)|$$

and

$$|f(x) - f_m(x)|^2 \leq 9\max\left\{|f(x)|^2, |f_1(x)|^2, |g_\infty(x)|^2\right\},$$

and again by the Lebesgue Dominated Convergence Theorem

$$\lim_{m\to\infty} \|f - f_m\|^2 = \lim_{m\to\infty} \int |f - f_m|^2 = \int \lim_{m\to\infty} |f - f_m|^2 = 0.$$

The result follows. ∎

We comment that there is nothing special about the set $[a, b]$ and, in fact, the arguments used in Theorem 5.3.1 are still valid if we choose to integrate over any measurable set $E \subset \mathbb{R}$. Therefore, we can consider

$$L^2(E) = \left\{ f \ \middle| \ \int_E |f|^2 < \infty \right\}$$

and we find that this is also a Hilbert Space.

Next, we introduce complex-valued functions of a real variable. That is, we are now interested in $f : \mathbb{R} \mapsto \mathbb{C}$ where the real and imaginary parts of f are measurable. In particular, for such an f and a measurable set E, we define

$$L^2(E) = \left\{ f \ \middle| \ \int_E f\bar{f} = \int_E |f|^2 < \infty \right\}.$$

In this case, we define the inner product of f and g in $L^2(E)$ as

$$\langle f, g \rangle = \int_E f\bar{g},$$

and hence $L^2(E)$ has the norm $\|f\| = \sqrt{\int |f|^2}$ (here $|f|$ refers to the *modulus* of f). Again, we can verify that $L^2(E)$ is in fact a Hilbert space. In Chapter 6, we will be especially interested in $L^2([-\pi, \pi])$. We will see that the set $\{e^{inx}/\sqrt{2\pi} \mid n \in \mathbb{Z}\}$ is an orthonormal set (in fact, a basis) in this space.

We now explore one slightly different example of a Hilbert space. Consider the collection of all square summable sequences of real numbers:

$$l^2 = \left\{ (x_1, x_2, x_3, \dots) \ \middle| \ x_i \in \mathbb{R} \text{ for all } i \in \mathbb{N}, \sum_{j=1}^{\infty} |x_j|^2 < \infty \right\}.$$

We define the inner product $\langle \mathbf{x}, \mathbf{y} \rangle = \sum_{j=1}^{\infty} x_j y_j$. Quite clearly, this is an inner product space. We leave the proof of the completeness as an exercise. We will show in the next section that $L^2([a, b])$ (and every infinite-dimensional Hilbert space with a particular type of basis) is isomorphic to l^2.

We now briefly step aside to informally discuss the geometry of l^2. At this stage, we are all quite comfortable with the geometry of finite-dimensional real vector spaces (which are, by the Fundamental Theorem of Finite-Dimensional Vector Spaces [Theorem 5.1.2], isomorphic to $\mathbb{R}^n$). In two or three dimensions, we are accustomed to representing vectors as arrows and discussing (especially in the presence of physicists and engineers) the ideas of "direction" and "magnitude." By analogy, we extend the idea of dot products (and hence orthogonality) to these higher dimensions. The higher dimensional spaces are hard to visualize (except by analogy), but projections (discussed in the next section in the Hilbert space setting) and algebraic manipulations are straightforward. What if we wish to extend all these nice properties to an infinite-dimensional space? Naively, we might jump to the conclusion that $\mathbb{R}^\infty$ (where vectors are infinite sequences) is a good candidate for an infinite-dimensional space. True, it is infinite dimensional, but we have some obvious problems extending all of the "nice" properties of $\mathbb{R}^n$. If we try to extend the inner product from $\mathbb{R}^n$ to $\mathbb{R}^\infty$ simply by exchanging (finite) sums for (infinite) series, then there is a problem. For example, the vector of all 1's has infinite length! As is always the case, the transition from the finite to the infinite raises a concern over divergence. If we address this *single* concern and take the subset of $\mathbb{R}^\infty$ of all vectors of finite length (i.e., all vectors from $\mathbb{R}^\infty$ whose components are square summable), then we *can* preserve all the nice properties of $\mathbb{R}^n$. As we'll see in the next section, l^2 will share with $\mathbb{R}^n$ such properties as the Pythagorean Theorem (which, for geometry fans, indicates that l^2, like $\mathbb{R}^n$, is flat — it has Euclidean geometry), the Parallelogram Rule, orthogonality of vectors, the Gram–Schmidt process, and the possession of an orthonormal basis.

Exercises

1. Use the completeness of $L^2([a, b])$ to show that $L^2(\mathbb{R})$ is complete.
2. Prove that l^2 is complete.
3. The *Legendre polynomials* are defined as

$$P_n(x) = \frac{1}{2^n n!} \frac{d^n}{dx^n}(x^2 - 1)^n \text{ for } n \in \mathbb{N}$$

and $P_0(x) = 1$. In the space $L^2([-1, 1])$, show that $\langle P_n, x^m \rangle = 0$ for $m \in \mathbb{N}$ and $m < n$, and then show that the set of Legendre polynomials is an orthogonal set. Show that $\|P_n\| = 1/\sqrt{n + \frac{1}{2}}$ and hence the set $\left\{ \sqrt{n + \frac{1}{2}} P_n(x) \mid n = 0, 1, 2, \ldots \right\}$ forms an orthonormal set.

4. We will see in Chapter 6 that the set

$$\left\{ \frac{1}{\sqrt{2\pi}}, \frac{\cos x}{\sqrt{\pi}}, \frac{\sin x}{\sqrt{\pi}}, \frac{\cos 2x}{\sqrt{\pi}}, \frac{\sin 2x}{\sqrt{\pi}}, \ldots \right\}$$

is an orthonormal basis for $L^2([-\pi, \pi])$. For now, show that this is in fact an orthonormal set.

5. A set D is said to be *dense* in l^2 if the topological closure of D is l^2. That is, if $x \in l^2$ then for all $\epsilon > 0$ there exists $d \in D$ such that $\|x - d\|_2 < \epsilon$. Find a countable dense subset of l^2.

4 Projections and Hilbert Space Isomorphisms

Since Hilbert spaces are endowed with an inner product, they have much of the associated geometry of familiar vector spaces. We have already defined orthogonality using the inner product and, of course, the inner product induces a norm. In this section, we take many of the ideas from $\mathbb{R}^n$ and extend them to Hilbert spaces.

Recall that the *projection* of a vector $\mathbf{x} \in \mathbb{R}^n$ onto a nonzero vector $\mathbf{a} \in \mathbb{R}^n$ is $\text{proj}_{\mathbf{a}}(\mathbf{x}) = \dfrac{\mathbf{x} \cdot \mathbf{a}}{\mathbf{a} \cdot \mathbf{a}} \mathbf{a}$. Therefore, in an inner product space we define the *projection of f onto nonzero g* as $\text{proj}_g(f) = \dfrac{\langle f, g \rangle}{\langle g, g \rangle} g$.

For a nonempty set S in a Hilbert space H, we say that $h \in H$ is *orthogonal* to S if $\langle h, s \rangle = 0$ for all $s \in S$. The *orthogonal complement* of S is

$$S^{\perp} = \{h \in H \mid \langle h, s \rangle = 0 \text{ for all } s \in S\}.$$

($S^{\perp}$ is pronounced "S perp" and $S^{\perp}$ is sometimes called the "perp space" of S.) In fact, $S^{\perp}$ is itself a Hilbert space:

Theorem 5.4.1 For any nonempty set S in a Hilbert space H, the set $S^{\perp}$ is a Hilbert space.

Proof: Clearly, $S^{\perp}$ is a vector space. We only need to show that it is complete. Let (s_n) be a Cauchy sequence in $S^{\perp}$. Then, since H is complete, there exists $h \in H$ such that $\lim\limits_{n \to \infty} s_n = h$. Now for all $s \in S$ we have

$$\langle h, s \rangle = \left\langle h, \lim_{n \to \infty} s_n \right\rangle = \lim_{n \to \infty} \langle h, s_n \rangle = 0,$$

since the inner product is continuous (Exercise 6 of Section 5.2). So $h \in S^{\perp}$ and (s_n) converges in $S^{\perp}$. Therefore $S^{\perp}$ is complete. ■

We now use the idea of an orthogonal complement to decompose a Hilbert space into subspaces. As we will see, the decomposition is an algebraic and not a set theoretic decomposition (it will involve vector addition as opposed to set union).

Theorem 5.4.2 Let S be a subspace of a Hilbert space H (that is, the set of vectors in S is a subset of the set of vectors in H, and S itself is a Hilbert space). Then for any $h \in H$, there exists a unique $t \in S$ such that $\inf\limits_{s \in S} \|h - s\| = \|h - t\|$.

Proof: Let $d = \inf\limits_{s \in S} \|h - s\|$ and choose a sequence $(s_n) \subset S$ such that $\lim\limits_{n \to \infty} \|h - s_n\| = d$. Then

$$
\begin{aligned}
\|s_m - s_n\|^2 &= \|(s_m - h) - (s_n - h)\|^2 \\
&= 2\|s_m - h\|^2 + 2\|s_n - h\|^2 - \|s_m + s_n - 2h\|^2 \\
&\quad \text{by the Parallelogram Law (Exercise 3 of Section 5.2)} \\
&= 2\|s_m - h\|^2 + 2\|s_n - h\|^2 - 4\left\|h - \frac{1}{2}(s_m + s_n)\right\|^2 \\
&\leq 2\|s_m - h\|^2 + 2\|s_n - h\|^2 - 4d^2 \text{ since } \frac{1}{2}(s_m + s_n) \in S.
\end{aligned}
$$

Now the fact that $\lim\limits_{n \to \infty} \|h - s_n\| = d$, implies that as $m, n \to \infty$, $\|s_m - s_n\| \to 0$ and so (s_n) is Cauchy and hence convergent to some $t_1 \in S$, where $\lim\limits_{n \to \infty} \|h - s_n\| = \|h - t_1\|$. For uniqueness, suppose for some $t_2 \in S$ we also have $\lim\limits_{n \to \infty} \|h - s_n\| = \|h - t_2\|$. Then

$$
\begin{aligned}
\|t_1 - t_2\|^2 &= 2\|h - t_1\|^2 + 2\|h - t_2\|^2 - 4\left\|h - \frac{1}{2}(t_1 + t_2)\right\|^2 \text{ (as above)} \\
&= 4d^2 - 4\left\|h - \frac{1}{2}(t_1 + t_2)\right\|^2.
\end{aligned}
$$

Next, $\frac{1}{2}(t_1 + t_2) \in S$ and so $\left\|h - \frac{1}{2}(t_1 + t_2)\right\| \geq \inf\limits_{s \in S} \|h - s\| = d$. Therefore $\|t_1 - t_2\|^2 = 0$ and $t_1 = t_2$. $\blacksquare$

In fact, Theorem 5.4.2 can be generalized somewhat. We say a set S, a subset of a Hilbert space (or a vector space), is *convex* if for each $s_1, s_2 \in S$, the vector $ts_1 + (1 - t)s_2$ is in S for all $t \in [0, 1]$. Clearly subspaces of a Hilbert space are convex. Theorem 5.4.2 can be generalized by requiring S to be a closed convex set instead of being a subspace (see the exercises). Subset S is a *cone* if for all $s \in S$ we have the vector $ts \in S$ for all $t > 0$. A cone which is convex is a *convex cone*.

EXAMPLE 5.4.1: Consider the set C of all vectors in $\mathbb{R}^3$ of the form (x, y, z) where $x^2 + y^2 \leq z^2$. C is a cone since $x^2 + y^2 \leq z^2$ implies $(tx)^2 + (ty)^2 \leq (tz)^2$, for all $t > 0$, but t is not convex since $(1, 1, \sqrt{2}), (1, 1, -\sqrt{2}) \in C$, but $\frac{1}{2}(1, 1, \sqrt{2}) + \frac{1}{2}(1, 1, -\sqrt{2}) = (1, 1, 0) \notin C$.

We now use Theorem 5.4.2 to uniquely decompose elements of H into a sum of an element of S and an element of $S^\perp$.

Theorem 5.4.3 Let S be a subspace of a Hilbert space H. Then for all $h \in H$, there exists a unique decomposition of the form $h = s + s'$ where $s \in S$ and $s' \in S^\perp$.

Proof: For $h \in H$, let t be as defined in Theorem 5.4.2. Let $r = h - t$. Now for any $s_1 \in S$ and any scalar a,

$$\|r\|^2 = \|h - t\|^2 \leq \|h - (t + as_1)\|^2 = \|r - as_1\|^2$$

$$= \langle r - as_1, r - as_1 \rangle = \|r\|^2 - \langle as_1, r \rangle - \langle r, as_1 \rangle + |a|^2 \|s_1\|^2.$$

Therefore $0 \leq |a|^2 \|s_1\|^2 - \langle as_1, r \rangle - \langle r, as_1 \rangle$. If the inner product is complex valued and a is real, then we have $0 \leq |a|^2 \|s_1\|^2 - 2|a|\text{Re}\langle r, s_1 \rangle$. Therefore by letting $a \to 0$, we see that $\text{Re}\langle r, s_1 \rangle = 0$. If a is purely imaginary, say $a = ib$ where $b \in \mathbb{R}$, then

$$\begin{aligned} 0 &\leq b^2 \|s_1\|^2 - \langle ibs_1, r \rangle - \langle r, ibs_1 \rangle \\ &= b^2 \|s_1\|^2 - ib\langle s_1, r \rangle + ib\langle r, s_1 \rangle \\ &= b^2 \|s_1\|^2 + ib \left(2i\text{Im}\langle r, s_1 \rangle \right) \\ &= b^2 \|s_1\|^2 - 2b\text{Im}\langle r, s_1 \rangle. \end{aligned}$$

Therefore, $b\|s_1\|^2 - 2\text{Im}\langle r, s_1 \rangle \geq 0$ and upon letting $b \to 0^+$, we see that $\text{Im}\langle r, s_1 \rangle = 0$. Hence, $\langle r, s_1 \rangle = 0$ for all $s_1 \in S$, and therefore $r \in S^\perp$. So we have written h as $h = t + r$ where $t = s \in S$ and $r = s' \in S^\perp$.

Now suppose that $h = t_1 + r_1 = t_2 + r_2$ where $t_1, t_2 \in S$ and $r_1, r_2 \in S^\perp$. Then $t_1 - t_2 = r_2 - r_1$ where $t_1 - t_2 \in S$ and $r_2 - r_1 \in S^\perp$. Since the only element common to S and $S^\perp$ is 0, then $t_1 - t_2 = r_2 - r_1 = 0$ and $t_1 = t_2$ and $r_1 = r_2$. Therefore the decomposition of h is unique. ∎

Theorem 5.4.3 allows us now to define the projection of a vector onto a subspace. With the notation of Theorem 5.4.3, define the *projection of vector $h \in H$ onto subspace S* as the unique $s \in S$ such that $h = s + s'$ for $s' \in S^\perp$, denoted $\text{proj}_S(h) = s$. We also say that H can be written as the *direct sum* of S and $S^\perp$, denoted $H = S \oplus S^\perp$, and that H has this as an (orthogonal) decomposition. Theorems 5.4.2 and 5.4.3 combine to show that the best approximation of $h \in H$ by elements of a subspace S is $\text{proj}_S(h)$.

Recall that the *standard basis* for $\mathbb{R}^n$ is $\{(1, 0, 0, \dots, 0), (0, 1, 0, \dots, 0), \dots, (0, 0, 0, \dots, 0, 1)\}$. This basis has two desirable properties, namely that each vector is a unit vector and the vectors are pairwise orthogonal. In an inner product space, a set of nonzero vectors is said to be *orthogonal* if the vectors are pairwise orthogonal, and an orthogonal set of vectors in which each vector is a unit vector is an *orthonormal set*. We are interested in Hilbert spaces with a basis in the sense of Schauder which is also an orthonormal set.

Definition 5.4.1 A Schauder basis of a Hilbert space which is also an orthonormal set is called an *orthonormal basis*.

Theorem 5.4.4 A Hilbert space with a Schauder basis has an orthonormal basis.

Proof: We start with a Schauder basis $S = \{s_1, s_2, \dots\}$ and construct an orthonormal basis $R = \{r_1, r_2, \dots\}$ using a method called the *Gram–Schmidt process*.

First define $r_1 = s_1/\|s_1\|$. Now for $k \geq 2$ define $R_k = \text{span}\{r_1, r_2, \dots, r_{k-1}\}$ and

$$r_k = \frac{s_k - \text{proj}_{R_k}(s_k)}{\|s_k - \text{proj}_{R_k}(s_k)\|}.$$

Then for $i \neq j$,

$$\langle r_i, r_j \rangle = \frac{\left\langle s_i - \text{proj}_{R_i}(s_i), s_j - \text{proj}_{R_j}(s_j) \right\rangle}{\|s_i - \text{proj}_{R_i}(s_i)\| \|s_j - \text{proj}_{R_j}(s_j)\|}.$$

Clearly the r_i's are unit vectors. We leave as an exercise the proof that the r_i's are pairwise orthogonal. ∎

The Gram–Schmidt process introduced in the proof of the previous theorem is very geometric. By removing from s_k its projection onto the space R_k, we construct r_k such that $r_k \in R_k^{\perp}$. Since $r_k \in R_{k+1}^{\perp}$, we are assured that the r_i's are orthogonal.

Theorem 5.4.5 If $R = \{r_1, r_2, \dots\}$ is an orthonormal basis for a Hilbert space H and if $h \in H$, then

$$h = \sum_{k=1}^{\infty} \langle h, r_k \rangle r_k.$$

Proof: We know from Bessel's inequality (Corollary 5.2.1) that for all $n \in \mathbb{N}$,

$$\sum_{k=1}^{n} |\langle h, r_k \rangle|^2 \leq \|h\|^2.$$

Therefore $s_n = \sum_{k=1}^{n} |\langle h, r_k \rangle|^2$ is a monotone bounded sequence of real numbers and hence converges (by Corollary 1.4.1) and is Cauchy. Define $h_n = \sum_{k=1}^{n} \langle h, r_k \rangle r_k$. Then for $n > m$

$$\|h_n - h_m\|^2 = \left\| \sum_{k=m}^{n} \langle h, r_k \rangle r_k \right\|^2 = \sum_{k=m}^{n} |\langle h, r_k \rangle|^2,$$

and as a consequence, the sequence (h_n) is a Cauchy sequence in H and so converges to some $h' \in H$. So for each $i \in \mathbb{N}$,

$$\langle h - h', r_i \rangle = \left\langle h - \lim_{n \to \infty} \sum_{k=1}^{n} \langle h, r_k \rangle r_k, r_i \right\rangle$$

$$= \lim_{n \to \infty} \left\langle h - \sum_{k=1}^{n} \langle h, r_k \rangle r_k, r_i \right\rangle$$

$$= \langle h, r_i \rangle - \langle h, r_i \rangle = 0$$

Therefore by Exercise 8, $h - h' = 0$ and $h = \sum_{k=1}^{\infty} \langle h, r_k \rangle r_k$. ∎

Theorem 5.4.6 Let $\{r_1, r_2, \dots\}$ be an orthonormal basis for Hilbert space H, let $R_k = \mathrm{span}\{r_1, r_2, \dots, r_{k-1}\}$, and let $h \in H$. Then $\inf_{s \in R_k} \|h - s\| = \|h - t\|$ where $t = \sum_{i=1}^{k-1} \langle h, r_i \rangle r_i$. That is, best approximations of h are given by partial sums of the orthonormal series of h.

Proof: Suppose to the contrary that there exists $t' \in R_k$ where $\inf_{s \in R_k} \|h - s\| = \|h - t'\| < \|h - t\|$ (we know the infimum holds for a unique element of R_k by Theorem 5.4.3). Then $t' = \sum_{i=1}^{k-1} t_i' r_i$ for some $t_1', t_2', \dots, t_{k-1}'$.

Now $\|h - t\|^2 = \left\|\sum_{i=1}^{\infty}\langle h, r_i\rangle r_i\right\|^2 = \sum_{i=k}^{\infty}|\langle h, r_i\rangle|^2$ and $\|h - t'\|^2 = \left\|\sum_{i=1}^{k-1}(\langle h, r_i\rangle - \right.$

$\left. t_i'\rangle r_i + \sum_{i=k}^{\infty}\langle h, r_i\rangle r_i\right\|^2 = \sum_{i=1}^{k-1}|\langle h, r_i\rangle - t_i'|^2 + \sum_{i=k}^{\infty}|\langle h, r_i\rangle|^2$. Clearly this contradicts

$\|h - t'\| < \|h - t\|$.   ∎

Notice that the previous theorem is consistent with our experience in $\mathbb{R}^n$. It merely states that every element h of a Hilbert space can be written as an (infinite) linear combination of the projections of h onto the elements of an orthonormal basis R:

$$h = \sum_{k=1}^{\infty} \text{proj}_{r_k}(h).$$

Though we don't usually speak of vectors in Hilbert spaces as having magnitude and direction, there is still some validity to this idea. Just as there are n "fundamental directions" $(1, 0, 0, \dots, 0), (0, 1, 0, \dots, 0), \dots, (0, 0, 0, \dots, 0, 1)$ in $\mathbb{R}^n$ (the "fundamental" property being given by the fact that every "direction" [i.e. nonzero vector] is a linear combination of these "directions" [i.e., vectors]), there are a countable number of "fundamental directions" in a Hilbert space with an orthonormal basis. This is a particularly tangible idea when we consider the Hilbert space

$$l^2 = \left\{(x_1, x_2, \dots) \,\middle|\, x_k \in \mathbb{R}, \sum_{k=1}^{\infty}|x_k|^2 < \infty\right\}$$

with orthonormal basis $R = \{(1, 0, 0, \dots), (0, 1, 0, \dots), (0, 0, 1, 0, \dots), \dots\}$.

Theorem 5.4.7 If $R = \{r_1, r_2, \dots\}$ is an orthonormal basis for a Hilbert space H and for $h \in H$, $h = \sum_{k=1}^{\infty} a_k r_k$, then $\|h\| = \sum_{k=1}^{\infty}|a_k|^2$.

We leave the proof of Theorem 5.4.7 as an exercise.

We state the following definition for Hilbert spaces, though it is valid for any space with a metric (see Exercise 26 of Section 1.4).

Definition 5.4.2 A Hilbert space with a countable dense subset is *separable*. That is, a separable Hilbert space H has a subset $D = \{d_1, d_2, \dots\}$ such that for any

$h \in H$ and for all $\epsilon > 0$, there exists $d_k \in D$ with $\|h - d_k\| < \epsilon$. Therefore the (topological) closure of D is H.

Most texts concentrate on a study of separable Hilbert spaces. We prefer to focus on Hilbert spaces with orthonormal bases. The following result resolves our apparently novel approach!

Theorem 5.4.8 A Hilbert space with scalar field $\mathbb{R}$ or $\mathbb{C}$ is separable if and only if it has an orthonormal basis.

   Proof: Suppose H is separable and $\{d_1, d_2, \dots\}$ is dense in H. For $k \geq 2$ define $D_k = \text{span}\{d_1, d_2, \dots, d_{k-1}\}$ and $e_k = d_k - \text{proj}_{D_k}(d_k)$. Then the set $E = \{e_1, e_2, \dots\} \setminus \{0\}$ is linearly independent (in the sense of Schauder) and dense in H. Applying the Gram–Schmidt process to E yields an orthonormal basis of H.
   Next, suppose $R = \{r_1, r_2 \dots\}$ is an orthonormal basis for H. Then for each $h \in H$, $h = \sum_{k=1}^{\infty} \langle h, r_k \rangle r_k$, by Theorem 5.4.5. Let $\epsilon > 0$ be given. For each k, there exists a rational number (or a rational complex number if $\langle h, r_k \rangle$ is complex) a_k such that $|\langle h, r_k \rangle - a_k| < \frac{\epsilon}{2^k}$. Then $\left| h - \sum_{k=1}^{\infty} a_k r_k \right| < \epsilon$ and the set

$$D = \left\{ \sum_{k=1}^{\infty} a_k r_k \ \middle| \ a_k \text{ is rational (or complex rational)} \right\}$$

is a countable dense subset of H.   ∎

   We want to study Hilbert spaces with orthonormal bases. Fortunately, there are not very many such spaces! In fact, there is (up to isomorphism) only one such space.

Definition 5.4.3 Let H_1 and H_2 be Hilbert spaces. If there exists a one-to-one and onto mapping $\pi : H_1 \mapsto H_2$ such that inner products are preserved: $(h, h') = (\pi(h), \pi(h'))$ for all $h, h' \in H_1$, then π is a *Hilbert space isomorphism* and H_1 and H_2 are *isomorphic*.

We are now ready to extend the Fundamental Theorem of Finite-Dimensional Vector Spaces (Theorem 5.1.2) to the infinite-dimensional case.

Theorem 5.4.9 [Fundamental Theorem of Infinite-Dimensional Vector Spaces]
Let H be a Hilbert space with an infinite orthonormal basis. Then H is isomorphic to l^2.

Proof: Let the orthonormal basis of H be $R = \{r_1, r_2, \dots\}$. Then for $h \in H$, define $\pi(h)$ to be the sequence of inner products of h with the elements of R:

$$\pi(h) = (\langle h, r_1 \rangle, \langle h, r_2 \rangle, \dots).$$

By Theorem 5.4.5, $h = \sum_{k=1}^{\infty} \langle h, r_k \rangle r_k$ and by Theorem 5.4.7 $\pi(h) \in l^2$. Now the representation of h in terms of the basis elements is unique, so π is one-to-one. Next, let $\sum_{k=1}^{\infty} |a_k|^2 \in l^2$, and consider the partial sums, s_n, of $\sum_{k=1}^{\infty} a_k r_k$. Then for $m < n$

$$\|s_n - s_m\|^2 = \left\| \sum_{k=m+1}^{n} a_k r_k \right\|^2 = \sum_{k=m+1}^{n} |a_k|^2.$$

Since $\sum_{k=1}^{\infty} |a_k|^2$ converges (its associated sequence of partial sums is a monotone, bounded sequence of real numbers), then the sequence of partial sums of this series is convergent and hence Cauchy. Therefore, (s_n) is a Cauchy sequence in H and hence is convergent. Therefore π is onto.

Now consider $h, h' \in H$ where $h = \sum_{k=1}^{\infty} a_k r_k$ and $h' = \sum_{k=1}^{\infty} a'_k r_k$. Then

$$\langle h, h' \rangle = \left\langle \sum_{k=1}^{\infty} a_k r_k, \sum_{k=1}^{\infty} a'_k r_k \right\rangle = \sum_{k=1}^{\infty} |a_k|^2$$

$$= \langle (a_1, a_2, \dots), (a'_1, a'_2, \dots) \rangle = \langle \pi(h), \pi(h') \rangle.$$

Therefore π is a Hilbert space isomorphism. ∎

As in Definition 5.1.7, we define a transformation T from one Hilbert space H_1 to another Hilbert space H_2 (both with the same scalar field) as *linear* if for all $h, h' \in H_1$ and for all scalars f and f', we have $T(fh + f'h') = fT(h) + f'T(h')$. We can use Theorem 5.4.9 to find a representation of a linear transformation from one Hilbert space with an infinite orthonormal basis to another such space.

Theorem 5.4.10 If $T : H_1 \mapsto H_2$ is a linear transformation where both H_1 and H_2 are Hilbert spaces (over the same field) with infinite orthonormal bases, then T is equivalent to the action of an infinite matrix $(A_{ij})_{i,j \in \mathbb{N}}$.

The proof parallels the proof of Theorem 5.1.3, and we leave it as an exercise.

Exercises

1. Find the projection of $f(x) = x^2$ onto $\sin x$ in $L^2([-\pi, \pi])$.
2. Prove that if S is an orthonormal set in an inner product space, then any finite subset of S is linearly independent.
3. Prove that Theorem 5.4.2 holds for S a closed convex set.
4. (a) Prove that if S is a subspace of an inner product space H and S has orthonormal basis $\{s_1, s_2, \ldots, s_n\}$ then for any $h \in H$, we have
$$\text{proj}_S(h) = \sum_{k=1}^{n} \langle h, s_k \rangle s_k.$$
   (b) Use induction to show that $\langle r_i, r_j \rangle = 0$ for $i \neq j$, where the r_i's are as defined in Theorem 5.4.4.
5. Suppose Hilbert space H has orthonormal basis $R = \{r_1, r_2, \ldots\}$. Let $h \in H$. Prove that $\|h\|^2 = \sum_{k=1}^{\infty} |\langle h, r_i \rangle|^2$.
6. Show that the natural orthonormal basis $R = \{(1, 0, 0, \ldots), (0, 1, 0, \ldots), \ldots\}$ for l^2 is a (topologically) closed set and a bounded set, but not a compact set. Recall that the Heine–Borel Theorem (Theorem 1.4.11) states that a set in $\mathbb{R}^n$ is compact if and only if it is closed and bounded. The example given here shows that the familiar Heine–Borel Theorem does not hold in all metric spaces. Also notice that R is an infinite bounded set with no limit points, indicating that the Bolzano–Weierstrass Theorem (see Exercise 3 of Section 1.4) does not hold in l^2.
7. Prove Theorem 5.4.7.
8. Prove that if $\{r_1, r_2, \ldots\}$ is an orthonormal basis for Hilbert space H, and for $h \in H$ we have $\langle h, r_i \rangle = 0$ for all $i \in \mathbb{N}$, then $h = 0$.
9. Prove Theorem 5.4.10.
10. Apply the Gram–Schmidt process to the set $\{1, x, x^2, x^3\}$ in the space $L^2([-1, 1])$. Verify that this yields the first four Legendre polynomials (see Exercise 3 of Section 5.3).

5 Banach Spaces

In our study of Hilbert spaces, we had the luxury of a norm with which to measure distances and an inner product with which to measure angles. In this section, we consider vector spaces with a norm, but will omit the requirement of an inner product.

Definition 5.5.1 A vector space with a norm is a *normed linear space*. A normed linear space which is complete with respect to the norm is a *Banach space*. That is, a Banach space is a normed linear space in which Cauchy sequences converge.

Since an inner product induces a norm, we see that every Hilbert space is a Banach space.

> **EXAMPLE 5.5.1:** Consider the linear space $C([a, b])$ of all continuous functions on $[a, b]$. Define $\|f\| = \max_{x \in [a,b]} |f(x)|$. Then we claim $C([a, b])$ is a Banach space with norm $\| \cdot \|$. First, it is easy to verify that $\| \cdot \|$ is a norm. Let (f_n) be a Cauchy sequence in $C([a, b])$. That is, for any $\epsilon_1 > 0$ there exists $N_1 \in \mathbb{N}$ such that for all $m, n > N_1$ we have $\|f_n - f_m\| < \epsilon_1$. Therefore for each $x \in [a, b]$, the sequence of real numbers $(f_n(x))$ is Cauchy. Since $\mathbb{R}$ is complete (or, if $f : \mathbb{R} \mapsto \mathbb{C}$, since $\mathbb{C}$ is complete), $(f_n(x))$ converges to some real number, call it $f(x)$. With ϵ_1, n, and N_1 as above and letting $m \to \infty$, we see that $|f_n(x) - f(x)| < \epsilon$ for all $x \in [a, b]$. Therefore f is the limit of (f_n) under the norm $\| \cdot \|$. We now only need to show that $f \in C([a, b])$.
>
> Let $\epsilon_2 > 0$. Then there exists $N_2 \in \mathbb{N}$ such that $|f_{N_2}(x_0) - f(x_0)| < \epsilon_2/3$ for all $x_0 \in [a, b]$. Since f_{N_2} is continuous, there exists $\delta > 0$ such that $|f_{N_2}(x_0) - f_{N_2}(y)| < \epsilon_2/3$ whenever $|x_0 - y| < \delta$ and $x_0, y \in [a, b]$. Hence for such y,

$$|f(x_0) - f(y)| \leq |f(x_0) - f_{N_2}(x_0)| + |f_{N_2}(x_0) - f_{N_2}(y)| + |f_{N_2}(y) - f(y)|$$
$$< \frac{\epsilon_2}{3} + \frac{\epsilon_2}{3} + \frac{\epsilon_2}{3} = \epsilon_2.$$

Therefore f is continuous at x_0 and since x_0 is arbitrary, f is continuous on $[a, b]$. Therefore $C([a, b])$ is complete and so is a Banach space.

For $p \geq 1$ and measurable set E, define

$$L^p(E) = \left\{ f \, \middle| \, \int_E |f|^p < \infty \right\}.$$

We partition $L^p(E)$ into equivalence classes by defining the equivalence relation $f \equiv g$ if and only if $f = g$ a.e. on E. As a consequence, we will not draw a distinction between two functions, which are equal almost everywhere. We define $\|f\|_p = \left\{ \int_E |f|^p \right\}^{1/p}$ and therefore $\|f\|_p = 0$ if and only if $f \equiv 0$. Also, for any scalar a, $\|af\|_p = |a| \|f\|_p$. For the case $p = \infty$, we define $L^\infty(E)$ to be the set of essentially bounded functions on E and let $\|f\|_\infty = \text{ess sup}\{|f(x)| \, | \, x \in E\}$. We now go to the task of showing that $L^p(E)$ (or simply L^p) is a Banach space. We establish the triangle inequality in the following theorem.

Theorem 5.5.1 [Minkowski's inequality] Let $p \geq 1$ and $f, g \in L^p$. Then for $1 \leq p \leq \infty$ we have $\|f + g\|_p \leq \|f\|_p + \|g\|_p$.

Proof: Suppose $1 \leq p < \infty$. If $\|f\|_p = 0$ or $\|g\|_p = 0$ then the result is obvious. Now for $x \in E$

$$|f(x) + g(x)|^p \leq (|f(x)| + |g(x)|)^p = \left(\|f\|_p \frac{|f(x)|}{\|f\|_p} + \|g\|_p \frac{|g(x)|}{\|g\|_p} \right)^p$$

$$= (\|f\|_p + \|g\|_p)^p \left(\frac{\|f\|_p}{\|f\|_p + \|g\|_p} \frac{|f(x)|}{\|f\|_p} + \frac{\|g\|_p}{\|f\|_p + \|g\|_p} \frac{|g(x)|}{\|g\|_p} \right)^p$$

$$= (\|f\|_p + \|g\|_p)^p \left(\lambda \frac{|f(x)|}{\|f\|_p} + (1 - \lambda) \frac{|g(x)|}{\|g\|_p} \right)^p$$

where $\lambda = \dfrac{\|f\|_p}{\|f\|_p + \|g\|_p}$. Now the function $h(y) = y^p$ is concave up on $y \in [0, \infty)$ for $p \geq 1$. Since $\lambda \in (0, 1)$; then

$$\lambda \frac{|f(x)|}{\|f\|_p} + (1 - \lambda) \frac{|g(x)|}{\|g\|_p} \in \left(\frac{|f(x)|}{\|f\|_p}, \frac{|g(x)|}{\|g\|_p} \right)$$

and

$$\left(\lambda \frac{|f(x)|}{\|f\|_p} + (1 - \lambda) \frac{|g(x)|}{\|g\|_p} \right)^p \leq \lambda \left(\frac{|f(x)|}{\|f\|_p} \right)^p + (1 - \lambda) \left(\frac{|g(x)|}{\|g\|_p} \right)^p.$$

Therefore for all $x \in E$,

$$|f(x) + g(x)|^p \leq (\|f\|_p + \|g\|_p)^p \left\{ \lambda \left(\frac{|f(x)|}{\|f\|_p} \right)^p + (1 - \lambda) \left(\frac{|g(x)|}{\|g\|_p} \right)^p \right\}$$

and integrating over E, we get the desired result. We leave the case $p = \infty$ as an exercise. ∎

Minkowski's inequality accomplishes two things — it guarantees that L^p is closed under addition and that $\|\cdot\|_p$ is a norm on L^p. Therefore the L^p spaces are normed linear spaces (at least for $1 \le p \le \infty$). If we try to extend the above arguments to L^p spaces where $0 < p < 1$, then there is a problem. In particular, we can modify the proof of Minkowski's inequality to show that for $f, g \in L^p$ with $0 < p < 1$, we have $\|f + g\|_p \ge \|f\|_p + \|g\|_p$ (see the exercises). Therefore the L^p spaces are not normed when $0 < p < 1$.

Theorem 5.5.2 [Hölder's inequality] If p and q are nonnegative (extended) real numbers such that $\dfrac{1}{p} + \dfrac{1}{q} = 1$, and if $f \in L^p$ and $g \in L^q$, then $fg \in L^1$ and $\|fg\|_1 \le \|f\|_p \|g\|_q$.

Proof: Suppose $p = 1$ and $q = \infty$. Then

$$\int |fg| \le \int |f| \|g\|_\infty = \|g\|_\infty \|f\|_1.$$

Therefore $fg \in L^1$ and $\|fg\|_1 \le \|f\|_1 \|g\|_\infty$.

Now suppose $1 < p < \infty$ and $1 < q < \infty$. Since L^p norms are calculated from absolute values, we may assume without loss of generality that f and g are nonnegative. First we show the result holds for $\|f\|_p = \|g\|_q = 1$. Define $\Phi(x) = \dfrac{x}{p} + \dfrac{1}{q} - x^{1/p}$. Then $\Phi(0) = \dfrac{1}{q} > 0$ and

$$\Phi'(x) = \frac{1}{p} - \frac{1}{p}x^{1/p-1} = \frac{1}{p} - \frac{1}{p}x^{-1/q} = \frac{1}{p}\left(1 - x^{-1/q}\right) = \frac{1}{p}\left(1 - \frac{1}{\sqrt[q]{x}}\right).$$

Therefore Φ has a minimum at $x = 1$ and $\Phi(1) = 0$. So for $x > 0$, $x^{1/p} \le \dfrac{x}{p} + \dfrac{1}{q}$.
Let $x = \dfrac{f^p}{g^q}$. Then

$$\Phi(x) = \Phi\left(\frac{f^p}{g^q}\right) = \frac{-f}{g^{q/p}} + \frac{f^p}{pg^q} + \frac{1}{q} \ge 0$$

and $gf \le \dfrac{g^q}{q} + \dfrac{f^p}{p}$. Hence

$$\|fg\|_1 \le \frac{\|f\|^p}{p} + \frac{\|g\|^q}{q} = 1 = \|f\|_p \|g\|_q. \tag{5.1}$$

Now for general $f \in L^p$ and $g \in L^q$, replace f and g of (5.1) with $\frac{f}{\|f\|_p}$ and $\frac{g}{\|g\|_q}$, respectively. This yields the result.   ∎

When $p = q = 2$, Hölder's inequality is sometimes called the Schwarz inequality.

We now want to establish the completeness of the L^p spaces for $p \geq 1$. To do so, we will study the summability of series.

Definition 5.5.2 For a sequence (f_n) in a normed linear space, we define the series $\sum\limits_{n=1}^{\infty} f_n$. The *nth partial sum* of this series is $s_n = \sum\limits_{k=1}^{n} f_k$, and we say that the series is *summable* if the sequence (s_n) converges. That is, $\sum\limits_{n=1}^{\infty} f_n$ is summable if there exists f such that

$$\lim_{n \to \infty} \left\| \sum_{k=1}^{n} f_k - f \right\| = 0.$$

In this case, f is called the *sum* of the series. The series $\sum\limits_{n=1}^{\infty} f_n$ is said to be *absolutely summable* if the series $\sum\limits_{n=1}^{\infty} \|f_n\|$ is summable in $\mathbb{R}$.

Before we explore the completeness of L^p, we need the following lemma:

Lemma 5.5.1 A normed linear space is complete if and only if every absolutely summable series is summable.

Proof: Suppose the normed linear space is complete (that is, is a Banach space) and let the series $\sum\limits_{k=1}^{\infty} f_k$ be absolutely summable. Then $\sum\limits_{k=1}^{\infty} \|f_k\|$ is summable and hence for all $\epsilon > 0$, there exists $N \in \mathbb{N}$ such that $\sum_{k=N}^{\infty} \|f_k\| < \epsilon$. Let the nth partial sum of the series $\sum\limits_{k=1}^{\infty} f_k$ be $s_n = \sum\limits_{k=1}^{n} f_k$. Then for $n > m > N$ we have

$$\|s_n - s_m\| = \left\| \sum_{k=m+1}^{n} f_k \right\| \leq \sum_{k=m+1}^{n} \|f_k\| \leq \sum_{k=m+1}^{\infty} \|f_k\| < \epsilon.$$

Therefore the sequence (s_n) is Cauchy and hence converges. That is, the series $\sum_{k=1}^{\infty} f_k$ converges.

Next, suppose that every absolutely summable series is summable, and let (f_k) be a Cauchy sequence. Then for all $l \in \mathbb{N}$, there exists $N_l \in \mathbb{N}$ such that $\|f_m - f_n\| < 1/2^l$ whenever $m > n > N_l$. Notice that we may choose the N_l to be an increasing sequence. Now consider the series $\sum_{l=1}^{\infty} \|f_{N_{l+1}} - f_{N_l}\|$. This series is absolutely summable and so, by our hypothesis, is summable. Say $f = \sum_{l=1}^{\infty} (f_{N_{l+1}} - f_{N_l})$. Notice that the nth partial sum of this series is $\sum_{l=1}^{n} (f_{N_{l+1}} - f_{N_l}) = f_{N_{n+1}} - f_{N_1}$, and so $\lim_{n \to \infty} f_{N_{n+1}} = f$ for some f. We will see that $f = \lim_{k \to \infty} f_k$. Let $\epsilon > 0$. Since (f_k) is Cauchy, there exists $N \in \mathbb{N}$ such that for all $m, n > N$ we have $\|f_n - f_m\| < \epsilon/2$. Since $F_{N_l} \to f$, there exists $L \in \mathbb{N}$ such that for all $l > L$ we have $\|f_{N_l} - f\| < \epsilon/2$. By choosing l such that both $l > L$ and $N_l > N$ then we have for all $n > N$

$$\|f_n - f\| = \|f_n - f_{N_l} + f_{N_l} - f\| \leq \|f_n - f_{N_l}\| + \|f_{N_l} - f\| < \frac{\epsilon}{2} + \frac{\epsilon}{2} = \epsilon.$$

Hence $\lim_{k \to \infty} f_k = f$ and the space is complete. ∎

We are now in a position to establish the completeness of L^p.

Theorem 5.5.3 [Riesz–Fisher Theorem] For $1 \leq p \leq \infty$, $L^p(E)$ is complete.

Proof: First suppose $1 \leq p < \infty$. Let (f_k) be an absolutely summable series: $\sum_{k=1}^{\infty} \|f_k\| \leq M$ for some $M \in \mathbb{R}$. Consider for each $x \in E$ the partial sum $g_n(x) = \sum_{k=1}^{n} |f_k(x)|$. This defines a sequence of functions (g_n) defined on E. Then by Minkowski's inequality (Theorem 5.5.1) we have $\|g_n\| \leq \sum_{k=1}^{n} \|f_k\| \leq M$, and so $\int_E (g_n)^p \leq M^p$. For each $x \in E$, the sequence $(g_n(x))$ is monotone increasing, and so $g = \lim_{n \to \infty} g_n$ on E. Then g is measurable and by Fatou's Lemma (Theorem 3.3.4) we have $\int_e g^p \leq \underline{\lim} \int_E g_n^p \leq M^p$. Therefore $g < \infty$ on $E \setminus I$ where

$m(I) = 0$. Now for each $x \in E \setminus I$, $\sum_{k=1}^{\infty} f_k(x)$ is absolutely summable and hence summable to some number $h(x)$. Extend $h(x)$ to E by defining $h(x) = 0$ for $x \in I$. Then $h(x) = \sum_{k=1}^{\infty} f_k(x)$ a.e. on E and so h is measurable. Now $\left| \sum_{k=1}^{n} f_k(x) \right| \leq g(x)$ for all $n \in \mathbb{N}$ and so $|h(x)| = |\sum_{k=1}^{\infty} f_k(x)| \leq \sum_{k=1}^{\infty} |f_k(x)| \leq g(x)$. Hence

$$\|h(x)\|^p = \int_E |h|^p \leq \int_E |g|^p = \int_E g^p \leq M^p$$

and so $h \in L^p$. We now show that h is the limit of $\sum_{k=1}^{\infty} f_k$ (in the L^p sense). Now for each $x \in E$

$$\left| h(x) - \sum_{k=1}^{n} f_k(x) \right|^p \leq \left(|h(x)| + \left| \sum_{k=1}^{n} f_k(x) \right| \right)^p \leq (2g(x))^p,$$

and since $2^p g^p$ is integrable over E, then

$$\lim_{n \to \infty} \int_E \left| h(x) - \sum_{k=1}^{n} f_k(x) \right|^p = \int_E \lim_{n \to \infty} \left| h(x) - \sum_{k=1}^{n} f_k(x) \right|^p = 0$$

by the Lebesgue Dominated Convergence Theorem (Theorem 3.3.1). That is, $\lim_{n \to \infty} \left\| h(x) - \sum_{k=1}^{n} f_k(x) \right\| = 0$ and so $\sum_{k=1}^{\infty} f_k(x) = h(x)$ (in the L^p sense). Therefore, the absolutely summable series $\sum_{k=1}^{\infty} f_k(x)$ is also summable, and by Lemma 5.5.1, L^p is complete. We leave the proof that L^∞ is complete as an exercise. ∎

We now see that, with an eye toward completeness, we can modify the hierarchy of the spaces mentioned in Section 5.2 to the following:

$$\left(\begin{array}{c} \text{vector} \\ \text{spaces} \end{array} \right) \supset \left(\begin{array}{c} \text{normed} \\ \text{vector} \\ \text{spaces} \end{array} \right) \supset \left(\begin{array}{c} \text{Banach} \\ \text{spaces} \end{array} \right) \supset \left(\begin{array}{c} \text{Hilbert} \\ \text{spaces} \end{array} \right).$$

We now turn our attention to a slightly different topic.

Definition 5.5.3 A *linear functional* T on a normed linear space $\langle V, \mathbb{F} \rangle$ is a mapping $T : V \mapsto \mathbb{F}$ such that for all scalars $f_1, f_2 \in \mathbb{F}$ and all vectors $\mathbf{v}_1, \mathbf{v}_2 \in V$

we have $T(f_1 v_1 + f_2 v_2) = f_1 T(v_1) + f_2 T(v_2)$. Linear functional T is *bounded* if there exists some $M \in \mathbb{R}$ such that $\|T(v)\| \leq M\|v\|$ for all $v \in V$. The collection of all bounded linear functionals from $\langle V, \mathbb{F} \rangle$ to $\mathbb{F}$ make up the *dual space* of $\langle V, \mathbb{F} \rangle$, denoted $\langle V, \mathbb{F} \rangle^*$ or simply V^*.

Notice that if $S, T \in V^*$, then $f_1 S + f_2 T \in V^*$ for all $f_1, f_2 \in \mathbb{F}$. Therefore V^* is itself a linear space.

Definition 5.5.4 For $T \in V^*$ where the scalar field of V is $\mathbb{R}$ or $\mathbb{C}$, define the *functional norm* $\|T\| = \sup\limits_{v \neq 0} \dfrac{|T(v)|}{\|v\|}$. (Notice that this equation actually involves three norms: the functional norm, the absolute value on $\mathbb{R}$ or $\mathbb{C}$, and the norm on the linear space $\langle V, \mathbb{F} \rangle$.)

We now want to investigate the structure of the dual space of L^p. We present a suggestive lemma.

Lemma 5.5.2 If $t \in L^q$ then $T(s) = \int st$ is a bounded linear functional on L^p where $\frac{1}{p} + \frac{1}{q} = 1, 1 \leq p \leq \infty$. In addition $\|T\| = \|t\|_q$.

Proof: By Hölder's inequality (Theorem 5.5.2),

$$|T(s)| = \left| \int st \right| \leq \|st\|_1 \leq \|s\|_p \|t\|_q$$

and hence for all $s \in L^p$, $\dfrac{|T(s)|}{\|s\|_q} \leq \|t\|_p$. Therefore $\|T\| \leq \|t\|_p$, and T is bounded. We leave the proof of the equality as an exercise. ∎

In fact, the converse of Lemma 5.5.2 holds for $1 \leq p < \infty$. This is the well-known Riesz Representation Theorem (the proof of which is rather long, so we omit it):

Theorem 5.5.4 [Riesz Representation Theorem] Let T be a bounded linear functional on L^p, $1 \leq p < \infty$. Then there is a function $t \in L^q$, where $\frac{1}{p} + \frac{1}{q} = 1$, such that $T(s) = \int st$. In addition, $\|T\| = \|t\|_q$.

In slightly different terminology, the Riesz Representation Theorem states that the dual space of L^p is L^q. In particular, we see that L^2 is *self dual* and bounded linear functionals on L^2 can be represented using the inner product:

Corollary 5.5.1 Let T be a bounded linear functional on L^2. Then there is $t \in L^2$ such that $T(s) = \int st = \langle s, t \rangle$. In addition, $\|T\| = \|t\|_2$.

Exercises

1. (a) Prove that Minkowski's inequality, $\|f + g\|_p \le \|f\|_p + \|g\|_p$, reduces to equality if and only if $g = af$ for some scalar a.
   (b) Prove Minkowski's inequality for the case $p = \infty$.
2. Modify the proof of Minkowski's inequality to show that for $f, g \in L^p$ with $0 < p < 1$, we have $\|f + g\|_p \ge \|f\|_p + \|g\|_p$.
3. Consider the vector space $\mathcal{P}_n$ of all complex-valued polynomials of degree n or less. For $1 \le p < \infty$ we define the L^p norm of $P \in L^p$ as

$$\|P\|_p = \frac{1}{2\pi} \left(\int_0^{2\pi} |P(e^{i\theta})|^p \, d\theta \right)^{1/p}.$$

   (a) Verify that $\lim_{p \to \infty} \|P\|_p = \max_{|z|=1} |P(z)|$.
   (b) For $0 < p < 1$, define $\|P\|_p$ as above (though for these values of p, this is not a norm because it does not satisfy the triangle inequality). Define $\|P\|_0 = \exp\left(\frac{1}{2\pi} \int_0^{2\pi} \log |P(e^{i\theta})| \, d\theta \right)$ and verify that $\lim_{p \to 0^+} \|P\|_p = \|P\|_0$. $\|P\|_0$ is called the *Mahler measure* of P [41].
4. Show that a normed linear space is an inner product space if and only if the norm satisfies the Parallelogram Law (see Exercise 3 of Section 5.2).
5. Show that, for $1 \le p \le \infty$, the only space $L^p([a, b])$ which is an inner product space is the space $L^2([a, b])$.
6. Show that $L^p([a, b]) \subset L^q([a, b])$ for $p \ge q$. By example, show that the subset inclusion is proper.
7. Prove that for T a bounded linear functional, $\|T\|$ is the infimum over all M such that $|T(v)| \le M\|v\|$ and also $\|T\| = \sup_{\|v\|=1} |T(v)|$.
8. Complete the Riesz–Fisher Theorem by showing that L^∞ is complete.
9. In Lemma 5.5.2, prove $\|T\| = \|t\|_q$.
10. For $1 \le p < \infty$, define

$$l^p = \left\{ x = (x_1, x_2, \dots) \,\middle|\, x_i \in \mathbb{R} \text{ for } i \in \mathbb{N}, \sum_{i=1}^{\infty} |x_i|^p < \infty \right\}$$

   and $\|x\|_p = \|(x_1, x_2, \dots)\|_p = \{\sum_{i=1}^{\infty} |x_i|^p\}^{1/p}$.
   (a) Prove Minkowski's inequality for l^p: $\|x + y\|_p \le \|x\|_p + \|y\|_p$.

   (b) Prove Hölder's inequality for l^p: $\sum_{i=1}^{\infty} |x_i y_i| \leq \|\mathbf{x}\|_p \|\mathbf{y}\|_p$ where $\frac{1}{p} + \frac{1}{q} = 1$.

   (c) Prove l^p is complete.

11. Define $l^{\infty} = \{\mathbf{x} = (x_1, x_2, \dots) \mid \sup_{i \in \mathbb{N}} |x_i| < \infty\}$. Prove l^{∞} is complete.

12. State a Riesz Representation Theorem for bounded linear functionals on l^p, $1 \leq p < \infty$.

Chapter 6

Fourier Analysis

Many real world phenomena can be described by periodic functions. For example, common household electrical current is alternating current. The voltage V in a typical 115-volt outlet with frequency 60 hertz can be expressed as the function $V = 163 \sin 60(2\pi t) = 163 \sin 120\pi t$. A basic component of music is a pure tone, which is a single sine wave. A musical instrument usually creates several different tones at the same time, which are linear combinations of sine and cosine functions with different frequencies. An electrocardiograph is nearly a periodic function. Joseph Fourier was the first person who created the idea of representing functions in the "frequency" domain to reveal "wave" properties of the function. Thus, he opened a new area in mathematics called *Fourier analysis*.

Joseph Fourier was born in 1768 in Auxserre, a small town in France. In 1807, he presented a memoir to the Institute of France (which was published in 1822 in his book *The Analytic Theory of Heat*), where he claimed that any periodic function could be expressed as the sum of sine and cosine functions. Fourier was motivated by the study of heat diffusion. He wanted to represent the solution of a heat equation in a practical way that would yield a deeper understanding of the physical problem. He found the series of harmonically related sinusoids to be a good representation. His ideas dominated mathematical analysis for 100 years. Outside mathematics, Fourier analysis is used in many areas. It contains all of the central ideas of electrical engineering. Crystallography, the telephone, the x-ray machine, and many other devices use Fourier's ideas.

In Fourier analysis, the complex exponential function e^{ix} is often used. We have by the *Euler formula*,

$$e^{ix} = \cos x + i \sin x, \tag{0.1}$$

that

$$\sin x = \frac{e^{ix} - e^{-ix}}{2i} \qquad \cos x = \frac{e^{ix} + e^{-ix}}{2}. \tag{0.2}$$

Hence, e^{ix} can be considered as a complex sinusoid. In general, a complex-valued function (with a real independent variable) has the form

$$f(x) = a(x) + ib(x), \quad x \in \mathbb{R},$$

where $a(x)$ is the *real part* of $f(x)$ and $b(x)$ is the *imaginary part* of $f(x)$, and both a and b are real-valued functions. The *complex conjugate* of $f(x)$ is

$$\overline{f(x)} = a(x) - ib(x), \quad x \in \mathbb{R}.$$

We have $f'(x) = a'(x) + ib'(x)$ and $\int f(x)\,dx = \int a(x)\,dx + i \int b(x)\,dx$. Hence, the calculus of complex-valued functions is similar to the calculus of real-valued functions. In this chapter, we assume functions are complex valued.

I Fourier Series

Following Fourier's original ideas, we discuss how to represent a periodic function as a series of sine and cosine functions. We first consider 2π-periodic functions. Each element of the set

$$\{1, \sin x, \cos x, \sin 2x, \cos 2x, \cdots\} \tag{1.1}$$

is 2π-periodic. They are "fundamental tones" for 2π-periodic functions: a linear combination of them represents a 2π-periodic function. We call $\frac{1}{2}a_0 + \sum_{k=1}^{n} a_k \cos kx + b_k \sin kx$ a *trigonometric polynomial* and call the series $\frac{1}{2}a_0 + \sum_{k=1}^{\infty} a_k \cos kx + b_k \sin kx$ a *trigonometric series*. The set (1.1) has an important property called *orthogonality*, which is stated in the following.

Theorem 6.1.1 The set (1.1) satisfies

(1) $\int_{-\pi}^{\pi} \sin nx \cos mx \, dx = 0, \quad (n, m) \in \mathbb{Z}^+ \times \mathbb{Z}^+,$

(2) $\frac{1}{\pi} \int_{-\pi}^{\pi} \sin nx \sin kx \, dx = \delta_{n,k}, \quad (n, k) \in \mathbb{N}^2,$

(3) $\frac{1}{\pi} \int_{-\pi}^{\pi} \cos mx \cos lx \, dx = \delta_{m,l}, \quad (m, l) \in (\mathbb{Z}^+ \times \mathbb{Z}^+) \setminus \{(0, 0)\},$

where $\delta_{n,k}$ is Kronecker's symbol: $\delta_{n,k} = \begin{cases} 1, & \text{if } n = k, \\ 0, & \text{otherwise.} \end{cases}$

Proof: We leave the proof as an exercise. ∎

Orthogonality is extremely useful when we represent a periodic function as a trigonometric series. For example, if

$$f(x) = \frac{1}{2}a_0 + \sum_{k=1}^{\infty} a_k \cos kx + b_k \sin kx,$$

where the series is uniformly convergent on $[-\pi, \pi]$, then

$$a_k = \frac{1}{\pi} \int_{-\pi}^{\pi} f(x) \cos kx \, dx, \quad b_k = \frac{1}{\pi} \int_{-\pi}^{\pi} f(x) \sin kx \, dx. \qquad (1.2)$$

Conversely, if $f(x)$ is integrable on $[-\pi, \pi]$, then the integrals (1.2) exist. In this case, we can make a trigonometric series $\frac{1}{2}a_0 + \sum_{k=1}^{\infty} a_k \cos kx + b_k \sin kx$ for f; although we do not know whether the series is convergent. The use of this series to study f was Fourier's original idea. We now put it on a solid foundation.

1.1 The Finite Fourier Transform and Its Properties

Let

$$\tilde{L}^1 = \left\{ f \, \middle| \, f \text{ is } 2\pi\text{-periodic and } \int_{-\pi}^{\pi} |f(x)| \, dx < \infty \right\}$$

Definition 6.1.1 For $f \in \tilde{L}^1$, define

$$\begin{aligned} f_c^{\wedge}(k) &= \frac{1}{\pi} \int_{-\pi}^{\pi} f(x) \cos kx \, dx, \\ f_s^{\wedge}(k) &= \frac{1}{\pi} \int_{-\pi}^{\pi} f(x) \sin kx \, dx, \end{aligned} \qquad k \in \mathbb{Z}^+. \qquad (1.3)$$

Then $f_c^{\wedge}(k)$ and $f_s^{\wedge}(k)$ are called *real Fourier coefficients* of f, and the trigonometric series

$$\frac{1}{2} f_c^{\wedge}(0) + \sum_{k=1}^{\infty} f_c^{\wedge}(k) \cos kx + f_s^{\wedge}(k) \sin kx \qquad (1.4)$$

is called the *real Fourier series* of f.

Note that we do not know whether the series (1.4) converges to $f(x)$. Hence, we use the notation

$$f(x) \sim \frac{1}{2}f_c{}^\wedge(0) + \sum_{k=1}^{\infty} f_c{}^\wedge(k) \cos kx + f_s{}^\wedge(k) \sin kx \qquad (1.5)$$

to indicate that (1.4) is the Fourier series of $f(x)$.

When we multiply two Fourier series together, it is easier to use the complex exponential representation instead of the sine and cosine representation. Hence, we apply the Euler formula (0.2) to change the form of (1.5):

$$\frac{1}{2}f_c{}^\wedge(0) + \sum_{k=1}^{\infty} f_c{}^\wedge(k) \cos kx + f_s{}^\wedge(k) \sin kx$$

$$= \frac{1}{2}f_c{}^\wedge(0) + \sum_{k=1}^{\infty} f_c{}^\wedge(k) \frac{e^{ikx} + e^{-ikx}}{2} + f_s{}^\wedge(k) \frac{e^{ikx} - e^{-ikx}}{2i}$$

$$= \frac{1}{2}f_c{}^\wedge(0) + \sum_{k=1}^{\infty} \frac{f_c{}^\wedge(k) - if_s{}^\wedge(k)}{2} e^{ikx} + \sum_{k=1}^{\infty} \frac{f_c{}^\wedge(k) + if_s{}^\wedge(k)}{2} e^{-ikx}.$$

Write

$$f^\wedge(k) = \frac{f_c{}^\wedge(k) - if_s{}^\wedge(k)}{2}, \quad f^\wedge(-k) = \frac{f_c{}^\wedge(k) + if_s{}^\wedge(k)}{2}, \quad k \in \mathbb{Z}^+;$$

then

$$\frac{1}{2}f_c{}^\wedge(0) + \sum_{k=1}^{\infty} f_c{}^\wedge(k) \cos kx + f_s{}^\wedge(k) \sin kx = \sum_{k \in \mathbb{Z}} f^\wedge(k) e^{ikx}.$$

Definition 6.1.2 The series

$$\sum_{k \in \mathbb{Z}} f^\wedge(k) e^{ikx} \qquad (1.6)$$

is called the (*complex*) *Fourier series* of f, and $f^\wedge(k)$ is called the kth *Fourier coefficient* of f.

It is easy to verify that $\{e^{ikx} \mid k \in \mathbb{Z}\}$ is also an orthogonal system on $[-\pi, \pi]$ in the sense that

$$\frac{1}{2\pi} \int_{-\pi}^{\pi} e^{ikx} \overline{e^{ilx}} \, dx = \delta_{k,l}. \qquad (1.7)$$

Equation (1.7) allows us to compute $f^\wedge(k)$ directly as

$$f^\wedge(k) = \frac{1}{2\pi} \int_{-\pi}^{\pi} f(x) e^{-ikx} \, dx, \quad k \in \mathbb{Z}. \qquad (1.8)$$

Formula (1.8) maps a function f to a sequence $(f\hat{}(k))_{k\in\mathbb{Z}}$. We call it the *finite Fourier transform*.

EXAMPLE 6.1.1: Let $f(x) \in \tilde{L}^1$ be defined by $f(x) = x, x \in [-\pi, \pi)$. Then

$$f(x) \sim 2\sum_{n=1}^{\infty} \frac{(-1)^n}{n} \sin nx.$$

EXAMPLE 6.1.2: Let $f(x) = |\sin x|$. Then

$$f(x) \sim \frac{2}{\pi}\left(1 + \sum_{n=1}^{\infty} \frac{2}{1 - (2n)^2} \cos 2nx\right).$$

EXAMPLE 6.1.3: Let $f(x) \in \tilde{L}^1$ be defined by

$$f(x) = \begin{cases} x(\pi + x), & -\pi \le x \le 0; \\ x(\pi - x), & 0 < x < \pi. \end{cases}$$

Then

$$f(x) \sim \frac{8}{\pi}\sum_{n=1}^{\infty} \frac{1}{(2n-1)^3} \sin(2n-1)x.$$

By definition, if $f(x) = 0$ almost everywhere, then its Fourier series is 0. The converse is also true. In fact, we have the following uniqueness theorem.

Theorem 6.1.2 The Fourier series of $f(x)$ is 0 if and only if $f(x) = 0$ almost everywhere.

The proof of the theorem needs knowledge of Fourier summation. We will not prove it here.

Fourier series are a useful tool for analyzing the frequency properties of a function. If you consider the function $f(t)$ as a record of a piece of music, then its Fourier series exhibits the tones in the music and shows the strength of each tone: The larger (the magnitude of) the kth coefficient, the stronger the kth tone.

We now discuss the basic properties of the finite Fourier transform. First, the finite Fourier transform is linear, i.e.,

$$[f + g]\hat{}(k) = f\hat{}(k) + g\hat{}(k), \quad [cf]\hat{}(k) = cf\hat{}(k).$$

Second, it has the following operation properties.

Lemma 6.1.1 Let $f \in \tilde{L}^1$. For $h \in \mathbb{R}, k \in \mathbb{Z}$, we have

(1) $[f(\cdot + h)]\hat{}(k) = e^{ihk}f\hat{}(k);$

(2) $[e^{-ij\cdot}f(\cdot)]\hat{}(k) = f\hat{}(k + j);$

(3) $\overline{[f(-\cdot)]}\hat{}(k) = \overline{f\hat{}(k)}.$

Proof: We have

$$[f(\cdot + h)]\hat{}(k) = \frac{1}{2\pi} \int_{-\pi}^{\pi} f(x + h)e^{-ikx}\, dx$$

$$= \frac{1}{2\pi} \int_{-\pi-h}^{\pi-h} f(x)e^{-ik(x-h)}\, dx$$

$$= \frac{1}{2\pi} \int_{-\pi}^{\pi} f(x)e^{-ikx}e^{ikh}\, dx = e^{ihk}f\hat{}(k)$$

and

$$[e^{-ij\cdot}f(\cdot)]\hat{}(k) = \frac{1}{2\pi} \int_{-\pi}^{\pi} e^{-ijx}f(x)e^{-ikx}\, dx$$

$$= \frac{1}{2\pi} \int_{-\pi}^{\pi} f(x)e^{-i(k+j)x}\, dx = f\hat{}(k + j).$$

We leave (3) as an exercise.   ■

The following theorem shows that the high-frequency components in a Fourier series are near 0.

Theorem 6.1.3 For $f \in \tilde{L}^1$,

$$\lim_{|k|\to\infty} f\hat{}(k) = 0, \quad \lim_{k\to\infty} f_s\hat{}(k) = 0, \quad \lim_{k\to\infty} f_c\hat{}(k) = 0. \qquad (1.9)$$

This result is a consequence of the famous Riemann–Lebesgue Lemma.

Lemma 6.1.2 [Riemann–Lebesgue] If f is integrable on the interval (a, b), then

$$\lim_{\rho\to\infty} \int_a^b f(t)\sin \rho t\, dt = 0, \quad \lim_{\rho\to\infty} \int_a^b f(t)\cos \rho t\, dt = 0. \qquad (1.10)$$

Proof: We only prove the first limit in (1.10). The proof of the second limit is similar, and we leave it as an exercise. First, we assume f' is continuous. Then

$$\lim_{\rho \to \infty} \int_a^b f(t) \sin \rho t \, dt = \lim_{\rho \to \infty} \left(-\frac{1}{\rho} f(t) \cos \rho t \big|_a^b + \frac{1}{\rho} \int_a^b f'(t) \cos \rho t \, dt \right) = 0.$$

If f is now only integrable on $[a, b]$, then for any $\epsilon > 0$, there is a continuously differentiable function g such that

$$\int_a^b |f(x) - g(x)| \, dx < \epsilon/2.$$

Since (1.10) holds for g, there is an $N > 0$ such that

$$\left| \int_a^b g(x) \sin \rho x \, dx \right| < \epsilon/2, \quad \text{for all } \rho \geq N.$$

Thus, for $\rho \geq N$,

$$\left| \int_a^b f(x) \sin \rho x \, dx \right| = \left| \int_a^b g(x) \sin \rho x \, dx + \int_a^b (f(x) - g(x)) \sin \rho x \, dx \right|$$

$$\leq \left| \int_a^b g(x) \sin \rho x \, dx \right| + \int_a^b |f(x) - g(x)| \, dx < \epsilon.$$

The lemma is proved. ∎

1.2 Convergence of Fourier Series

Is the Fourier series of $f(t)$ always convergent to $f(t)$? In Fourier's time, people thought so, but they were wrong! The convergence of Fourier series is much harder to handle than expected. It took one and a half centuries to fully analyze. We have no plan to discuss the convergence of Fourier series in detail. We only give a very brief introduction here.

Let

$$S_n(f, t) = \sum_{k=-n}^{n} f^{\wedge}(k) e^{ikt},$$

which is called the *Fourier partial sum* (of order n) of f. We consider the pointwise convergence of Fourier series. That is, we address the questions: "For a fixed

$t \in [-\pi, \pi]$, does $\lim_{n \to \infty} S_n(f, t)$ exist? If it exists, is the limit equal to $f(t)$?" The following result is one of the most important results in this direction.

Theorem 6.1.4 If $f(x) \in \tilde{L}^1$ has bounded variation, then at each point $x \in [-\pi, \pi]$, its Fourier series is convergent to

$$\frac{f(x^-) + f(x^+)}{2}.$$

The proof of this theorem is beyond the scope of this book. We omit it. Readers can refer to [2]. By Theorem 6.1.4, the Fourier series of a piecewise differentiable function is convergent everywhere. But be aware that it is convergent to the function value only at points of continuity.

> **EXAMPLE 6.1.4:** We consider the convergence of the Fourier series in Examples 6.1.1, 6.1.2, and 6.1.3. All of them are functions of bounded variation. $f(x)$ in Example 6.1.1 is discontinuous at $x = \pm\pi$, the other two are continuous everywhere. By Theorem 6.1.4, we have
>
> $$2\sum_{n=1}^{\infty} \frac{(-1)^n}{n} \sin nx = x, \quad x \in (-\pi, \pi),$$
>
> $$\frac{2}{\pi}\left(1 + \sum_{n=1}^{\infty} \frac{2}{1 - (2n)^2} \cos 2nx\right) = |\sin x|, \quad x \in [-\pi, \pi],$$
>
> $$\frac{8}{\pi}\sum_{n=1}^{\infty} \frac{1}{(2n-1)^3} \sin(2n-1)x = \begin{cases} x(\pi + x), & -\pi \le x \le 0; \\ x(\pi - x), & 0 < x \le \pi. \end{cases}$$

Note that for $f(x)$ in Example 6.1.1, $f(\pi^-) = \pi$ and $f(\pi^+) = -\pi$. Hence,

$$0 = \frac{f(\pi^+) + f(\pi^-)}{2} = 2\sum_{n=1}^{\infty} \frac{(-1)^n}{n} \sin n\pi.$$

> **EXAMPLE 6.1.5:** Let
>
> $$f(x) = \begin{cases} \frac{\pi}{2}, & x \in (0, \pi), \\ 0, & x = 0, \\ -\frac{\pi}{2}, & x \in (-\pi, 0). \end{cases}$$
>
> Its Fourier series is
>
> $$2\sum_{n=1}^{\infty} \frac{\sin(2n-1)x}{2n-1}.$$

Hence,

$$f(x) = 2 \sum_{n=1}^{\infty} \frac{\sin(2n-1)x}{2n-1}, \quad x \in (-\pi, \pi) \setminus \{0\}.$$

Since $2 \sum_{n=1}^{\infty} \frac{\sin(2n-1)x}{2n-1}\Big|_{x=0} = 0$, it is also convergent to $f(x)$ at $x = 0$.
We now consider the jumps of $f(x)$ and the amplitude of $S_n(f; x)$ at $x = 0$ respectively. We have

$$f(0^+) - f(0^-) = \pi.$$

Let $\xi_n^+ = \frac{\pi}{2n}$ and $\xi_n^- = -\frac{\pi}{2n}$. Then

$$\lim_{n \to \infty} S_n(f; \xi_n^+) = \int_0^{\pi} \frac{\sin t}{t} dt \approx 1.85193706 \cdots$$

and

$$\lim_{n \to \infty} S_n(f; \xi_n^-) = -\int_0^{\pi} \frac{\sin t}{t} dt \approx -1.85193706 \cdots,$$

which yields

$$\lim_{n \to \infty} \left| S_n(f; \xi_n^+) - S_n(f; \xi_n^-) \right| \approx 3.70387412 \cdots$$

$$\approx 1.179\pi,$$

i.e., the limit of the amplitudes of $S_n(f, x)$ around 0 is at least a 1.179 multiple of the jump of f at 0. This is called *Gibbs' phenomenon*.

1.3 The Study of Functions Using Fourier Series

A 2π-periodic function f can be represented in two ways. It can be represented in the time domain in the form $f(t)$, or it can be represented as a Fourier series. Since the components in a Fourier series are sinusoids, this is a representation in the frequency domain. (If f is represented as $f(x)$, we say it is in the spatial domain. However, there is no difference between $f(t)$ and $f(x)$ mathematically.) We represent f in the time domain when we want to study the motion of the object. In this case, we are interested in its differential properties, called smoothness. When we want to study the frequency properties of an object, we represent f in the frequency domain using its Fourier series. In this case, the frequency range and the strength of each wave component in the series will be our concern. In this section, we shall show the relationship between the smoothness of f and the magnitude of $f^{\wedge}(k)$. Note that $\max \left| \frac{d^n}{dt^n} e^{ikt} \right| = |k|^n$.

Hence, strong high-frequency components in a Fourier series will destroy the smoothness of the function.

We now study the behavior of the Fourier series of functions which have high-order derivatives. Let $\tilde{W}_1^r$ be the set of all 2π-periodic functions that have derivatives up to order r and their rth derivative is in $\tilde{L}^1$:

$$\tilde{W}_1^1 = \{f \in \tilde{L}_1 \mid f \text{ is absolutely continuous}\},$$
$$\tilde{W}_1^r = \{f \in \tilde{L}_1 \mid f' \in \tilde{W}_1^{r-1}\}, \quad r > 1.$$

The integer r indicates the smoothness of the functions in $\tilde{W}_1^r$— the larger the number r the smoother the functions.

EXAMPLE 6.1.6:   Consider the 2π-periodic functions defined as

(1) $f(x) = x^2, \quad x \in [0, 2\pi)$;
(2) $g(x) = x^2, \quad x \in [-\pi, \pi)$; and
(3) $h(x) = \begin{cases} x(\pi + x), & -\pi \le x \le 0, \\ x(\pi - x), & 0 < x \le \pi. \end{cases}$

Then $f \in \tilde{L}^1$, but not in $\tilde{W}_1^1$; $g \in \tilde{W}_1^1$, but not in $\tilde{W}_1^2$; $h \in \tilde{W}_1^2$, but not in $\tilde{W}_1^3$.

A function f is *of smaller order* than function g as $x \to \infty$ if $\lim\limits_{x \to \infty} \dfrac{f(x)}{g(x)} = 0$. We denote this as $f(x) = o(g(x))$, pronounced "f is little-oh of g."

Theorem 6.1.5 If $f \in \tilde{W}_1^r$, then

$$f^\wedge(k) = o\left(\frac{1}{|k|^r}\right) \quad \text{as } |k| \to \infty.$$

Proof: Assume $f \in \tilde{W}_1^r, r \ge 1$. Then

$$[f']^\wedge(k) = \frac{1}{2\pi} \int_{-\pi}^{\pi} f'(t)e^{-ikt}\, dt$$

$$= \frac{1}{2\pi}f(t)e^{-ikt}\big|_{-\pi}^{\pi} + ik\frac{1}{2\pi}\int_{-\pi}^{\pi} f(t)e^{-ikt}\, dt$$

$$= ikf^\wedge(k),$$

which, by induction, yields

$$f^\wedge(k) = \frac{[f^{(r)}]^\wedge(k)}{(ik)^r}.$$

Since $f^{(r)} \in \tilde{L}^1$, we have $\lim_{|k| \to \infty} [f^{(r)}]\,\hat{}\,(k) = 0$, which implies

$$\lim_{|k| \to \infty} |k^r f\hat{}\,(k)| = \lim_{|k| \to \infty} |[f^{(r)}]\,\hat{}\,(k)| = 0.$$

The theorem is proved. ∎

EXAMPLE 6.1.7: The function in Example 6.1.3 is in $\tilde{W}_1^2$ since $f'' \in \tilde{L}^1$. Hence, $f\hat{}\,(k) = o(\frac{1}{k^2})$. In fact, we have

$$f(x) = \frac{8}{\pi} \sum_{n=1}^{\infty} \frac{1}{(2n-1)^3} \sin(2n-1)x,$$

where $|f\hat{}\,(n)| = \frac{4}{\pi(2|n|-1)^3} = o\left(\frac{1}{(2|n|-1)^2}\right)$.

Note that if $f(t) = g(t)$ almost everywhere, then they have the same Fourier series. Hence, Theorem 6.1.5 is also true for the function that is almost everywhere equal to a function in $\tilde{W}_1^r$.

EXAMPLE 6.1.8: Let the 2π-periodic function $g(x)$ be defined as

$$g(x) = \begin{cases} f(x), & x \text{ is an irrational number,} \\ 0, & x \text{ is a rational number,} \end{cases}$$

where $f(x)$ is the function in Example 6.1.3. Then

$$g(x) \sim \frac{8}{\pi} \sum_{n=1}^{\infty} \frac{1}{(2n-1)^3} \sin(2n-1)x$$

and the series converges to $g(x)$ at each irrational point.

EXAMPLE 6.1.9: The function $f(x)$ in Example 6.1.3 is discontinuous at $x = \pm\pi$. Hence $f \notin \tilde{W}_1^1$. Thus, we cannot expect $f\hat{}\,(k) = o\left(\frac{1}{|k|}\right)$. In fact, we have

$$f\hat{}\,(k) = \frac{(-1)^{k-1} i}{k}, \quad k \in \mathbb{Z} \setminus \{0\},$$

which yields

$$\lim_{|k| \to \infty} \left| \frac{f\hat{}\,(k)}{k} \right| = 1 \neq 0.$$

The result in Theorem 6.1.5 can be partially reversed. First, we have the following.

Theorem 6.1.6 If $f(t) = \sum_{k \in \mathbb{Z}} c_k e^{ikt}$ and $\sum_{k \in \mathbb{Z}} |c_k|$ is convergent, then $f(t)$ is a continuous function and $f^{\wedge}(k) = c_k$.

Proof: Since $\sum_{k \in \mathbb{Z}} |c_k|$ is convergent, the series $\sum_{k \in \mathbb{Z}} c_k e^{ikt}$ is uniformly convergent to $f(t)$. Note that all functions $e^{ikx}, k \in \mathbb{Z}$, are continuous, hence the series is convergent to a continuous function. ∎

By Theorem 6.1.6, if $|c_k| \le \frac{M}{|k|^{1+\epsilon}}, k \in \mathbb{N}$, holds for an $M > 0$ and an $\epsilon > 0$, then $f(t) = \sum_{k \in \mathbb{Z}} c_k e^{ikt}$ is continuous. Assume now that c_k decays faster. Then f becomes smoother. This fact is shown in the following theorem.

Theorem 6.1.7 If

$$|c_k| \le \frac{M}{|k|^{1+n+\epsilon}}, \quad k \in \mathbb{N},$$

holds for an $M > 0$, an $\epsilon > 0$, and an integer $n \in \mathbb{Z}^+$, then $f(t) = \sum_{k \in \mathbb{Z}} c_k e^{ikt}$ has continuous derivatives up to order n.

Proof: We leave the proof as an exercise. ∎

Exercises

1. Prove Theorem 6.1.1: The set $\{1, \sin x, \cos x, \sin 2x, \cos 2x, \cdots\}$ satisfies the following.
   (a) $\int_{-\pi}^{\pi} \sin nx \cos mx \, dx = 0$, $(n, m) \in \mathbb{Z}^+ \times \mathbb{Z}^+$.
   (b) $\frac{1}{\pi} \int_{-\pi}^{\pi} \sin nx \sin kx \, dx = \delta_{n,k}$, $(1)(n, k) \in \mathbb{N}^2$.
   (c) $\frac{1}{\pi} \int_{-\pi}^{\pi} \cos mx \cos lx \, dx = \delta_{m,l}$, $(m, l) \in (\mathbb{Z}^+ \times \mathbb{Z}^+) \setminus \{(0, 0)\}$, where
   $$\delta_{n,k} = \begin{cases} 1, & \text{if } n = k, \\ 0, & \text{otherwise.} \end{cases}$$
2. Prove $\frac{1}{2\pi} \int_{-\pi}^{\pi} e^{ikt} \overline{e^{ilt}} \, dt = \delta_{k,l}$.
3. Prove $|e^{ih} - 1| \le |h|$, where h is a real number.
4. Assume $f \in \tilde{L}^1$ and $f^{\wedge}(k) = 0$, $|k| > n$. Prove that $f(x)$ is almost everywhere equal to a trigonometric polynomial of degree less than or equal to n.
5. Prove Lemma 6.1.1 (3): $\overline{[f(-\cdot)]}^{\wedge}(k) = \overline{f^{\wedge}(k)}$.

6. Find the real and complex Fourier series for the 2π-periodic piecewise constant function

$$f(x) = \begin{cases} -\frac{\theta}{2}, & \theta < |x| < \pi, \\ \frac{\pi-\theta}{2}, & |x| \leq \theta, \end{cases}$$

where θ is a constant such that $0 < \theta < \pi$.

7. Find the real and complex Fourier series for the 2π-periodic piecewise linear function

$$f(x) = \begin{cases} -x, & 0 \leq x < \theta, \\ \pi - x, & \theta \leq x < 2\pi, \end{cases}$$

where θ is a constant such that $0 < \theta < 2\pi$.

8. Find the real and complex Fourier series for the 2π-periodic continuous function $f(x) = |x|, |x| \leq \pi$.

9. Prove that if $f(x)$ is even, then $f_s\hat{}(k) = 0$ for all $k \in \mathbb{N}$; else if $f(x)$ is odd, then $f_c\hat{}(k) = 0$ for all $k \in \mathbb{N}$.

10. Show that if f is a real 2π-periodic function, then $f\hat{}(-k) = \overline{f\hat{}(k)}$.

11. Assume $f(x)$ is a real 2π-periodic function. Prove that all $f\hat{}(k), k \in \mathbb{Z}$, are real if and only if $f(x)$ is an even function.

12. Prove that $|f\hat{}(k)| \leq \frac{1}{2\pi} \int_{-\pi}^{\pi} |f(x)| \, dx$, for all $k \in \mathbb{Z}$.

13. Apply Lemma 6.1.1 to represent the Fourier series of the following functions, by $\{f\hat{}(k)\}$ assuming f is a 2π-periodic function.
    (a) $f(x - \frac{\pi}{2}) + f(x + \frac{\pi}{2})$.
    (b) $(\sin x)f(x)$.
    (c) $(\cos x)f(x)$.
    (d) $f(x) + f(-x)$.

14. Prove the second limit in the Riemann–Lebesgue Lemma: If f is integrable on the interval (a, b), then $\lim_{\rho \to \infty} \int_a^b f(t) \cos \rho t \, dt = 0$.

15. Let $f_1(x)$ and $f_2(x)$ be two 2π-periodic functions such that

$$f_1(x) = x^2 + x, \quad -\pi < x \leq \pi,$$

$$f_2(x) = x^2 + x, \quad 0 < x \leq 2\pi.$$

(a) Find the Fourier series for these two functions.
(b) Explain why the Fourier series for them are not equal.
(c) Does the Fourier series of $f_1(x)$ converge? Is the Fourier series of $f_1(x)$ convergent to $f_1(x)$ everywhere? To what numbers is its Fourier series convergent at $x = 0$, $x = \frac{\pi}{2}$, and $x = \pi$?
(d) Does the Fourier series of $f_2(x)$ converge? Is the Fourier series of $f_2(x)$ convergent to $f_2(x)$ everywhere? To what numbers is its Fourier series convergent at $x = 0$, $x = \frac{\pi}{2}$, and $x = \pi$?

16. Let $f_a(x)$ be a 2π-periodic function defined by

$$f_a(x) = \begin{cases} \frac{1}{\sqrt{2a}}, & |x| \leq a, \\ 0, & a < |x| \leq \pi, \end{cases}$$

where a is a positive constant less than π.
   (a) Find the Fourier series of $f_a(x)$.
   (b) Find the value of $\lim_{a \to 0^+} [f_a]\hat{}(k)$ for each $k \in \mathbb{Z}$.

17. Use mathematical induction to prove Theorem 6.1.7: If $|c_k| \leq \frac{M}{|k|^{1+n+\epsilon}}$, $k \in \mathbb{N}$, holds for an $M > 0$, an $\epsilon > 0$, and an integer $n \in \mathbb{Z}^+$, then $f(t) = \sum_{k \in \mathbb{Z}} c_k e^{ikt}$ has continuous derivatives up to order n.

18. Assume f is a 2π-periodic, absolutely continuous function. Prove $\lim_{|k| \to \infty} k f\hat{}(k) = 0$.

19. Assume the sequence $(c_k)_{k \in \mathbb{Z}}$ exponentially decays, i.e.,

$$|c_k| \leq C e^{-\lambda |k|}, \quad k \in \mathbb{Z},$$

for a $\lambda > 0$ and a $C > 0$. Prove that $f(t) = \sum_{k \in \mathbb{Z}} c_k e^{ikt}$ is infinitely differentiable.

2 Parseval's Formula

The Fourier series of a function in $\tilde{L}^1$ is not always pointwise convergent. This fact restricts applications of the finite Fourier transform in $\tilde{L}^1$. Hence, we want to find a different space that overcomes this restriction. We already know that $\{e^{ikt}\}_{k \in \mathbb{Z}}$ is an orthonormal system on $[-\pi, \pi]$. This provides a geometric explanation of Fourier series: They form an orthonormal decomposition of a function. Hence, in this section, we introduce the space of square integrable periodic functions and show that $\{e^{ikt}\}_{k \in \mathbb{Z}}$ is an orthonormal basis of this space. Therefore, Fourier partial sums are best approximations for functions (recall Theorem 5.4.6).

2.1 The Space of Square Integrable Periodic Functions

We define the linear space

$$\tilde{L}^2 = \left\{ f \in \tilde{L}^1 \,\middle|\, \frac{1}{2\pi} \int_{-\pi}^{\pi} |f(t)|^2 \, dt < \infty \right\}$$

and equip $\tilde{L}^2$ with the inner product

$$\langle f, g \rangle = \frac{1}{2\pi} \int_{-\pi}^{\pi} f(x)\overline{g(x)}\, dx \text{ for all } f, g \in \tilde{L}^2.$$

Then the norm of a function $f \in \tilde{L}^2$ is $\|f\|_{\tilde{L}^2} = \sqrt{\langle f, f \rangle}$, the distance between functions $f, g \in \tilde{L}^2$ is $\|f - g\|_{\tilde{L}^2}$, and the angle between f and g is

$$\cos^{-1} \frac{\langle f, g \rangle}{\|f\| \|g\|}.$$

We now discuss the partial Fourier sum of a function $f \in L^2$. We already know that $\{e^{ikt}\}_{k \in \mathbb{Z}}$ is an orthonormal system in $\tilde{L}^2$. Let V_n be the subspace spanned by $\{e^{ikt}\}_{k=-n}^{n}$. Then each function in V_n is of the form $\sum_{k=-n}^{n} c_k e^{ikt}$, and the partial Fourier sum of f is the orthogonal projection of $f \in \tilde{L}^2$ onto V_n:

$$\text{proj}_{V_n} f = \sum_{k=-n}^{n} \langle f, e^{ikt} \rangle e^{ikt} = \sum_{k=-n}^{n} f^\wedge(k) e^{ikt} = S_n(f).$$

From Theorem 5.4.6, we know

$$\|S_n(f) - f\|_{\tilde{L}^2} \leq \|t_n - f\|_{\tilde{L}^2} \text{ for all } t_n \in V_n. \tag{2.1}$$

Note that $\|S_n(f)\|_{\tilde{L}^2} = \sqrt{\sum_{k=-n}^{n} |f^\wedge(k)|^2}$ and the distance from f to V_n is $\|f - S_n(f)\|_{\tilde{L}^2}$. Hence,

$$\sum_{k=-n}^{n} |f^\wedge(k)|^2 + \|f - S_n(f)\|_{\tilde{L}^2}^2 = \|f\|_{\tilde{L}^2}^2, \text{ for all } n \in \mathbb{N},$$

which yields

$$\sqrt{\sum_{k \in \mathbb{Z}} |f^\wedge(k)|^2} \leq \|f\|_{\tilde{L}^2}. \tag{2.2}$$

Let

$$l^2 = \left\{ \mathbf{a} := (a_k)_{k \in \mathbb{Z}} \,\middle|\, \sum_{k \in \mathbb{Z}} |a_k|^2 < \infty \right\}$$

in which the inner product is defined by $\langle \mathbf{a}, \mathbf{b} \rangle = \sum_{k \in \mathbb{Z}} a_k \overline{b_k}$. Then the norm of $\mathbf{a}$ is $\sqrt{\sum_{k \in \mathbb{Z}} |a_k|^2}$. Let $\mathbf{e}^k$ be the sequence such that its kth term $e_j^k = \delta_{jk}$. Then $\{\mathbf{e}^k\}_{k \in \mathbb{Z}}$ is an orthonormal basis of l^2. Inequality (2.2) shows that $(f^\wedge(k)) \in l^2$. Hence, the finite Fourier transform is a mapping from $\tilde{L}^2$ to l^2. To prove $\{e^{ikt}\}_{k \in \mathbb{Z}}$ is a basis of $\tilde{L}^2$, we have to confirm $\sqrt{\sum_{k \in \mathbb{Z}} |f^\wedge(k)|^2} = \|f\|_{\tilde{L}^2}$. We use convolution to complete the task.

2.2 The Convolution Theorem

In Section 4 of Chapter 3, we introduced the convolution for $L(\mathbb{R})$. We now extend it to $\tilde{L}^1$.

Definition 6.2.1 Let $f, g \in \tilde{L}^1$. Then the 2π-periodic function

$$(f * g)(x) = \frac{1}{2\pi} \int_{-\pi}^{\pi} f(x - t) g(t) \, dt$$

is called the *convolution* of f and g.

Convolution has many properties similar to multiplication. We have

$$f * g = g * f,$$
$$f * (g + h) = f * g + f * h,$$
$$(f * g) * h = f * (g * h).$$

An important property of convolution is the following.

Theorem 6.2.1 [Convolution Theorem] If $f, g \in \tilde{L}^2$, then

$$[f * g](t) = \sum_{k \in \mathbb{Z}} f^{\wedge}(k) g^{\wedge}(k) e^{ikt}. \tag{2.3}$$

Proof: We have

$$
\begin{aligned}
[f * g]^{\wedge}(k) &= \frac{1}{2\pi} \int_{-\pi}^{\pi} \left(\frac{1}{2\pi} \int_{-\pi}^{\pi} f(x - t) g(t) dt \right) e^{-ikx} \, dx \\
&= \frac{1}{2\pi} \int_{-\pi}^{\pi} \left(\frac{1}{2\pi} \int_{-\pi}^{\pi} f(x - t) g(t) e^{-i(x-t)k} dx \right) e^{-itk} \, dt \\
&= \frac{1}{2\pi} \int_{-\pi}^{\pi} \left(\frac{1}{2\pi} \int_{-\pi}^{\pi} f(x - t) e^{-i(x-t)k} dx \right) g(t) e^{-itk} \, dt \\
&= f^{\wedge}(k) g^{\wedge}(k).
\end{aligned}
$$

By Schwarz's inequality (Theorem 5.2.1) and (2.2),

$$\sum_{k \in \mathbb{Z}} |f^{\wedge}(k) g^{\wedge}(k)| \leq \sqrt{\sum_{k \in \mathbb{Z}} |f^{\wedge}(k)|^2} \sqrt{\sum_{k \in \mathbb{Z}} |g^{\wedge}(k)|^2} \leq \|f\|_{\tilde{L}^2} \|g\|_{\tilde{L}^2},$$

which implies the series $\sum_{k \in \mathbb{Z}} |[f * g]^{\wedge}(k)|$ is convergent. By Theorem 6.1.6 and Theorem 6.2.1, we obtain (2.3). ■

2.3 Parseval's Formula

We are ready to prove that the finite Fourier transform is an orthonormal mapping from $\tilde{L}$ to l^2.

Theorem 6.2.2 For $f, g \in \tilde{L}^2$,

$$\langle f, g \rangle = \langle (f\hat{}(k)) , (g\hat{}(k)) \rangle \tag{2.4}$$

and therefore

$$\|f\|_{\tilde{L}^2} = \sqrt{\sum_{k \in \mathbb{Z}} |f\hat{}(k)|^2}. \tag{2.5}$$

Proof: In (2.3), letting $t = 0$, we have

$$\frac{1}{2\pi} \int_{-\pi}^{\pi} f(x)g(-x)\, dx = \sum_{k \in \mathbb{Z}} f\hat{}(k)g\hat{}(k).$$

Let $g(-x) = \overline{h(x)}$. By (3) in Lemma 6.1.1,

$$g\hat{}(k) = \overline{[h(-\cdot)]}\hat{}(k) = \overline{h\hat{}(k)}.$$

Hence,

$$\langle f, h \rangle = \frac{1}{2\pi} \int_{-\pi}^{\pi} f(x)\overline{h(x)}\, dx = \frac{1}{2\pi} \int_{-\pi}^{\pi} f(x)g(-x)\, dx$$
$$= \sum_{k \in \mathbb{Z}} f\hat{}(k)g\hat{}(k) = \sum_{k \in \mathbb{Z}} f\hat{}(k)\overline{h\hat{}(k)},$$

which yields (2.4). Letting $g = f$ in (2.4), we get (2.5). ∎

Formula (2.5) is called the *Parseval formula* and (2.4) is called the *general Parseval formula*. From the Parseval formula, we get the following.

Corollary 6.2.1 For $f \in \tilde{L}^2$,

$$\lim_{n \to \infty} \|S_n(f) - f\|_{\tilde{L}^2} = 0. \tag{2.6}$$

Therefore, $\{e^{ikt}\}_{k\in\mathbb{Z}}$ is an orthonormal basis of $\tilde{L}^2$.

We have shown that the finite Fourier transform is an orthonormal transform from $\tilde{L}^2$ to l^2. We now show that it is an onto mapping from $\tilde{L}^2$ to l^2.

Theorem 6.2.3 For each $(a_k) \in l^2$, there is a function $f \in \tilde{L}^2$ such that $f^\wedge(k) = a_k$, $k \in \mathbb{Z}$.

Proof: Let $s_n(x) = \sum_{k=-n}^{n} a_k e^{ikx}$. For $m \geq n$, we have

$$\|s_m - s_n\|_{L^2} = \sum_{n+1\leq|k|\leq m} |a_k|^2,$$

where the limit of the right hand side is 0 as $n, m \to \infty$. Hence, (s_n) is a Cauchy sequence in $\tilde{L}^2$. Therefore, there is a function $f \in \tilde{L}^2$ such that

$$\lim_{n\to\infty} \|s_n - f\|_{L^2} = 0. \tag{2.7}$$

We now prove $f^\wedge(k) = a_k$. For an arbitrary fixed k and any $n > |k|$, we have

$$|f^\wedge(k) - a_k| = \left|\frac{1}{2\pi}\int_{-\pi}^{\pi}(f(t) - s_n(t))e^{-ikt}dt\right| \leq \|f - s_n\|_{L^2}.$$

By (2.7), we get $f^\wedge(k) = a_k$. The theorem is proved.   ∎

We summarize Theorem 6.2.2 and Theorem 6.2.3 in the following.

Theorem 6.2.4 The finite Fourier transform $f \mapsto f^\wedge$ is an orthonormal isomorphism from $\tilde{L}^2$ to l^2.

EXAMPLE 6.2.1:   The Fourier series of

$$f(x) = \begin{cases} \frac{\pi}{2}, & x \in (0, \pi), \\ 0, & x = 0, \\ -\frac{\pi}{2}, & x \in (-\pi, 0), \end{cases}$$

is

$$2\sum_{n=1}^{\infty} \frac{\sin(2n-1)x}{2n-1} = \sum_{n=1}^{\infty} \frac{e^{i(2n-1)x} - e^{-i(2n-1)x}}{i(2n-1)}.$$

It is obvious that $f \in \tilde{L}^2$. We have

$$\|f^\wedge\|_2^2 = 2\sum_{n=1}^{\infty} \frac{1}{(2n-1)^2}$$

and

$$\|f\|_{L^2}^2 = \frac{1}{2\pi} \int_{-\pi}^{\pi} f(x)\,dx = \frac{\pi^2}{4}.$$

By (2.5), we have $\|f\|_{L^2} = \|f^{\wedge}\|_2$. Hence,

$$\sum_{n=1}^{\infty} \frac{1}{(2n-1)^2} = \frac{\pi^2}{8}. \tag{2.8}$$

From (2.8), we can derive the sum $\sum_{n=1}^{\infty} \frac{1}{n^2}$. Assume

$$\sum_{n=1}^{\infty} \frac{1}{n^2} = s.$$

Then

$$\sum_{n=1}^{\infty} \frac{1}{(2n)^2} = \frac{1}{4}s.$$

Note that

$$\frac{3}{4}s = \sum_{n=1}^{\infty} \frac{1}{n^2} - \sum_{n=1}^{\infty} \frac{1}{(2n)^2} = \sum_{n=1}^{\infty} \frac{1}{(2n-1)^2} = \frac{\pi^2}{8}.$$

Hence,

$$\sum_{n=1}^{\infty} \frac{1}{n^2} = \frac{\pi^2}{6}.$$

2.4 The Finite Fourier Transform on General Intervals

Before we end this section, we briefly discuss the finite Fourier transforms of the functions with periods other than 2π. Let $\tilde{L}_{2\sigma}^1$ be the space of all 2σ-periodic integrable functions and $\tilde{L}_{2\sigma}^2$ be the space defined by

$$\tilde{L}_{2\sigma}^2 = \left\{ f \in \tilde{L}_{2\sigma}^1 \,\middle|\, \int_{-\sigma}^{\sigma} |f(x)|^2\,dx < \infty \right\}.$$

In $\tilde{L}_{2\sigma}^2$ the inner product is defined by

$$\langle f, g \rangle = \frac{1}{2\sigma} \int_{-\sigma}^{\sigma} f(x)\overline{g(x)}\,dx.$$

The Fourier series of $f \in \tilde{L}_{2\sigma}^1$ is

$$\sum_{k=-\infty}^{\infty} f_\sigma^{\wedge}(k)e^{ik\frac{\pi}{\sigma}x},$$

where

$$f_\sigma\hat{}(k) = \frac{1}{2\sigma} \int_{-\sigma}^{\sigma} f(x)e^{-ik\frac{\pi}{\sigma}x}\, dx. \tag{2.9}$$

Similarly, the real Fourier series is

$$\frac{1}{2}a_0 + \sum_{k=1}^{\infty} a_k \cos k\frac{\pi}{\sigma}x + b_k \sin k\frac{\pi}{\sigma}x,$$

where

$$a_k = \frac{1}{\sigma} \int_{-\sigma}^{\sigma} f(x) \cos k\frac{\pi}{\sigma}x\, dx, \quad k \in \mathbb{Z}^+,$$

$$b_k = \frac{1}{\sigma} \int_{-\sigma}^{\sigma} f(x) \sin k\frac{\pi}{\sigma}x\, dx, \quad k \in \mathbb{N}.$$

Thus, all results we have obtained for $\tilde{L}^1$ and $\tilde{L}^2$ also hold for $\tilde{L}^1_{2\sigma}$ and $\tilde{L}^2_{2\sigma}$.

When a function f is defined on a finite interval $[a, b)$, we can use the following method to make f periodic on $\mathbb{R}$. Let $\sigma = \frac{b-a}{2}$. Define

$$\tilde{f}(x) = f(x - 2k\sigma), \quad x \in [a + 2k\sigma, a + 2(k + 1)\sigma), \quad k \in \mathbb{Z}. \tag{2.10}$$

Then $\tilde{f}(x)$ has period 2σ. If its Fourier series converges to $\tilde{f}(x)$, then in particular the series converges to $f(x)$ on $[a, b)$. However, the periodization (2.10) may destroy the continuity of $\tilde{f}(x)$ at the endpoints a and b if $f(a) \neq f(b)$. In order to keep the continuity at the ends, we adopt the following method. For simplicity, we assume $a = 0$ and $b = \sigma$. We first extend f to an even function on $[-\sigma, \sigma]$ by defining

$$f(-x) = f(x), \quad x \in [0, \sigma].$$

Thus, $f(x)$ is well defined on $[-\sigma, \sigma]$ and $f(\sigma) = f(-\sigma)$. Then we make periodic the extended function using (2.10).

EXAMPLE 6.2.2: Let $f(x) = |x|, x \in [-l, l]$. Then $f(x)$ is continuous and $f'(x)$ is integrable on $[-l, l]$. Hence, its Fourier series converges to f. We have

$$f(x) = \frac{l}{2} - \frac{4l}{\pi^2} \sum_{n=1}^{\infty} \frac{1}{(2n - 1)^2} \cos \frac{(2n - 1)\pi x}{l}.$$

EXAMPLE 6.2.3: Assume $0 < a < l$. Let $f(x)$ be a $2l$-periodic function defined by

$$f(x) = \begin{cases} 1 - \frac{|x|}{a}, & |x| \le a, \\ 0, & a < |x| \le l \end{cases}.$$

Then we have

$$f(x) = \frac{a}{2l} + \frac{2l}{\pi^2 a} \sum_{n=1}^{\infty} \frac{1}{n^2} \left(1 - \cos\frac{n\pi a}{l}\right) \cos\frac{n\pi x}{l}.$$

For $\tilde{L}_{2\sigma}^2$, the orthonormal basis used is the Fourier series under $\{e^{i\frac{\pi}{\sigma}kt}\}_{k\in\mathbb{Z}}$. The frequency "spectrum" of this system is $\{\frac{k}{2\sigma}\}_{k\in\mathbb{Z}}$, which becomes more dense as σ gets larger. When we consider the nonperiodic function defined on the whole real line, the frequency spectrum becomes continuous and Fourier integrals are introduced.

Exercises

1. Prove that, if $f, g \in \tilde{L}^1$, then $f * g \in \tilde{L}^1$ and $[f * g]\hat{}(k) = f\hat{}(k)g\hat{}(k)$.
2. Prove that if $f, g \in \tilde{L}^1$ and the series $\sum |g\hat{}(k)|$ is convergent, then

$$f * g(t) = \sum_{k\in\mathbb{Z}} f\hat{}(k)g\hat{}(k)e^{ikt}.$$

3. Let $f, g \in \tilde{L}^2$. Prove $fg \in \tilde{L}^1$ and $[fg]\hat{}(k) = \sum_{j\in\mathbb{Z}} f\hat{}(j)g\hat{}(k-j)$.
4. Let $f(x) = \sum_{k=1}^{n} a_k \cos kx$ and $g(x) = \sum_{k=1}^{n} b_k \cos kx$. Find $\|f - g\|_{L^2}$.
5. Let $t_n \in V_n$ be a trigonometric polynomial of order n. Prove that $\min_{t_n \in V_n} \| \sin(n+1)x - t_n(x)\|_{L^2}$ is obtained at $t_n = 0$.
6. Let $f(t) = \sum_{k\in\mathbb{Z}} \frac{1}{2^{|k|}} e^{ikt}$. Find $\int_{-\pi}^{\pi} |f(t)|^2 \, dt$.
7. Assume $f, g \in \tilde{L}^1$. Prove that $\|f * g - f\|_{L^2}^2 = \sum_{k\in\mathbb{Z}} |f\hat{}(k)|^2 |g\hat{}(k) - 1|^2$.
8. Use the Fourier series in Example 6.1.1 in Section 6.1 to derive $\sum_{n=1}^{\infty} \frac{1}{n^2} = \frac{\pi^2}{6}$.
9. Let $f(x)$ be a 2π-periodic function defined by $f(x) = |x|(\pi - |x|), |x| \leq \pi$.
   (a) Find the Fourier series of f.
   (b) Use the result in (a) to prove $\sum_{n=1}^{\infty} \frac{1}{n^4} = \frac{\pi^4}{90}$.
10. Let $f(x)$ be a 2π-periodic function defined by

$$f(x) = \begin{cases} (\pi - \theta)x, & |x| \leq \theta, \\ \theta(\pi - x), & \theta < x \leq 2\pi - \theta. \end{cases}$$

   (a) Prove $f(x) = 2\sum_{n=1}^{\infty} \frac{\sin n\theta}{n^2} \sin nx$.
   (b) Use the result in (a) to find the sum $\sum_{n=1}^{\infty} \frac{\sin^2 n\theta}{n^4}$.

11. Assume $f \in \tilde{L}^1$ and $|f^{\wedge}(k)| \le \frac{M}{|k|^{n+1/2+\epsilon}}$, for an $M > 0$ and an $\epsilon > 0$. Prove that f has derivatives up to order n and $f^{(r)} \in \tilde{L}^2$ for $r \le n$.

12. Assume $f \in \tilde{L}^2_{2\sigma}$. Prove

$$\sum_{k=-\infty}^{\infty} |f_\sigma^{\wedge}(k)|^2 = \frac{1}{2\sigma} \int_{-\sigma}^{\sigma} |f(x)|^2 \, dx,$$

where $f_\sigma^{\wedge}(k)$ is defined by (2.9).

13. Let $f(x)$ be a π-periodic function defined by $f(x) = |x|, -\pi/2 \le x < \pi/2$.
    (a) Find its Fourier series $\sum f_\pi^{\wedge}(k)e^{i2kt}$.
    (b) Considering it as a 2π-periodic function, find its Fourier series $\sum f^{\wedge}(k)e^{ikt}$.
    (c) Are they the same? What is the general conclusion?

14. Let $f(x)$ be a $2l$-periodic function defined by $f(x) = x, -l \le x < l$. Find its Fourier series.

15. Prove that $x^2 = \frac{l^2}{3} + \frac{4l^2}{\pi^2} \sum_{n=1}^{\infty} \frac{(-1)^n}{n^2} \cos \frac{n\pi x}{l}, \quad |x| \le l$.

3 The Fourier Transform of Integrable Functions

Let L^1 be the space of all (complex-valued) functions integrable on $\mathbb{R}$. The norm of $f \in L^1$, is defined by

$$\|f\|_1 = \int_{\mathbb{R}} |f(t)| \, dt.$$

We have the following:
   (1) $\|f\|_1 = 0$ if and only if $f(x) = 0$ a.e. on $\mathbb{R}$.
   (2) $\|cf\|_1 = |c| \|f\|_1$, for a complex number c.
   (3) $\|f + g\|_1 \le \|f\|_1 + \|g\|_1$.

We now study the Fourier transform on L^1.

3.1 Definition and Properties

Definition 6.3.1   For a function $f \in L^1$, the complex-valued function

$$\hat{f}(\omega) = \int_{\mathbb{R}} f(t)e^{-it\omega} \, dt, \quad \omega \in \mathbb{R}, \tag{3.1}$$

is called the *Fourier transform* of f.

The Fourier transform is linear:

$$\widehat{f+g} = \hat{f} + \hat{g} \text{ and } \widehat{cf} = c\hat{f},$$

where c is a complex constant.

EXAMPLE 6.3.1: Let $f(t) = e^{-at^2}, a > 0$. Then $\hat{f}(\omega) = \sqrt{\frac{\pi}{a}} e^{-\frac{\omega^2}{4a}}$.

EXAMPLE 6.3.2: Let $g(t) = \frac{1}{t^2+a^2}$. Then $\hat{g}(\omega) = \frac{\pi e^{-a|\omega|}}{a}$.

EXAMPLE 6.3.3:

Let $f(x) = \frac{\sin^2 x}{x^2}$. Then $\hat{f}(\omega) = \begin{cases} \pi\left(1 - \frac{|\omega|}{2}\right), & |\omega| < 2, \\ 0 & |\omega| \geq 2. \end{cases}$

Similar to Lemma 6.1.1, we have the following.

Lemma 6.3.1 For $f \in L^1$, we have

(1) $[f(\cdot + h)]\hat{}(\omega) = e^{ih\omega}\hat{f}(\omega), \quad h, \omega \in \mathbb{R}$;
(2) $[e^{-ih\cdot}f(\cdot)]\hat{}(\omega) = \hat{f}(\omega + h), \quad h, \omega \in \mathbb{R}$;
(3) $[af(a\cdot)]\hat{}(\omega) = \hat{f}(\omega/a), \quad a > 0, \omega \in \mathbb{R}$;
(4) $[f(-\cdot)]\hat{}(\omega) = \overline{\hat{f}(\omega)}, \quad \omega \in \mathbb{R}$.

Proof: We have

$$[f(\cdot + h)]\hat{}(\omega) = \int_{\mathbb{R}} f(t + h)e^{-it\omega}\, dt$$

$$= \int_{\mathbb{R}} f(t + h)e^{-i(t+h)\omega}e^{ih\omega}\, d(t + h)$$

$$= e^{ih\omega}\int_{\mathbb{R}} f(x)e^{-ix\omega}\, dx = e^{ih\omega}\hat{f}(\omega).$$

(1) is proved. The proofs of the remaining parts are left as exercises. ∎

The Fourier transform represents a function in the frequency domain. It is the continuous version of the Fourier series. Many results for the Fourier transform are parallel to those for Fourier series. Readers can compare the results in this section with those in Section 6.1. The following is a result similar to Theorem 6.1.3.

Theorem 6.3.1 If $f \in L^1$, then $\hat{f}(\omega)$ is continuous on $\mathbb{R}$ and

$$\lim_{|\omega|\to\infty} \hat{f}(\omega) = 0.$$

Proof: We have

$$\left|\hat{f}(\omega+h) - \hat{f}(\omega)\right| \leq \int_{-\infty}^{\infty} \left|e^{iht} - 1\right| |f(t)|\, dt.$$

Recall that

$$\left|e^{iht} - 1\right| |f(t)| \leq 2|f(t)|$$

and

$$\lim_{h\to 0} \left|e^{iht} - 1\right| |f(t)| = 0.$$

By the Lebesgue Dominated Convergence Theorem (Theorem 3.3.1), $\hat{f}(\omega)$ is uniformly continuous on $\mathbb{R}$. Besides, for $\omega \neq 0$ we have

$$\begin{aligned}
\hat{f}(\omega) &= \int_{-\infty}^{\infty} f(t) e^{-it\omega}\, dt \\
&= \int_{-\infty}^{\infty} f\left(t + \frac{\pi}{\omega}\right) e^{-i(t+\frac{\pi}{\omega})\omega}\, d\left(t + \frac{\pi}{\omega}\right) \\
&= -\int_{-\infty}^{\infty} f\left(t + \frac{\pi}{\omega}\right) e^{-it\omega}\, dt,
\end{aligned}$$

which implies that

$$\hat{f}(\omega) = \frac{1}{2} \int_{-\infty}^{\infty} \left(f(t) - f\left(t + \frac{\pi}{\omega}\right)\right) e^{-it\omega}\, dt.$$

We know that, for $f \in L^1$,

$$\lim_{h\to 0} \|f(\cdot + h) - f(\cdot)\|_1 = 0.$$

Hence,

$$0 \leq \lim_{|\omega|\to\infty} \left|\hat{f}(\omega)\right| \leq \lim_{|\frac{\pi}{\omega}|\to 0} \frac{1}{2} \int_{-\infty}^{\infty} \left|f(t) - f\left(t + \frac{\pi}{\omega}\right)\right|\, dt = 0,$$

i.e., $\lim_{|\omega|\to\infty} \hat{f}(\omega) = 0$. The theorem is proved. ∎

For convenience, we define

$$C_0 = \left\{ f \in C \,\middle|\, \lim_{|t|\to\infty} f(t) = 0 \right\},$$

which is a linear subspace of C. So Theorem 6.3.1 states that the Fourier transform is a linear transform from L^1 to C_0.

3.2 The Convolution Theorem

As we have seen in Section 6.2, convolution is a useful tool in the study of relationships between functions. The convolution of two functions in L^1 is defined in Section 3.4. For convenience, we restate it here.

Definition 6.3.2 Let $f, g \in L^1$. The *convolution* of f and g is defined by

$$f * g(x) = \int_{-\infty}^{\infty} f(x - t)g(t) \, dt. \qquad (3.2)$$

EXAMPLE 6.3.4:   For a function $f \in L^1$, define the *first moving average* of f as

$$A(f; x, h) = h^{-1} \int_{x-h/2}^{x+h/2} f(t) \, dt = h^{-1} \int_{-h/2}^{h/2} f(x - t) \, dt.$$

We define $\chi_h(x) = \begin{cases} 1, & |x| < h/2, \\ 0, & |x| \geq h/2. \end{cases}$ Then we can write the moving average $A(f; x, h)$ in a convolution form: $A(f; x, h) = f * \chi_h(x)$. Note that if $f(t)$ is continuous at x, then $\lim_{h \to 0} A(f; x, h) = f(x)$. Hence, we can use $A(f; x, h)$ to approximate $f(x)$. Note also that $A(f; x, h)$ is absolutely continuous. Hence, $A(f; x, h)$ is smoother than f.

Similar to the convolution of periodic functions, for L^1 functions we have

$$f * g = g * f,$$
$$f * (g + h) = f * g + f * h,$$
$$(f * g) * h = f * (g * h).$$

We also have the following.

Theorem 6.3.2 If $f, g \in L^1$ then $f * g \in L^1$ and

$$\int_{\mathbb{R}} |f * g(t)| \, dx \leq \left(\int_{\mathbb{R}} |f(t)| \, dt \right) \left(\int_{\mathbb{R}} |g(t)| \, dt \right). \qquad (3.3)$$

If $f, g \in L^2$ then $f * g \in C_0$ and

$$\sup_{t \in \mathbb{R}} |f * g(t)| \leq \|f\|_2 \|g\|_2. \qquad (3.4)$$

Proof: Equation (3.3) is established in Theorem 3.4.4 If $f, g \in L^2$, then by Schwarz's inequality for L^2 (Theorem 5.5.2 with $p = q = 2$) we have

$$|f * g(x)| = \left| \int_{\mathbb{R}} f(x - t)g(t)\, dt \right|$$

$$\leq \left(\int_{\mathbb{R}} |f(x - t)|^2\, dt \right)^{1/2} \left(\int_{\mathbb{R}} |g(t)|^2\, dt \right)^{1/2}$$

$$= \|f\|_2 \|g\|_2.$$

Besides,

$$|f * g(x + h) - f * g(x)| = \left| \int_{\mathbb{R}} \left(f(x + h - t) - f(x - t) \right) g(t)\, dt \right|$$

$$\leq \|f(\cdot + h) - f(\cdot)\|_2 \|g\|_2.$$

Since $f \in L^2$,

$$\lim_{h \to 0} \|f(\cdot + h) - f(\cdot)\|_2 = 0,$$

which implies

$$\lim_{h \to 0} |f * g(x + h) - f * g(x)| = 0,$$

i.e., $f * g \in C$. We now prove $\lim_{|x| \to \infty} f * g(x) = 0$. Since $f, g \in L^2$, for any $\epsilon > 0$, there is an $N > 0$ such that

$$\left(\int_{|t| \geq N} |f(t)|^2\, dt \right)^{1/2} < \epsilon \quad \text{and} \quad \left(\int_{|t| \geq N} |g(t)|^2\, dt \right)^{1/2} < \epsilon.$$

Let $x \in (-\infty, -2N) \cup (2N, \infty)$. Then $(x - N, x + N) \subset (-\infty, -N) \cup (N, \infty)$. For this x, we have

$$|f * g(x)| \leq \left(\int_{|t| \leq N} + \int_{|t| \geq N} \right) |f(x - t)g(t)|\, dt$$

$$\leq \left(\int_{|t| \leq N} |f(x - t)|^2\, dt \right)^{1/2} \|g\|_2$$

$$+ \|f\|_2 \left(\int_{|t| \geq N} |g(t)|^2\, dt \right)^{1/2}$$

$$\leq (\|g\|_2 + \|f\|_2)\, \epsilon,$$

which yields $\lim_{|x| \to \infty} |f * g(x)| = 0$. The lemma is proved. ■

We now give the convolution theorem for L^1.

Theorem 6.3.3 If $f, g \in L^1$, then

$$[f * g]\hat{}(\omega) = \hat{f}(\omega)\hat{g}(\omega), \quad \omega \in \mathbb{R}.$$

Proof: We have

$$
\begin{aligned}
[f * g]\hat{}(\omega) &= \int_{-\infty}^{\infty} \left(\int_{-\infty}^{\infty} f(x - t)g(t)\, dt \right) e^{-ix\omega}\, dx \\
&= \int_{-\infty}^{\infty} \left(\int_{-\infty}^{\infty} f(x - t)g(t)e^{-i(x-t)\omega}\, dx \right) e^{-it\omega}\, dt \\
&= \int_{-\infty}^{\infty} \left(\int_{-\infty}^{\infty} f(x - t)e^{-i(x-t)\omega}\, dx \right) g(t)e^{-it\omega}\, dt \\
&= \hat{f}(\omega)\hat{g}(\omega).
\end{aligned}
$$

This completes the proof. ■

3.3 The Inverse Fourier Transform

The convergence theorem for Fourier series confirms that a periodic function can be identified with its Fourier series. Similarly, we expect that a function on $\mathbb{R}$ can be totally recovered from its Fourier transform. Unfortunately, $f \in L^1$ does not imply $\hat{f} \in L^1$. Hence, the recovery involves a limiting process.

Theorem 6.3.4 Assume $f \in L^1$. For $\rho > 0$, define

$$A_\rho(t) = \frac{1}{2\pi} \int_{\mathbb{R}} e^{-\frac{|\omega|}{\rho}} \hat{f}(\omega)e^{it\omega}\, d\omega. \tag{3.5}$$

Then

$$\|A_\rho\|_1 \leq \|f\|_1 \tag{3.6}$$

and

$$\lim_{\rho \to \infty} A_\rho(t) = f(t) \quad a.e. \tag{3.7}$$

Furthermore, if f is also continuous, then (3.7) holds everywhere.

The function $e^{-\frac{|\omega|}{\rho}}$ in (3.5) is called a *convergence factor*, which accelerates the decay of the integrand in (3.5) so that the improper integral exists. We put the proof of the theorem at the end of the section. A direct consequence of Theorem 6.3.4 is the following uniqueness theorem for the Fourier transform.

Theorem 6.3.5 For $f \in L^1$, if $\hat{f} = 0$, then $f(t) = 0$ a.e.

Proof: If $\hat{f} = 0$, then $A_\rho(t) = 0$ for each $\rho > 0$. By (3.7), $f(t) = 0$ a.e. ∎

If the function $\hat{f}(\omega)e^{it\omega}$ is integrable on $\mathbb{R}$, then the convergence factor $e^{-\frac{|\omega|}{\rho}}$ in the integral of (3.5) can be removed. Thus, we have the following.

Theorem 6.3.6 For function f, if $\hat{f} \in L^1$, then

$$\frac{1}{2\pi} \int_{-\infty}^{\infty} \hat{f}(\omega)e^{it\omega}\, d\omega = f(t), \quad a.e. \tag{3.8}$$

Furthermore, if f is also continuous, then (3.8) holds everywhere.

Proof: By Theorem 6.3.4, we have

$$\lim_{\rho \to \infty} \frac{1}{2\pi} \int_{-\infty}^{\infty} e^{-\frac{|\omega|}{\rho}} \hat{f}(\omega)e^{it\omega}\, d\omega = f(t) \quad a.e.$$

Note that

$$\left| e^{-\frac{|\omega|}{\rho}} \hat{f}(\omega)e^{it\omega} \right| \le |\hat{f}(\omega)| \in L^1$$

and

$$\lim_{\rho \to \infty} e^{-|\omega|/\rho}\hat{f}(\omega)e^{it\omega} = \hat{f}(\omega)e^{it\omega}.$$

By the Lebesgue Dominated Convergence Theorem (Theorem 3.3.1), we have

$$\lim_{\rho \to \infty} \frac{1}{2\pi} \int_{-\infty}^{\infty} e^{-\frac{|\omega|}{\rho}} \hat{f}(\omega)e^{it\omega}\, d\omega = \frac{1}{2\pi} \int_{-\infty}^{\infty} \hat{f}(\omega)e^{it\omega}\, d\omega, \text{ for all } t \in \mathbb{R}.$$

The theorem is proved. ∎

Formula (3.8) inspires the following definition.

Definition 6.3.3 Let $g \in L^1$. The integral $\frac{1}{2\pi} \int_{\mathbb{R}} g(\omega) e^{it\omega} \, d\omega$ is called the *inverse Fourier transform* of g and denoted by $g^{\vee}$.

Using normalization, we can define the Fourier transform and inverse Fourier transform in a more symmetric way. Let

$$\mathcal{F}(f)(\omega) = \frac{1}{\sqrt{2\pi}} \hat{f}(\omega) = \frac{1}{\sqrt{2\pi}} \int_{-\infty}^{\infty} f(\omega) e^{-it\omega} \, dt$$

and

$$\mathcal{F}^{-1}(g)(t) = \sqrt{2\pi} g^{\vee}(\omega) = \frac{1}{\sqrt{2\pi}} \int_{-\infty}^{\infty} g(\omega) e^{it\omega} \, d\omega.$$

Thus, if f and $\hat{f}$ both are in L^1, we have $\mathcal{F}^{-1}\mathcal{F}(f) = f$. In this book, we call $\mathcal{F}$ and $\mathcal{F}^{-1}$ the *normalized Fourier transform* and the *normalized inverse Fourier transform*, respectively.

3.4 The Study of Functions Using Fourier Transforms

As with Fourier series, there is a close relationship between the smoothness of f and the decay rate of $\hat{f}$. Let W_1^1 denote the space of all absolutely continuous functions on $\mathbb{R}$. Then $f \in W_1^1$ implies $f' \in L^1(\mathbb{R})$. We inductively define

$$W_1^r = \{ f \mid f' \in W_1^{r-1}, r \geq 2 \}.$$

The following theorem is similar to Theorem 6.1.5.

Theorem 6.3.7 If $f \in W_1^r$, then

$$[f^{(r)}]^{\wedge}(\omega) = (i\omega)^r \hat{f}(\omega) \tag{3.9}$$

and

$$\lim_{|\omega| \to \infty} |\omega^r \hat{f}(\omega)| = 0. \tag{3.10}$$

Conversely, if $f \in L^1$ and there is a function $g \in L^1$ such that

$$(i\omega)^r \hat{f}(\omega) = \hat{g}(\omega), \quad \omega \in \mathbb{R}, \quad r \in \mathbb{N}, \tag{3.11}$$

then $f \in W_1^r$.

Proof: We prove the first claim. Assume $f \in W_1^1$. We have

$$f(x) - f(x-1) = \int_{x-1}^{x} f'(t)\, dt = \int_{-\infty}^{\infty} \chi(x-t) f'(t)\, dt, \qquad (3.12)$$

where

$$\chi(x) = \begin{cases} 1, & x \in [0,1), \\ 0, & \text{otherwise.} \end{cases}$$

Taking the Fourier transform of (3.12), we have

$$(1 - e^{-i\omega})\hat{f}(\omega) = \hat{\chi}(\omega)\, (f')\,\hat{}(\omega)$$

$$= \frac{1 - e^{-i\omega}}{i\omega}\, (f')\,\hat{}(\omega).$$

Thus

$$(f')\,\hat{}(\omega) = i\omega \hat{f}(\omega), \qquad \omega \neq 2k\pi. \qquad (3.13)$$

However, if both $(f')\hat{}(\omega)$ and $i\omega\hat{f}(\omega)$ are continuous functions, then (3.13) holds for all $\omega \in \mathbb{R}$. Using mathematical induction we can prove the claim for a general r. We now prove the converse claim for $r = 1$. It can be done by reversing the steps above. First, $(i\omega)\hat{f}(\omega) = \hat{g}(\omega)$ implies

$$\frac{1 - e^{-ih\omega}}{i\omega}\hat{g}(\omega) = (1 - e^{-ih\omega})\hat{f}(\omega) = [f(\cdot) - f(\cdot - h)]\hat{}(\omega).$$

On the other hand, let $w(x) = \int_{x-h}^{x} g(t)\, dt$. Then

$$\frac{1 - e^{-ih\omega}}{i\omega}\hat{g}(\omega) = w\hat{}(\omega).$$

By the uniqueness of the Fourier transform, for each fixed $h > 0$,

$$f(x) - f(x-h) = \int_{x-h}^{x} g(t)\, dt \quad \text{a.e.,}$$

which implies

$$\int_{0}^{y} [f(x) - f(x-h)]\, dx = \int_{0}^{y} \int_{x-h}^{x} g(t)\, dt\, dx.$$

Since $f \in L^1$, for each fixed $y > 0$, we have

$$\lim_{h \to \infty} \int_0^y [f(x) - f(x - h)] \, dx$$

$$= \int_0^y f(x) \, dx - \lim_{h \to \infty} \int_0^y f(x - h) \, dx$$

$$= \int_0^y f(x) \, dx.$$

By the Lebesgue Dominated Convergence Theorem (Theorem 3.3.1), we also have

$$\lim_{h \to \infty} \int_0^y \int_{x-h}^x g(t) \, dt \, dx = \int_0^y \int_{-\infty}^x g(t) \, dt \, dx.$$

Hence,

$$\int_0^y f(x) \, dx = \int_0^y \int_{-\infty}^x g(t) \, dt \, dx, \quad \text{for all } y > 0,$$

which yields

$$f(x) = \int_{-\infty}^x g(t) \, dt \quad a.e.$$

The claim is proved for $r = 1$. The result for a general r can be proved by mathematical induction.   ∎

Limit (3.10) indicates $\hat{f}(\omega) = o\left(\frac{1}{|\omega|^r}\right)$ as $|\omega| \to \infty$. Again, we see that, the smoother the function f, the faster $\hat{f}$ is decaying as $|\omega| \to \infty$. Note that the Fourier transform and the inverse Fourier transform are defined in a similar way. Hence, we also expect that fast decay of f implies high smoothness of $\hat{f}$. In fact, we have the following.

Theorem 6.3.8 If $f \in L^1$ and $x^r f(x) \in L^1$ for an $r \in \mathbb{N}$, then $\hat{f}$ has continuous derivatives up to order r and $\lim_{|\omega| \to \infty} \hat{f}^{(r)}(\omega) = 0$. Besides,

$$(\hat{f})^{(k)}(\omega) = (-i)^k \int_{-\infty}^{\infty} x^k f(x) e^{-ix\omega} \, dx, \quad 0 \le k \le r.$$

In particular,

$$(\hat{f})^{(k)}(0) = (-i)^k \int_{-\infty}^{\infty} x^k f(x) \, dx, \quad 0 \le k \le r.$$

Proof: Assume $r = 1$. Then

$$\frac{\hat{f}(\omega + h) - \hat{f}(\omega)}{h} = \int_{-\infty}^{\infty} f(x) \frac{(e^{-ihx} - 1)}{h} e^{-ix\omega} \, dx.$$

Note that

$$\left| f(x) \frac{(e^{-ihx} - 1)}{h} e^{-ix\omega} \right| \le |xf(x)| \in L^1$$

and

$$\lim_{h \to 0} f(x) \frac{(e^{-ihx} - 1)}{h} e^{-ix\omega} = -ixf(x)e^{-ix\omega}.$$

By the Lebesgue Dominated Convergence Theorem (Theorem 3.3.1),

$$\left(\hat{f} \right)' (\omega) = \lim_{h \to 0} \frac{\hat{f}(\omega + h) - \hat{f}(\omega)}{h} = -i \int_{-\infty}^{\infty} xf(x)e^{-ix\omega} \, dx.$$

Hence, the theorem is true for $r = 1$. The result for a general r can be proved by mathematical induction. ∎

Corollary 6.3.1 If $f(x)$ decays faster than $\frac{1}{|x|^n}$ for all $n \in \mathbb{N}$, i.e., if $|f(x)| \le \frac{M}{|x|^n}$ for some $M \in \mathbb{R}$, then $\hat{f}(\omega)$ is infinitely differentiable.

Proof: We leave the proof as an exercise. ∎

EXAMPLE 6.3.5: Consider the function f in Example 6.3.3: $f(x) = \frac{\sin^2 x}{x^2}$. We have $f \in L^1$, but $xf(x) = \frac{\sin^2 x}{x}$ is not in L^1. We can see that the derivative of its Fourier transform $\hat{f}(\omega)$ is not continuous at 0 and ±2.

EXAMPLE 6.3.6: The function $e^{-ax^2}, a > 0$, decays faster than x^n for all $n \in \mathbb{N}$. Its Fourier transform $\hat{f}(\omega) = \sqrt{\frac{\pi}{a}} e^{-\frac{\omega^2}{4a}}$ is infinitely differentiable.

Formula (3.9) is useful in solving differential equations.

EXAMPLE 6.3.7: Find the solution of the heat equation

$$\frac{\partial u}{\partial t} = \frac{\partial^2 u}{\partial x^2} \qquad (3.14)$$

with the initial condition $u(x, 0) = f(x)$, where $f, \hat{f} \in L^1$. Taking the Fourier transform of Equation (3.14), we have

$$\frac{\partial}{\partial t} \hat{u}(\omega, t) = -\omega^2 \hat{u}(\omega, t).$$

Hence, $\hat{u}(\omega, t) = c_0(\omega)e^{-\omega^2 t}$. Note that $\hat{u}(\omega, 0) = \hat{f}(\omega)$. Hence,

$$\hat{u}(\omega, t) = \hat{f}(\omega)e^{-\omega^2 t}.$$

Taking the inverse Fourier transform, we get

$$u(x, t) = \frac{1}{2\sqrt{\pi t}} \int_{\mathbb{R}} f(y) e^{-\frac{(x-y)^2}{4t}} dt.$$

3.5 Proof of Theorem 6.3.4

We now prove Theorem 6.3.4. Recall that $A_\rho(t) = \frac{1}{2\pi} \int_{\mathbb{R}} e^{-\frac{|\omega|}{\rho}} \hat{f}(\omega) e^{it\omega} d\omega.$

Lemma 6.3.2 If $f, g \in L^1$, then

$$\int_{\mathbb{R}} \hat{f}(u) g(u) \, du = \int_{\mathbb{R}} f(u) \hat{g}(u) \, du. \tag{3.15}$$

Proof: Applying Fubini's Theorem (Theorem 3.4.2), we have the following:

$$\int_{\mathbb{R}} \hat{f}(u) g(u) \, du = \int_{\mathbb{R}} \left(\int_{\mathbb{R}} f(t) e^{-itu} dt \right) g(u) \, du$$

$$= \int_{\mathbb{R}} \left(\int_{\mathbb{R}} g(u) e^{-itu} du \right) f(t) \, dt$$

$$= \int_{\mathbb{R}} f(t) \hat{g}(t) \, dt.$$

The lemma is proved.   ∎

Lemma 6.3.3 We have the following:

$$\frac{1}{2\pi} \int_{-\infty}^{\infty} \frac{2\rho}{1 + (\rho u)^2} \, du = 1, \quad \rho > 0, \tag{3.16}$$

and

$$A_\rho(t) = \frac{1}{2\pi} \int_{-\infty}^{\infty} f(t - u) \frac{2\rho}{1 + (\rho u)^2} \, du. \tag{3.17}$$

Proof: We have

$$\left(e^{-|\cdot|} \right)^{\wedge}(v) = \int_{-\infty}^{\infty} e^{-|x|} e^{-ixv} \, dx = 2 \int_{0}^{\infty} e^{-x} \cos xv \, dx = \frac{2}{1 + v^2}.$$

Write

$$g_t(\cdot) = e^{-\frac{|\cdot|}{\rho}} e^{i \cdot t}.$$

By Lemma 6.3.1,

$$\hat{g}_t(v) = \frac{2\rho}{1 + [\rho(v - t)]^2}.$$

By Lemma 6.3.2,

$$A_\rho(t) = \frac{1}{2\pi} \int_{-\infty}^{\infty} e^{-\frac{|\omega|}{\rho}} \hat{f}(\omega) e^{it\omega} \, d\omega = \frac{1}{2\pi} \int_{-\infty}^{\infty} \hat{f}(\omega) g_t(\omega) d\omega$$

$$= \frac{1}{2\pi} \int_{-\infty}^{\infty} f(u) \hat{g}_t(u) \, du = \frac{1}{2\pi} \int_{-\infty}^{\infty} f(u) \frac{2\rho}{1 + [\rho(u - t)]^2} \, du$$

$$= \frac{1}{2\pi} \int_{-\infty}^{\infty} f(t - u) \frac{2\rho}{1 + (\rho u)^2} \, du.$$

Note that

$$\int_{-\infty}^{\infty} \frac{2\rho}{1 + (\rho u)^2} \, du = \int_{-\infty}^{\infty} \frac{2}{1 + u^2} \, du = 2\pi.$$

The lemma is proved.

We are ready to prove the theorem. By (3.17), we have

$$\|A_\rho\|_1 \le \frac{1}{2\pi} \int_{-\infty}^{\infty} \|f(\cdot + u)\|_1 \frac{2\rho}{1 + (\rho u)^2} \, du$$

$$\le \|f\|_1 \frac{1}{2\pi} \int_{-\infty}^{\infty} \frac{2\rho}{1 + (\rho u)^2} \, du = \|f\|_1.$$

This is (3.6). By (3.16) we have $f(t) = \frac{1}{2\pi} \int_{-\infty}^{\infty} f(t) \frac{2\rho}{1+(\rho u)^2} \, du$. Therefore,

$$A_\rho(t) - f(t) = \frac{1}{2\pi} \int_{-\infty}^{\infty} \left(f(t - u) - f(t) \right) \frac{2\rho}{1 + (\rho u)^2} \, du.$$

Since $f \in L^1$, for any $\epsilon > 0$, there is a $\delta > 0$ such that

$$\|f(\cdot - u) - f(\cdot)\|_1 < \epsilon \text{ whenever } |u| \le \delta. \tag{3.18}$$

We now have

$$\|A_\rho - f\|_1$$

$$\le \frac{1}{2\pi} \int_{-\infty}^{\infty} \|f(\cdot - u) - f\|_1 \frac{2\rho}{1 + (\rho u)^2} \, du$$

$$= \frac{1}{2\pi} \int_{-\delta}^{\delta} \|f(\cdot - u) - f\|_1 \frac{2\rho}{1 + (\rho u)^2} \, du$$

$$+ \frac{1}{2\pi} \int_{|u| \ge \delta} \|f(\cdot - u) - f\|_1 \frac{2\rho}{1 + (\rho u)^2} \, du$$

where

$$\frac{1}{2\pi} \int_{-\delta}^{\delta} \|f(\cdot - u) - f\|_1 \frac{2\rho}{1 + (\rho u)^2} \, du \leq \frac{1}{2\pi} \int_{-\infty}^{\infty} \frac{2\rho\epsilon}{1 + (\rho u)^2} \, du = \epsilon$$

and

$$0 \leq \frac{1}{2\pi} \int_{|u| \geq \delta} \|f(\cdot - u) - f\|_1 \frac{2\rho}{1 + (\rho u)^2} \, du$$

$$\leq 2\|f\|_1 \frac{1}{2\pi} \int_{|t| > \delta} \frac{2\rho}{1 + (\rho u)^2} \, du = \|f\|_1 \frac{1}{\pi} \int_{|t| > \rho\delta} \frac{1}{1 + u^2} \, du.$$

For a fixed $\delta > 0$,

$$\lim_{\rho \to \infty} \int_{|t| > \rho\delta} \frac{1}{1 + u^2} du = 0.$$

Hence,

$$\lim_{\rho \to \infty} \frac{1}{2\pi} \int_{|u| \geq \delta} \|f(\cdot - u) - f\|_1 \frac{2\rho}{1 + (\rho u)^2} \, du \to 0.$$

Therefore, $\lim_{\rho \to \infty} \|A_\rho - f\|_1 = 0$, which implies (3.7). ∎

Exercises

1. Let $f \in L^1$. Prove that $\hat{f}$ is odd (even) if and only if f is odd (even).
2. Find the Fourier transforms for the following functions.
   (a) $y = \chi_{[-1,1]}(x)$.
   (b) $y = \left(\frac{\sin x/2}{x/2}\right)^2$.
   (c) $y = \frac{1}{1+x^2}$.
   (d) $y = \begin{cases} e^{-x}, & 0 \leq x < \infty, \\ 0, & -\infty < x < 0. \end{cases}$
3. Prove Lemma 6.3.1 (2)–(4):
   (a) $[e^{-ih\cdot} f(\cdot)]\hat{}(\omega) = \hat{f}(\omega + h), \quad h, \omega \in \mathbb{R};$
   (b) $[af(a\cdot)]\hat{}(\omega) = \hat{f}(\omega/a), \quad a > 0, \omega \in \mathbb{R};$
   (c) $\overline{[f(-\cdot)]}\hat{}(\omega) = \hat{f}(\omega), \quad \omega \in \mathbb{R}.$
4. Assume $f, f_n \in L^1$ satisfy

$$\lim_{n \to \infty} \|f - f_n\|_1 = 0.$$

Prove

$$\lim_{n \to \infty} \hat{f}_n(\omega) = \hat{f}(\omega)$$

uniformly for all $\omega \in \mathbb{R}$.

5. For $f, g \in L^1$, the *correlation* of f and g is the function

$$[f, g](x) = \int_{-\infty}^{\infty} f(x + t)\overline{g(t)}\, dt,$$

and $[f, f]$ is called the *auto-correlation* of f. Prove
   (a) $[f, g] = f * \overline{g(-\cdot)}$.
   (b) $\widehat{[f, g]}(\omega) = \hat{f}(\omega)\overline{\hat{g}(\omega)}$.
   (c) $\widehat{[f, f]}(\omega) = |\hat{f}(\omega)|^2$.
6. Let $b_1(x) = \chi_{[0,1]}(x)$.
   (a) Find an explicit expression for $b_2(x) = b_1 * b_1(x)$.
   (b) Find $\hat{b}_2(\omega)$.
   (c) Let $b_n(x) = b_{n-1} * b_1(x)$, $n = 2, 3, \cdots$. Find $\hat{b}_n(\omega)$.
   (d) Use (3.9) (for $r = 1$) to prove that $b'_n(x) = b_{n-1}(x) - b_{n-1}(x - 1)$.
7. Assume $f \in L^1$ and $g(x) = e^{-x^2}$. Prove that $g * f$ is infinitely differentiable.
8. Prove Corollary 6.3.1: If $|f(x)| \le \frac{M}{|x|^n}$ for all $n \in \mathbb{N}$, where $M > 0$ is a constant independent of n, then $\hat{f}(\omega)$ is infinitely differentiable.
9. For $g \in L^1$, define $g^*(x) = \sum_{k \in \mathbb{Z}} g(x + 2k\pi)$. Prove that
   (a) $g^* \in \tilde{L}^1$ and $\int_{-\pi}^{\pi} g^*(t)\, dt = \int_{\mathbb{R}} g(t)\, dt$.
   (b) $(g^*)\,\hat{}\,(k) = \frac{1}{2\pi}\hat{g}(k)$.
   (c) If $\sum_{k \in \mathbb{Z}} g(x + 2k\pi)$ is uniformly convergent to $g^*(x)$, then $\sum_{k \in \mathbb{Z}} g(x + 2k\pi) = \frac{1}{2\pi} \sum_{k \in \mathbb{Z}} \hat{g}(k)e^{ikx}$.

4 Fourier Transforms of Square Integrable Functions

In Section 6.2, we saw that the finite Fourier transform is an orthonormal isomorphism from $\tilde{L}^2$ to l^2. In this section, we want to extend this result to the space

$$L^2 = \left\{ f \,\middle|\, \int_{\mathbb{R}} |f(t)|^2 dt < \infty \right\}.$$

As we stated in Section 5.3, this space is equipped with the inner product

$$\langle f, g \rangle = \int_{\mathbb{R}} f(t)\overline{g(t)}\, dt,$$

and L^2 is a Hilbert space. The norm of $f \in L^2$ is $\|f\|_2 = \sqrt{\langle f, f \rangle}$, the distance between f and g is $\|f - g\|_2$, and the angle between two functions $f, g \in L^2(f \neq 0, g \neq 0)$ is

$$\cos^{-1} \frac{\langle f, g \rangle}{\|f\|_2 \|g\|_2}.$$

We can see that the geometry in L^2 is quite similar to that in $\tilde{L}^2$. Hence, the results in this section are similar to those in Section 2. However, there is still a main difference between $\tilde{L}^2$ and $L^2 : \tilde{L}^2 \subset \tilde{L}^1$, but L^2 is not a subspace of L^1. Thus, for $f \in L^2$ the integral $\int_{-\infty}^{\infty} f(x) e^{-ix\omega} \, dx$ may not exist. Hence, we cannot use it to define the Fourier transform of $f \in L^2$. In order to overcome this difficulty, we adopt a limit approach to the Fourier transform for L^2. Recall that any function $f \in L^2$ can be considered as a limit of a sequence of functions in $L^2 \cap L^1$, say (f_n). Since the Fourier transform of f_n is well defined, the Fourier transform of f can be defined as the limit of $\hat{f}_n$. We now provide the details.

4.1 Definition and Properties

We start our discussion with $f \in L^1 \cap L^2$.

Lemma 6.4.1 If $f \in L^1 \cap L^2$, then $\hat{f} \in L^2$ and

$$\|f\|_2^2 = \frac{1}{2\pi} \|\hat{f}\|_2^2. \tag{4.1}$$

Proof: Let $f^*(x) = \overline{f(-x)}$. It is clear that $f^* \in L^1 \cap L^2$. We define $h = f * f^*$. Since both $f, f^* \in L^1 \cap L^2$, by Lemma 6.3.2 h is continuous and in L^1. We also have $\hat{h}(\omega) = |\hat{f}(\omega)|^2$ and $h(0) = \|f\|_2^2$. Applying Theorem 6.3.4 and setting $t = 0$, we have

$$\|f\|_2^2 = h(0) = \lim_{\rho \to \infty} \frac{1}{2\pi} \int_{-\infty}^{\infty} e^{-|\omega|/\rho} \hat{h}(\omega) \, d\omega$$

$$= \lim_{\rho \to \infty} \frac{1}{2\pi} \int_{-\infty}^{\infty} e^{-|\omega|/\rho} |\hat{f}(\omega)|^2 \, d\omega.$$

Note that $e^{-|\omega|/\rho} |\hat{f}(\omega)|^2$ is a positive, increasing function of $\rho > 0$ with

$$\lim_{\rho \to \infty} e^{-|\omega|/\rho} |\hat{f}(\omega)|^2 = |\hat{f}(\omega)|^2.$$

By the Monotone Convergence Theorem (Theorem 3.3.3),

$$\|f\|_2^2 = \lim_{\rho \to \infty} \frac{1}{2\pi} \int_{-\infty}^{\infty} e^{-|\omega|/\rho} |\hat{f}(\omega)|^2 \, d\omega$$

$$= \frac{1}{2\pi} \int_{-\infty}^{\infty} \lim_{\rho \to \infty} e^{-|\omega|/\rho} |\hat{f}(\omega)|^2 \, d\omega$$

$$= \frac{1}{2\pi} \int_{-\infty}^{\infty} |\hat{f}(\omega)|^2 \, d\omega = \frac{1}{2\pi} \|\hat{f}\|_2^2.$$

The lemma is proved.   ∎

The next step is to construct a sequence $f_N \in L^1 \cap L^2$ which has limit f as $N \to \infty$. Naturally, we choose the truncated functions of f :

$$f_N(t) = \begin{cases} f(t), & |t| \leq N, \\ 0, & \text{otherwise,} \end{cases} \quad N > 0.$$

It is clear that $f_N \in L^1 \cap L^2$ and

$$\lim_{N \to \infty} \|f_N - f\|_2 = 0. \tag{4.2}$$

We then use the limit of $\widehat{f_N}$ to define the Fourier transform of f. First, we confirm the existence of the limit.

Lemma 6.4.2 There is a function $\phi \in L^2$ such that

$$\lim_{N \to \infty} \|\widehat{f_N} - \phi\|_2 = 0. \tag{4.3}$$

Proof: By Lemma 6.4.1, for arbitrary $N, M > 0$,

$$\|\widehat{f_N} - \widehat{f_M}\|_2^2 = 2\pi \|f_N - f_M\|_2^2. \tag{4.4}$$

By (4.2), (f_N) is a Cauchy sequence in L^2. Then (4.4) implies $(\widehat{f_N})$ is a Cauchy sequence in L^2 as well. By the completeness of L^2, there is a function $\phi \in L^2$ such that (4.3) holds.   ∎

If $\lim_{n \to \infty} \|f_n - f\|_2 = 0$, then we denote it by

$$f(t) = L^2\text{-}\lim_{n \to \infty} f_n(t).$$

We are ready to define the Fourier transform for a function $f \in L^2$.

Definition 6.4.1 For $f \in L^2$, the Fourier transform is defined by

$$\hat{f}(\omega) = L^2\text{-}\lim_{N\to\infty} \int_{-N}^{N} f(x)e^{-ix\omega}\, dx. \qquad (4.5)$$

By Lemma 6.4.1, for a function $f \in L^2 \cap L^1$,

$$\int_{-\infty}^{\infty} f(x)e^{-ix\omega}\, dx = L^2\text{-}\lim_{N\to\infty} \int_{-N}^{N} f(x)e^{-ix\omega}\, dx, \quad a.e.$$

Hence this definition is consistent with Definition 6.3.1.

Note the difference between the pointwise limit

$$\lim_{n\to\infty} f_n(t) = f(t)$$

and the limit L^2-limit

$$L^2\text{-}\lim_{n\to\infty} f_n(t) = f(t).$$

For example, the function

$$f_n(t) = \begin{cases} n, & t \in (0, \tfrac{1}{n}), \\ 0, & \text{otherwise} \end{cases}$$

converges to 0 everywhere: $\lim_{n\to\infty} f_n(t) = 0$, for all $t \in \mathbb{R}$, but

$$\lim_{n\to\infty} \|f_n - 0\|_2 = \lim_{n\to\infty} \sqrt{n} = \infty.$$

Hence $L^2\text{-}\lim_{n\to\infty} f_n(t)$ does not exist. Conversely, there exists a sequence (f_n) such that its L^2-limit exists but the piecewise limit does not exist. For example, let

$$g_{n,m}(t) = \begin{cases} 1, & t \in (\tfrac{m}{n+1}, \tfrac{m+1}{n+1}) \\ 0, & \text{otherwise} \end{cases} \quad \text{for all } n \in \mathbb{Z}^+, \quad m = 0, 1, \cdots, n.$$

We now arrange $g_{n,m}$ to a biindex sequence

$$g_{0,0}, g_{1,0}, g_{1,1}, g_{2,0}, g_{2,1}, g_{2,2}, \cdots, \qquad (4.6)$$

and rewrite it as the sequence $f_1, f_2, \cdots$. That is, $f_{\tau(n,m)} = g_{n,m}$, where $\tau(n, m) = \tfrac{1}{2}(n+1)n+m$. Then $L^2\text{-}\lim_{k\to\infty} f_k(t) = 0$, but $f_k(t)$ is divergent at any point $t \in (0, 1)$.

Similar to the case for L^1, the Fourier transform of $f \in L^2$ has the following properties.

Lemma 6.4.3 Lemma 6.3.1 also holds for each $f \in L^2$.

Theorem 6.4.1 [Convolution Theorem] If $f \in L^2$ and $g \in L^1$, then $[f * g]\hat{}(\omega) = \hat{f}(\omega)\hat{g}(\omega)$ a.e.

Theorem 6.4.2 [Parseval's Formula] If $f, g \in L^2$, then

$$\int_{-\infty}^{\infty} \hat{f}(u)g(u)\,du = \int_{-\infty}^{\infty} f(u)\hat{g}(u)\,du.$$

We leave the proofs of these results to the reader.

4.2 Plancherel's Theorem

We have already seen that the Fourier transform is a linear transform on L^2. We now study its geometric properties. From Lemma 6.4.1 and 6.4.2, we can see that the normalized Fourier transform is orthonormal.

Theorem 6.4.3 If $f \in L^2$, then $\hat{f} \in L^2$ and

$$\|\mathcal{F}(f)\|_2^2 = \|f\|_2^2. \tag{4.7}$$

Proof: By Lemma 6.4.1, we have

$$\frac{1}{2\pi}\|\hat{f}_N\|_2^2 = \|f_N\|_2^2, \quad N \geq 0.$$

Letting $N \to \infty$, we have $\frac{1}{2\pi}\|\hat{f}\|_2^2 = \|f\|_2^2$ which implies (4.7). ∎

As a consequence of Theorem 6.4.3, we have:

Corollary 6.4.1 Assume $f \in L^2$ and $\hat{f}(\omega) = 0$ a.e. Then $f(t) = 0$ a.e.

Proof: If $\hat{f}(\omega) = 0$ a.e., then $\|\hat{f}\|_2 = 0$, which yields $\|f\|_2 = 0$. Hence, $f(t) = 0$ a.e. ∎

We also have the following.

Corollary 6.4.2 If $f, g \in L^2$, then

$$\langle f, g \rangle = \frac{1}{2\pi} \langle \hat{f}, \hat{g} \rangle. \tag{4.8}$$

Proof: We leave the proof as an exercise. ∎

We now prove that the Fourier transform is also an onto mapping on L^2.

Theorem 6.4.4 For $f \in L^2$, set $g(x) = \frac{1}{2\pi}\hat{f}(-x)$. Then $\hat{g}(t) = f(t)$ a.e. and

$$f(t) = L^2 - \lim_{N \to \infty} \frac{1}{2\pi} \int_{\mathbb{R}} \hat{f}(\omega) e^{i\omega t} \, d\omega. \tag{4.9}$$

Proof: By Lemma 6.4.3, we have $\hat{g}(x) = [g(-\cdot)]\hat{\ }(x)$. Hence,

$$\langle f, \hat{g} \rangle = \int_{\mathbb{R}} f(u)\overline{\hat{g}(u)} \, du = \int_{\mathbb{R}} f(u)\overline{[g(-\cdot)]\hat{\ }(u)} \, du.$$

By Theorem 6.4.2,

$$\int_{\mathbb{R}} f(u)\overline{[g(-\cdot)]\hat{\ }(u)} \, du = \int_{\mathbb{R}} \hat{f}(u)\overline{g(-u)} \, du$$

$$= \frac{1}{2\pi} \int_{\mathbb{R}} \hat{f}(u)\overline{\hat{f}(u)} \, du = \frac{1}{2\pi}\|\hat{f}\|_2^2 = \|f\|_2^2.$$

Hence, $\langle f, \hat{g} \rangle = \|f\|_2^2$ and $\langle \hat{g}, f \rangle = \overline{\langle f, \hat{g} \rangle} = \|f\|_2^2$. We also have

$$\langle \hat{g}, \hat{g} \rangle = 2\pi\langle g, g \rangle = \frac{1}{2\pi}\|\hat{f}(-\cdot)\|_2^2 = \|f\|_2^2.$$

Hence,

$$\|f - \hat{g}\|_2^2 = \langle f - \hat{g}, f - \hat{g} \rangle$$
$$= \langle f, f \rangle - \langle f, \hat{g} \rangle - \langle \hat{g}, f \rangle + \langle \hat{g}, \hat{g} \rangle$$
$$= 0,$$

which yields $\hat{g} = f$ a.e. Note that

$$\hat{g}(t) = L^2 - \lim_{N \to \infty} \int_{-N}^{N} g(u) e^{-iut} \, du$$

$$= L^2 - \lim_{N \to \infty} \frac{1}{2\pi} \int_{-N}^{N} \hat{f}(-u) e^{-iut} \, du$$

$$= L^2 - \lim_{N \to \infty} \frac{1}{2\pi} \int_{-N}^{N} \hat{f}(\omega) e^{i\omega t} \, d\omega.$$

Hence (4.9) holds.   ∎

We summarize Theorem 6.4.3 and Theorem 6.4.4 in the *Plancherel Theorem.*

Theorem 6.4.5 [Plancherel Theorem] The normalized Fourier transform $\mathcal{F}$ is an orthonormal automorphism on L^2.

Recall that the formula (4.9) recovers f from $\hat{f}$. Hence we use it to define the inverse Fourier transform for the functions in L^2.

For $g \in L^2$, its inverse Fourier transform is defined by

$$g^\vee(t) = L^2 - \lim_{N \to \infty} \frac{1}{2\pi} \int_{-N}^{N} \hat{g}(\omega) e^{i\omega t} \, d\omega.$$

Thus, if $f \in L^2$, then $[\hat{f}]^\vee = f$, or equivalently, $\mathcal{F}^{-1}\mathcal{F}(f) = f$.

4.3 The Fourier Transform of Derivatives

The result for Fourier transforms of derivatives in L^2 is similar to that for L^1. We state the result below and leave the proof as an exercise. Define

$$W_2^r = \left\{ f \in L^2 \,\middle|\, f^{(j)} \in L^2, 0 \le j \le r \right\}.$$

Theorem 6.4.6 If $f \in W_2^r$, then

$$[f^{(r)}]^\wedge(\omega) = (i\omega)^r \hat{f}(\omega) \quad \text{a.e.} \tag{4.10}$$

Conversely, if $f \in L^2$ and there is a function $g \in L^2$ such that

$$(i\omega)^r \hat{f}(\omega) = \hat{g}(\omega) \quad \text{a.e.,}$$

then $f \in W_2^r$.

A useful consequence of the theorem is the following.

Corollary 6.4.3 A function $f \in W_2^r$ if and only if

$$\int_{-\infty}^{\infty} (1 + \omega^{2r}) |\hat{f}(\omega)|^2 \, d\omega < \infty. \tag{4.11}$$

Proof: If $f \in W_2^r$, then $f^{(r)} \in L^2$, and then so is $[f^{(r)}]^\wedge$. By (4.10), $(i\omega)^r \hat{f}(\omega) \in L^2$. We now have

$$\int_{-\infty}^{\infty} (1 + \omega^{2r}) |\hat{f}(\omega)|^2 \, d\omega = \left(\int_{-1}^{1} + \int_{|\omega|>1} \right) (1 + \omega^{2r}) |\hat{f}(\omega)|^2 \, d\omega,$$

where

$$\int_{-1}^{1} (1 + \omega^{2r})|\hat{f}(\omega)|^2 \, d\omega \leq 2 \int_{-1}^{1} |\hat{f}(\omega)|^2 \, d\omega \leq 2\|\hat{f}\|_2^2$$

and

$$\int_{|\omega|>1} (1 + \omega^{2r})|\hat{f}(\omega)|^2 \, d\omega \leq 2 \int_{|\omega|>1} \omega^{2r}|\hat{f}(\omega)|^2 \, d\omega \leq 2\|[f^{(r)}]^\smallfrown\|_2.$$

Hence, (4.11) is true. Conversely, if (4.11) is true, then

$$\left|(i\omega)^r \hat{f}(\omega)\right| \leq \sqrt{(1 + \omega^{2r})}|\hat{f}(\omega)| \in L^2.$$

Hence, $(i\omega)^r \hat{f}(\omega) \in L^2$, which yields $f \in W_2^r$. ∎

Exercises

1. Let $f(x) = \frac{e^{ix}-1}{ix}$. Prove $f \in L^2, f \notin L^1$, and $\hat{f} \in L^1$.
2. Let $f(x) = \frac{\sin x}{x}$. Prove $\hat{f}(\omega) = \begin{cases} \pi, & |\omega| < 1, \\ 0, & |\omega| > 1. \end{cases}$
3. Prove Lemma 6.4.3: For $f \in L^2$, the following hold.
   (a) $[f(\cdot + h)]^\smallfrown(\omega) = e^{ih\omega}\hat{f}(\omega), \quad h, \omega \in \mathbb{R}.$
   (b) $[e^{-ih\cdot}f(\cdot)]^\smallfrown(\omega) = \hat{f}(\omega + h), \quad h, \omega \in \mathbb{R}.$
   (c) $[af(a\cdot)]^\smallfrown(\omega) = \hat{f}(\omega/a), \quad a > 0, \omega \in \mathbb{R}.$
   (d) $\overline{[f(-\cdot)]}^\smallfrown(\omega) = \hat{f}(\omega), \quad \omega \in \mathbb{R}.$
4. Prove Theorem 6.4.1: If $f \in L^2$ and $g \in L^1$, then $[f * g]^\smallfrown(\omega) = \hat{f}(\omega)\hat{g}(\omega)$ a.e.
5. Prove Theorem 6.4.2: If $f, g \in L^2$, then $\int_{-\infty}^{\infty} \hat{f}(u)g(u) \, du = \int_{-\infty}^{\infty} f(u)\hat{g}(u) \, du.$
6. Prove Corollary 6.4.2: If $f, g \in L^2$, then $\langle f, g \rangle = \frac{1}{2\pi}\langle \hat{f}, \hat{g} \rangle.$
7. Use mathematical induction to prove Theorem 6.4.6: If $f \in W_2^r$, then

$$[f^{(r)}]^\smallfrown(\omega) = (i\omega)^r \hat{f}(\omega) \quad \text{a.e.}$$

Conversely, if $f \in L^2$ and there is a function $g \in L^2$ such that

$$(i\omega)^r \hat{f}(\omega) = \hat{g}(\omega) \quad \text{a.e.}$$

Then $f \in W_2^r$.
8. Let $f \in L^2$. Prove the following. For $x, h \in \mathbb{R}$,

$$\int_{x}^{x+h} \hat{f}(v) \, dv = \int_{-\infty}^{\infty} \frac{e^{-ihu} - 1}{-iu} f(u)e^{-ixu} \, du;$$

in particular,

$$\hat{f}(\omega) = \frac{d}{d\omega}\left[\int_{-\infty}^{\infty}\frac{e^{-i\omega u} - 1}{-iu}f(u)\,du\right] \quad \text{a.e.}$$

9. Let $f \in L^2$. Prove the following. For $x, h \in \mathbb{R}$,

$$\int_x^{x+h} f(u)\,du = \frac{1}{2\pi}\int_{-\infty}^{\infty}\frac{e^{ih\omega} - 1}{i\omega}\hat{f}(\omega)e^{ix\omega}\,d\omega;$$

in particular,

$$f(x) = \frac{d}{dx}\left[\frac{1}{2\pi}\int_{-\infty}^{\infty}\frac{e^{ixu} - 1}{iu}\hat{f}(u)\,du\right] \quad \text{a.e.}$$

10. Use the result of Exercise 2 and the result of Example 6.3.3 to prove:
    (a) $\int_{-\infty}^{\infty}\frac{\sin^2 u}{u^2}\,du = \pi$.
    (b) $\int_{-\infty}^{\infty}\frac{\sin^4 u}{u^4}\,du = \frac{2\pi}{3}$.
11. Let $b_n(x)$ be the function defined in Exercise 6 in Section 6.3. Prove $b_n \in W_2^{n-1}$.
12. Assume $f \in L^2$ and $(ix)f(x) \in L^2$. Write $g(x) = (ix)f(x)$. Prove that $\frac{d}{d\omega}\hat{f} \in L^2$ and

$$\frac{d}{d\omega}\hat{f}(\omega) = -\hat{g}(\omega) \quad \text{a.e.}$$

5 The Poisson Summation Formula

At the end of this chapter, we give an important formula which links the sum of the shifts of a function to the sum of the shifts of its Fourier transform, called the Poisson Summation Formula.

5.1 The Poisson Summation Formula for L^1

The first elementary result in this direction is the following.

Lemma 6.5.1 For $g \in L^1$, we define

$$g^*(x) = \sum_{k=-\infty}^{\infty} g(x + 2k\pi). \tag{5.1}$$

Then $g^* \in \tilde{L}^1$,

$$\|g^*\|_{\tilde{L}^1} \le \frac{1}{2\pi}\|g\|_1, \tag{5.2}$$

and

$$\int_{-\pi}^{\pi} g^*(x)\,dx = \int_{-\infty}^{\infty} g(x)\,dx. \tag{5.3}$$

Assume $f \in \tilde{C}$. Then

$$\int_{-\pi}^{\pi} f(x-t)g^*(t)\,dt = \int_{-\infty}^{\infty} f(x-t)g(t)\,dt. \tag{5.4}$$

In particular,

$$\int_{-\pi}^{\pi} f(t)g^*(t)\,dt = \int_{-\infty}^{\infty} f(t)g(t)\,dt, \quad f \in \tilde{C}. \tag{5.5}$$

Proof: We have

$$\sum_{k=-\infty}^{\infty} \int_{-\pi}^{\pi} |g(x+2k\pi)|\,dx = \int_{-\pi}^{\pi} \sum_{k=-\infty}^{\infty} |g(x+2k\pi)|\,dx$$
$$= \int_{-\infty}^{\infty} |g(x)|\,dx = \|g\|_1 < \infty.$$

By the Monotone Convergence Theorem (Theorem 3.3.3), $g^*(x)$ exists and (5.3) holds. Equations (5.4) and (5.5) are directly derived from (5.3). We also have

$$\|g^*\|_{\tilde{L}^1} = \frac{1}{2\pi} \int_{-\pi}^{\pi} |g^*(x)|\,dx$$
$$= \frac{1}{2\pi} \int_{-\pi}^{\pi} \left| \sum_{k=-\infty}^{\infty} g(x+2k\pi) \right| dx$$
$$\le \frac{1}{2\pi} \int_{-\pi}^{\pi} \sum_{k=-\infty}^{\infty} |g(x+2k\pi)|\,dx$$
$$= \frac{1}{2\pi}\|g\|_1,$$

which implies (5.2). ∎

In (5.5), choosing $f(t) = \frac{1}{2\pi}e^{-ikt}$, we have the following.

Corollary 6.5.1 Let $g \in L^1$ and g^* be given by (5.1). Then the Fourier coefficients of the 2π-periodic function g^* are

$$[g^*]\hat{}(k) = \frac{1}{2\pi}\hat{g}(k), \quad k \in \mathbb{Z}.$$

From Corollary 6.5.1, the Fourier series of g^* is $\frac{1}{2\pi} \sum_{k=-\infty}^{\infty} \hat{g}(k)e^{-ikx}$. In general, the Fourier series of a function may not converge to the function. The theory of Fourier series provides many different conditions for the convergence of Fourier series; some of them ensure uniform convergence, others ensure almost everywhere convergence or other types of convergence. When the Fourier series of g^* is convergent to g^*, we have the formula

$$\sum_{k=-\infty}^{\infty} g(x + 2k\pi) = \frac{1}{2\pi} \sum_{k=-\infty}^{\infty} \hat{g}(k)e^{-ikx},$$

which holds in a certain sense. This formula is called the *Poisson summation formula*. According to the conditions which ensure the formula, we can derive many different versions of the Poisson summation formula. We now establish a few of them.

Theorem 6.5.1 Assume $g \in L^1 \cap C$ and the series $\sum_{k=-\infty}^{\infty} g(x + 2k\pi)$ is uniformly convergent to a function g^*. Assume also $\sum_{k \in \mathbb{Z}} |\hat{g}(k)| < \infty$. Then

$$\sum_{k=-\infty}^{\infty} g(x + 2k\pi) = \frac{1}{2\pi} \sum_{k=-\infty}^{\infty} \hat{g}(k)e^{ikx}, \quad x \in \mathbb{R}. \tag{5.6}$$

Particularly,

$$\sum_{k=-\infty}^{\infty} g(2k\pi) = \frac{1}{2\pi} \sum_{k=-\infty}^{\infty} \hat{g}(k). \tag{5.7}$$

Proof: Since g is continuous and $\sum_{k=-\infty}^{\infty} g(x + 2k\pi)$ is uniformly convergent to $g^*(x)$, g^* is a continuous function and $g^*(x) = \sum_{k=-\infty}^{\infty} g(x + 2k\pi)$ holds everywhere. The condition $\sum_{k \in \mathbb{Z}} |\hat{g}(k)| < \infty$ implies $[g^*]\hat{} \in l^1$ and therefore $\sum_{k \in \mathbb{Z}} [g^*]\hat{}(k)e^{ikx} = g^*(x)$ everywhere. This yields (5.6). Formula (5.7) is obtained by setting $x = 0$ in (5.6). The theorem is proved. ∎

Corollary 6.5.2 If $g \in C$ satisfies

$$g(x) = O\left(\frac{1}{1 + |x|^\alpha}\right) \tag{5.8}$$

and

$$\hat{g}(\omega) = O\left(\frac{1}{1 + |\omega|^\alpha}\right) \tag{5.9}$$

for some $\alpha > 1$, then (5.6) holds.

Proof: By (5.8), for any $x \in [-\pi, \pi]$,

$$\sum_{k=-\infty}^{\infty} |g(x + 2k\pi)| \le C \sum_{k=-\infty}^{\infty} \frac{1}{1 + |x + 2k\pi|^\alpha} < M,$$

where M is independent of x. Hence, $\sum_{k=-\infty}^{\infty} g(x+2k\pi)$ is uniformly convergent to a function $g^* \in \check{C}$. Similarly, by (5.9),

$$\sum_{k \in \mathbb{Z}} |\hat{g}(k)| \le C \sum_{k=-\infty}^{\infty} \frac{1}{1 + |k|^\alpha} < \infty.$$

By Theorem 6.5.1, (5.6) holds. ∎

By changing the period of the function, we obtain an alternative form of the Poisson summation formula.

Theorem 6.5.2 If $g \in L^1 \cap C$, the series $\sum_{k=-\infty}^{\infty} g(x + k)$ is uniformly convergent, and $\sum_{k \in \mathbb{Z}} |\hat{g}(2k\pi)| < \infty$, then

$$\sum_{k=-\infty}^{\infty} g(x + k) = \sum_{k=-\infty}^{\infty} \hat{g}(2k\pi)e^{i2k\pi x}, \quad x \in \mathbb{R}. \tag{5.10}$$

In particular,

$$\sum_{k=-\infty}^{\infty} g(k) = \sum_{k=-\infty}^{\infty} \hat{g}(2k\pi). \tag{5.11}$$

Proof: Let $g_{2\pi}(x) = g(\frac{x}{2\pi})$. Then $g_{2\pi}$ is a 2π-periodic function and $g_{2\pi}$ satisfies the conditions of Theorem 6.5.1. Hence,

$$\sum_{k=-\infty}^{\infty} g_{2\pi}(t + 2k\pi) = \frac{1}{2\pi} \sum_{k=-\infty}^{\infty} \hat{g}_{2\pi}(k)e^{ikt}, \quad t \in \mathbb{R}.$$

Setting $t = 2\pi x$, by $g_{2\pi}(x) = g(\frac{x}{2\pi})$ and $\hat{g}_{2\pi}(\omega) = \hat{g}(2\pi\omega)$, we have (5.10). ∎

Finally, we give the Poisson summation formula for the autocorrelation of f, which is often used in wavelet analysis. Recall that the autocorrelation of $f \in L^2$ is defined by (see Section 6.3, Exercise 5)

$$[f, f](x) = \int_{\mathbb{R}} f(t)\overline{f(t + x)}dt. \tag{5.12}$$

Theorem 6.5.3 If $f(x)$ satisfies

$$f(x) = O\left(\frac{1}{1 + |x|^\alpha}\right) \tag{5.13}$$

for some $\alpha > 1$ and

$$\hat{f}(x) = O\left(\frac{1}{1 + |x|^\beta}\right) \tag{5.14}$$

for some $\beta > 1/2$, then

$$\sum_{k=-\infty}^{\infty} \left|\hat{f}(x + 2k\pi)\right|^2 = \sum_{k=-\infty}^{\infty} [f,f](k)e^{-ikx}, \quad x \in \mathbb{R}. \tag{5.15}$$

Proof: By (5.13), $f \in L^1 \cap L^2$ and then $\hat{f} \in L^2 \cap C_0$, which implies $\widehat{[f,f]} = |\hat{f}|^2 \in L^1 \cap C_0$. Besides, by the Convolution Theorem (Theorem 6.3.2), $[f,f] \in L^1 \cap C_0$. We now write $g = |\hat{f}|^2$. Condition (5.14) yields the uniform convergence of the series $\sum_{k=-\infty}^{\infty} g(x + 2k\pi) = \sum_{k=-\infty}^{\infty} |\hat{f}(x + 2k\pi)|^2$. By Theorem 6.3.6, we also have

$$\frac{1}{2\pi}\hat{g}(k) = \frac{1}{2\pi} \int_{-\infty}^{\infty} \widehat{[f,f]}(t)e^{-itk} dt = [f,f](-k). \tag{5.16}$$

By (5.13), there is a $C > 0$ such that

$$\sum_{k\in\mathbb{Z}} |f(t + k)| \le C \sum_{k\in\mathbb{Z}} \frac{1}{1 + |t + k|^\alpha}.$$

The function $\sum_{k\in\mathbb{Z}} \frac{1}{1+|t+k|^\alpha}$ is a 1-periodic continuous function. Let $M = \max_{x\in[0,1]} \sum_{k\in\mathbb{Z}} \frac{1}{1+|t+k|^\alpha}$. Then $\max_{x\in[0,1]} \sum_{k\in\mathbb{Z}} |f(t + k)| \le CM$. By (5.16), we have

$$\frac{1}{2\pi} \sum_{k=-\infty}^{\infty} |\hat{g}(k)| = \sum_{k=-\infty}^{\infty} |[f,f](k)| = \sum_{k=-\infty}^{\infty} \left|\int_{\mathbb{R}} f(t)\overline{f(t + k)}\, dt\right|$$

$$\le \int_{\mathbb{R}} |f(t)| \sum_{k=-\infty}^{\infty} |\overline{f(t + k)}|\, dt \le CM\|f\|_1 < \infty.$$

This allows us to apply the Poisson summation formula (5.6) to g. Thus, we have $\sum_{k=-\infty}^{\infty} g(x + 2k\pi) = \frac{1}{2\pi} \sum_{k=-\infty}^{\infty} \hat{g}(k)e^{ikx}$, which implies (5.15). The theorem is proved. ∎

EXAMPLE 6.5.1: Let

$$F(x) = \frac{\sin^2(x/2)}{(x/2)^2}.$$

We have

$$\hat{F}(\omega) = \begin{cases} 2\pi(1 - |\omega|), & |\omega| \le 1, \\ 0 & |\omega| \ge 1. \end{cases}$$

The function satisfies the conditions of the Poisson summation formula and so does the function $\rho F(\rho x)$, $\rho > 0$. Let $g(x) = (n + 1)F((n + 1)x)$. Then $\hat{g}(\omega) = \hat{F}(\frac{\omega}{n+1})$. Applying the Poisson summation formula (5.6) to $g(x)$, we have

$$(n + 1) \sum_{k \in \mathbb{Z}} \frac{\sin^2((n + 1)(x/2 + k\pi))}{((n + 1)(x/2 + k\pi))^2} = \sum_{k=-n}^{n} \left(1 - \frac{|k|}{n + 1}\right) e^{ikx}.$$

Note that

$$(n + 1) \sum_{k \in \mathbb{Z}} \frac{\sin^2((n + 1)(x/2 + k\pi))}{((n + 1)(x/2 + k\pi))^2} = \frac{1}{n + 1} \sum_{k \in \mathbb{Z}} \frac{\sin^2((n + 1)x/2)}{(x/2 + k\pi)^2}.$$

We also have

$$\sum_{k=-n}^{n} \left(1 - \frac{|k|}{n + 1}\right) e^{ikx} = \frac{1}{n + 1} \frac{\sin^2((n + 1)x/2)}{\sin^2(x/2)}, \qquad (5.17)$$

(see Exercise 1), which yields

$$\frac{1}{n + 1} \frac{\sin^2((n + 1)x/2)}{\sin^2(x/2)} = \frac{1}{n + 1} \sum_{k \in \mathbb{Z}} \frac{\sin^2((n + 1)x/2)}{(x/2 + k\pi)^2},$$

and therefore

$$\frac{1}{\sin^2 u} = \sum_{k \in \mathbb{Z}} \frac{1}{(u + k\pi)^2}, \qquad u \in \mathbb{R} \setminus \pi\mathbb{Z}. \qquad (5.18)$$

5.2 Fourier Transforms of Compactly Supported Functions

In many applications, functions are actually defined on finite intervals, not on the whole real line. In this subsection we briefly discuss the properties of the Fourier transforms of these functions.

Definition 6.5.1 Let $f \in L^1_{loc}$. A point $x \in \mathbb{R}$ is called a *support point* of f if, for any $\delta > 0$, there is a function $g \in L^\infty_{loc}$ such that

$$\int_{x-\delta}^{x+\delta} f(x)g(x)\, dx \ne 0.$$

The set of all support points of f is called the *support* of f and denoted by supp f.

Note that
$$L^p \subset L^1_{loc}, 1 \leq p \leq \infty,$$
and $C \subset L^1_{loc}$. Hence, by Definition 6.5.1, the support of a function in L^p or in C is well defined. It is easy to verify that if $f \in C$, then

$$\text{supp} f = \text{closure } \{x \mid f(x) \neq 0\},$$

which is a closed set. In general, we have the following.

Lemma 6.5.2 The support of a function $f \in L^1_{loc}$ is a closed set.

Proof: Let $\bar{x}$ be in the closure of supp f. Then, for any $\delta > 0$, there is an $x \in \text{supp} f$ such that $x \in (\bar{x} - \delta, \bar{x} + \delta)$. Let $d = \min(x - \bar{x} + \delta, \bar{x} + \delta - x)$. Then $(x - d, x + d) \subset (\bar{x} - \delta, \bar{x} + \delta)$, and there is a function $g \in L^\infty_{loc}$ such that

$$\int_{x-d}^{x+d} f(x)g(x)\,dx \neq 0.$$

We now set $g^*(x) = g(x)\chi_{[x-d,x+d]}(x)$. Then

$$\int_{\bar{x}-\delta}^{\bar{x}+\delta} f(x)g^*(x)\,dx = \int_{x-d}^{x+d} f(x)g(x)\,dx \neq 0.$$

Hence, $\bar{x} \in \text{supp} f$. The lemma is proved. ∎

Definition 6.5.2 A function is said to be *compactly supported* if its support is a compact set.

For convenience, we denote the subspace of L^p that contains all compactly supported functions in L^p by L^p_0 and denote the subspace of C_0 that contains all compactly supported functions in C_0 by C_{00}. (Note that L^p_0 (or C_{00}) is not a closed subspace of L^p (or C_0).) Then we have the following relations:

$$C_{00} \subset L^p_0 \subset L^1_0, \quad 1 \leq p \leq \infty.$$

If $f \in L^1_0$, then, by Theorem 6.4.6, it is easy to see that $\hat{f}(\omega)$ is infinitely differentiable and

$$[\hat{f}]^{(r)}(\omega) = \int_{-\infty}^{\infty} (-ix)^r f(x)e^{-ix\omega}\,dx. \tag{5.19}$$

We now prove a stronger result.

Theorem 6.5.4 If $f \in L_0^1$, then $\hat{f}(\omega)$ is infinitely differentiable. Moreover, the Maclaurin series of $\hat{f}(\omega)$ converges to $\hat{f}(\omega)$ on $\mathbb{R}$.

Proof: Since $f \in L_0^1$, there is an $M > 0$ such that $\operatorname{supp} f \subset [-M, M]$. By (5.19)

$$\frac{\left|[\hat{f}]^{(n)}(\omega)\right|}{n!} \leq \frac{\int_{-M}^{M} \left|(-ix)^n f(x) e^{-ix\omega}\right| dx}{n!}$$

$$\leq \|f\|_1 \frac{M^n}{n!} \to 0, \quad \text{as } n \to \infty.$$

Hence,

$$\hat{f}(\omega) = \sum_{k=0}^{\infty} \frac{[\hat{f}]^{(k)}(0)}{k!} \omega^k, \quad \omega \in \mathbb{R}.$$

The theorem is proved. ∎

If $f \in C^\infty$ and the Maclaurin series of f converges to f on $\mathbb{R}$, then f is called an *entire function*. Theorem 6.5.4 confirms that the Fourier transform of a compactly supported function is an entire function.

Even though the Fourier transform of a compactly supported function f is very smooth, it may decay very slowly as $x \to \infty$. For instance, the Fourier transform of a compactly supported function may not be in L^1. It is the smoothness of f, not the compactness of f, that results in the fast decay of $\hat{f}$. The following is such a result.

Lemma 6.5.3 If $f \in C_{00}$ and $f' \in L^1$, then $\hat{f} \in L^1$ and

$$[\hat{f}]\,\hat{}(-x) = f(x).$$

Proof: We leave the proof as an exercise. ∎

5.3 The Poisson Summation Formula for Compactly Supported Functions

We now apply Lemma 6.5.3 and the Poisson summation formula to obtain a result analogous to Theorem 6.5.3, but for compactly supported functions.

Theorem 6.5.5 If $f \in C_{00}$ and $f' \in L^1$, then

$$\sum_{k=-\infty}^{\infty} \left| \hat{f}(\omega + 2k\pi) \right|^2 = \sum_{k=-\infty}^{\infty} [f,f](k) e^{-ik\omega}, \quad \omega \in \mathbb{R}. \tag{5.20}$$

Proof: We leave the proof as an exercise. ∎

Exercises

1. Prove (see (5.17))

$$\sum_{k=-n}^{n} \left(1 - \frac{|k|}{n+1} \right) e^{ikx} = \frac{1}{n+1} \frac{\sin^2((n+1)x/2)}{\sin^2(x/2)}.$$

by completing the following steps.

   (a) Let $D_n(x) = 1 + 2\sum_{k=1}^{n} \cos x$. Prove

$$D_n(x) = \sum_{k=-n}^{n} e^{ikx} = \frac{\sin \frac{(2n+1)x}{2}}{\sin \frac{x}{2}}, \quad x \neq 2j\pi, j \in \mathbb{Z}.$$

   (b) Prove

$$\sum_{k=-n}^{n} \left(1 - \frac{|k|}{n+1} \right) e^{ikx} = 1 + 2\sum_{k=1}^{n} \left(1 - \frac{k}{n+1} \right) \cos kx$$

$$= \frac{1}{n+1} \sum_{k=0}^{n} D_k(x).$$

   (c) Prove

$$\sum_{k=0}^{n} D_k(x) = \left(\frac{\sin \frac{(2n+1)x}{2}}{\sin \frac{x}{2}} \right)^2, \quad x \neq 2j\pi, j \in \mathbb{Z}.$$

   (d) Prove (5.17).

2. Prove (see (5.18))

$$\frac{1}{\sin^2 u} = \sum_{k \in \mathbb{Z}} \frac{1}{(u + k\pi)^2}, \quad u \in \mathbb{R} \setminus \pi\mathbb{Z}$$

by applying the Poisson summation formula (5.10) to the function
$$h(x) = \begin{cases} 1 - |x|, & x \in [-1, 1], \\ 0, & \text{otherwise} \end{cases} \cdot \left[\text{Hint: } \sum_{k=-\infty}^{\infty} h(x + k) = 1.\right]$$

3. Use the identity (5.18) to show that

$$\sum_{n=1}^{\infty} \frac{1}{(2n - 1)^2} = \frac{\pi^2}{8}$$

and then

$$\sum_{n=1}^{\infty} \frac{1}{n^2} = \frac{\pi^2}{6}.$$

[Hint: set $x = \frac{\pi}{2}$ in (5.18).]

4. By applying the Poisson summation formula (5.6) to the function $\frac{1}{\sqrt{2t}} e^{-\frac{x^2}{4t}}$, prove

$$\sqrt{\frac{\pi}{t}} \sum_{k=-\infty}^{\infty} e^{-\frac{(x + 2k\pi)^2}{4t}} = \sum_{k=-\infty}^{\infty} e^{-k^2 t} e^{ikx}, \quad t > 0.$$

5. By applying the Poisson summation formula (5.6) to the function $\sqrt{\frac{2}{\pi}} \frac{y}{x^2 + y^2}$, prove

$$2 \sum_{k=-\infty}^{\infty} \frac{y}{y^2 + (x + 2k\pi)^2} = \sum_{k=-\infty}^{\infty} e^{-|k|y} e^{ikx}, \quad y > 0.$$

6. Prove the following (weak) Poisson summation formula: If $f \in L^1$ and $(\hat{f}(k))_{k \in \mathbb{Z}} \in l^1$ then the Poisson Summation Formula (5.6) holds almost everywhere.

7. Prove Lemma 6.5.3: If $f \in C_{00}$ and $f' \in L^1$, then $\hat{f} \in L^1$ and $[\hat{f}]\,\hat{}\,(-x) = f(x)$.

8. Prove that if $f \in C_{00}$ and $f^{(n)} \in L^2$, then $\lim_{n \to \infty} \left|\frac{\hat{f}(\omega)}{\omega^n}\right| = 0$.

9. Complete the proof of Theorem 6.5.5: If $f \in C_{00}$ and $f' \in L^1$, then

$$\sum_{k=-\infty}^{\infty} |\hat{f}(\omega + 2k\pi)|^2 = \sum_{k=-\infty}^{\infty} [f, f](k) e^{-ik\omega}, \quad \omega \in \mathbb{R}.$$

10. Let $f \in L^2$ be compactly supported. Write $g = |\hat{f}|^2$. Prove that $\sum_{k \in \mathbb{Z}} g(x + k)$ is convergent almost everywhere.

11. Use the result in Exercise 10 to prove that if $f \in L^2$ is compactly supported, then

$$\sum_{k=-\infty}^{\infty} \left| \hat{f}(\omega + 2k\pi) \right|^2 = \sum_{k=-\infty}^{\infty} [f, f](k) e^{-ik\omega}, \quad \omega \in \mathbb{R},$$

holds almost everywhere for f.

Chapter 7

Orthonormal Wavelet Bases

In the previous chapter, we introduced Fourier series and the Fourier transform. We have seen that the basis used in Fourier series, $\{e^{-ikx}\}_{k \in \mathbb{Z}}$, is an orthonormal basis for the space $\tilde{L}^2$, and many important properties of a function can be described by its Fourier coefficients. This is one of the main points of interest in the theory of Fourier series. The Fourier series of a function defined on a finite interval other than $[-\pi, \pi]$ can be obtained by dilating the orthonormal basis $\{e^{-ikx}\}_{k \in \mathbb{Z}}$. For instance,

$$\left\{ e^{-ik(\frac{\pi}{\sigma}x)} \right\}_{k \in \mathbb{Z}}$$

is an orthonormal basis of the space $\tilde{L}^2_{2\sigma}$.

However, each element in the basis $\left\{ e^{-ik(\frac{\pi}{\sigma}x)} \right\}_{k \in \mathbb{Z}}$ is a *complex sinusoidal wave*, which is "global" in the x-domain. Hence, the coefficients of the Fourier transform of a function do not provide the "local" behavior of the function in the x-domain. For example, consider the 2π-periodic function $f(x)$ of Example 6.1.5. It is defined by

$$f(x) = \begin{cases} \pi/2, & x \in (0, \pi), \\ 0, & x = 0, \\ -\pi/2, & x \in (-\pi, 0). \end{cases}$$

f is essentially a constant in each of the "local" areas $(-\pi, 0)$ and $(0, \pi)$. We have $f(x) = 2 \sum_{n=1}^{\infty} \frac{\sin(2n-1)x}{2n-1}$. The Fourier coefficients of f do not provide direct information of the "local" behavior of $f(x)$.[1] Besides, the Gibbs phenomenon

---

[1]Here we use the word "local" in an intuitive sense. Roughly speaking, the local behavior of $f(x), x \in \mathbb{R}$, on a finite interval (a, b) is the behavior of the cutoff function $f_{loc}(x) = \begin{cases} f(x), & x \in (a, b), \\ 0, & \text{otherwise.} \end{cases}$

revealed in Example 6.1.5 indicates that the Fourier series does not provide a good approximation of $f(x)$ in a neighborhood of $x = 0$.

Another shortcoming of Fourier series involves convergence. In 1873, Paul Du Bois-Reymond constructed a continuous, 2π-periodic function, whose Fourier series diverged at a given point.

Therefore a question arises: "Is it possible to find other orthogonal systems for which the phenomenon discovered by Du Bois-Reymond cannot happen and the 'local' behavior of a function can be easily recognized from its coefficients?" To answer this question, Haar (1909) began with the function

$$H(x) = \begin{cases} 1, & 0 \leq x < \dfrac{1}{2}, \\ -1, & \dfrac{1}{2} \leq x < 1, \\ 0, & \text{otherwise} \end{cases}$$

and defined $H_n(x) = 2^{j/2}H(2^j x - k)$ for $n = 2^j + k$, where $j > 0$ and $0 \leq k < 2^j$. He added a function $H_0(x) = 1, x \in [0, 1]$, to the function sequence $H_1(x), H_2(x), \cdots$. Then the system $\{H_n(x)\}_{n=0}^{\infty}$ became an orthonormal basis for $L^2[0, 1]$.

The functions in this system are all locally supported: The support of $H_n(x)$ is the dyadic interval $I_n = [k2^{-j}, (k + 1)2^{-j}] \subset [0, 1]$. Besides, they are essentially generated by a single function $H_1(x)$ (except $H_0(x)$). It is also easy to see that any continuous function $f \in C[0, 1]$ can be represented by a series $\sum_{n=0}^{\infty} a_n H_n(x)$ and the series is uniformly convergent to $f(x)$ on $[0, 1]$. Since all functions in the system have their supports in $[0, 1]$, by taking integer translates, it is easy to extend the orthonormal basis $\{H_n(x)\}_{n=0}^{\infty}$ on $L^2[0, 1]$ to an orthonormal basis on $L^2(\mathbb{R})$. The functions $H_n(x), n = 0, 1, \cdots$, are not continuous, and this limits applications of the system. Haar's idea of using the translates and dilations of a locally supported function to construct an orthonormal basis opens a wide door to the construction of an orthonormal basis. This leads to wavelets. In this chapter, we shall introduce wavelet functions and show how they generate orthonormal bases of $L^2(\mathbb{R})$, which are called "wavelet bases." Briefly, a wavelet basis is the basis generated by the dilations and translates of a "wavelet" function ψ. More precisely, if

$$\left\{ 2^{j/2}\psi(2^j x - k) \right\}_{j,k \in \mathbb{Z}} \tag{0.1}$$

forms a basis of $L^2(\mathbb{R})$, then we call ψ a wavelet and call the system (0.1) a wavelet basis. Wavelets are usually compactly supported, or exponentially decay, i.e., they are "local" functions in the x-domain. Each element in the wavelet basis (0.1), then, is also a "local" function. Hence, wavelet bases are useful tools for analyzing the "local" behaviors of functions in the x-domain.

Wavelet analysis also considers many topics other than the construction and analysis of wavelet bases in $L^2(\mathbb{R})$. The contents of wavelet analysis include continuous wavelet transforms, wavelet bases in function spaces other than $L^2(\mathbb{R})$, wavelet frames, vector-valued wavelets, and their applications in many areas. Since our purpose is to introduce wavelets in a real analysis book, we have no desire to cover all topics. Readers can refer to [4] and [6] for more in-depth studies. We select the study of wavelet bases in $L^2(\mathbb{R})$ as our main topic, which seems a suitable topic in the landscape of this book.

I Haar Wavelet Basis

To understand the idea of the construction of orthonormal wavelet bases of L^2, we introduce in this section a simple orthonormal wavelet basis of L^2, called the Haar basis. Since in the remainder of the book we mainly discuss functions in L^2, we shall abbreviate $\|\cdot\|_2$ as $\|\cdot\|$.

I.I Approximation by Step Functions

We start from a well-known result: Any function in L^2 can be approximated by step functions. In particular, let

$$\chi_{n,k}(x) = \begin{cases} 1, & 2^{-n}k \le x < 2^{-n}(k+1), \\ 0, & \text{otherwise.} \end{cases}$$

Then for any function $f \in L^2$, there exist step functions

$$f_n(x) = \sum_{k \in \mathbb{Z}} c_{n,k} \chi_{n,k}(x), \quad n \in \mathbb{N},$$

such that

$$\lim_{n \to \infty} \|f_n - f\| = 0.$$

We define

$$V_n = \left\{ g_n \;\middle|\; g_n = \sum_{k \in \mathbb{Z}} a_{n,k} \chi_{n,k}, \; (a_{n,k})_{k \in \mathbb{Z}} \in l^2 \right\}. \tag{1.1}$$

Then $\{V_n\}$ is a sequence of subspaces of L^2, which approximates L^2 in the sense that, for any $f \in L^2$, there are functions $f_n \in V_n$ such that

$$\lim_{n \to \infty} \|f_n - f\| = 0.$$

We call $\{V_n\}$ an *approximation* of L^2. It is obvious that the larger n is, the "finer resolution" the space V_n has. Note that the subspaces $\{V_n\}$ are nested:

$$\cdots \subset V_{-1} \subset V_0 \subset V_1 \subset \cdots . \tag{1.2}$$

It is easy to see that

$$\bigcup_{n \in \mathbb{Z}} V_n = L^2 \tag{1.3}$$

and

$$\bigcap_{n \in \mathbb{Z}} V_n = \{0\}. \tag{1.4}$$

The basis of the space V_n has a simple structure. First we construct an orthonormal basis of V_0. Let $B(x)$ be the *box function* defined by

$$B(x) = \begin{cases} 1, & 0 \leq x < 1, \\ 0, & \text{otherwise,} \end{cases}$$

which is the *characteristic function* of the interval $[0, 1)$. It is clear that the function system $\{B(x-k)\}_{k \in \mathbb{Z}}$ forms an orthonormal basis of V_0. An orthonormal basis of V_n can be obtained by dilating the system $\{B(x - k)\}_{k \in \mathbb{Z}}$. For a function $f \in L^2$, we write

$$f_{n,k}(x) = 2^{n/2} f(2^n x - k), \quad k \in \mathbb{Z}, \quad n \in \mathbb{Z},$$

and abbreviate $f_{0,k}$ as f_k. The system $\{B_{n,k}\}_{k \in \mathbb{Z}}$ then forms an orthonormal basis of V_n.

1.2 The Haar Wavelet Basis

With the aid of the nested structure of (1.2), we can construct an orthonormal basis of L^2. Let W_n be the orthogonal complement of V_n with respect to V_{n+1}:

$$W_n \oplus V_n = V_{n+1}, \quad W_n \perp V_n.$$

By the nested structure of (1.3) and (1.4), we have

$$L^2 = \bigoplus_{n \in \mathbb{Z}} W_n, \quad W_n \perp W_{n'}, n \neq n'. \tag{1.5}$$

Since each subspace V_n is a 2^n-dilation of V_0, W_n is also a 2^n-dilation of W_0. Recall that dilation preserves orthogonality. Therefore, if $\{e_k\}_{k \in \mathbb{Z}}$ is an orthogonal basis of W_0, then its 2^n-dilation is an orthogonal basis of W_n. Thus, our task is reduced to finding an orthonormal basis of W_0. To do so, we define

$$H(x) = \begin{cases} 1, & 0 \le x < \frac{1}{2}, \\ -1, & \frac{1}{2} \le x < 1. \\ 0, & \text{otherwise}, \end{cases}$$

which is called the *Haar function*.

Lemma 7.1.1 Let

$$H_k(x) = H(x - k), \quad k \in \mathbb{Z}. \tag{1.6}$$

Then the system $\{H_k\}_{k\in\mathbb{Z}}$ is an orthonormal basis of W_0. Consequently, the system $\{H_{n,k}(x)\}_{k\in\mathbb{Z}}$ is an orthonormal basis of the space W_n.

Proof: We only need to prove that the functions in (1.6) form an orthonormal basis of W_0. It is clear that $\{H_k(x)\}_{k\in\mathbb{Z}}$ is an orthonormal system in W_0. We now claim that it is also a basis of W_0. Let g be a function in W_0. Then $g \in V_1$ and there is a sequence $(c_k) \in l^2$ such that

$$g = \sum_{k\in\mathbb{Z}} c_k B_{1,k} = \sum_{l\in\mathbb{Z}} (c_{2l}B_{1,2l} + c_{2l+1}B_{1,2l+1}).$$

Since $g \perp V_0$, we have $c_{2l+1} = -c_{2l}$. Note that $H_l = \frac{1}{\sqrt{2}} (B_{1,2l} - B_{1,2l+1})$. Hence,

$$g = \sqrt{2} \sum_{l\in\mathbb{Z}} c_{2l}H_l, \quad (c_{2l}) \in l^2.$$

The lemma is proved. ∎

From the lemma, we can obtain the following theorem.

Theorem 7.1.1 The system $\{H_{n,k}\}_{n,k\in\mathbb{Z}}$ forms an orthonormal basis of L^2.

Proof: It is obvious that $\{H_{n,k}\}_{n,k\in\mathbb{Z}}$ is an orthonormal system in L^2. The fact that it is a basis of L^2 is a consequence of (1.5). ∎

We usually call the space W_n the *Haar space* and call $\{H_{n,k}\}_{n,k\in\mathbb{Z}}$ the *Haar basis* of L^2. The geometric meaning of the Haar decomposition of a function can be explained as follows. Note that each element $H_{n,k}$ in the Haar basis represents a square wave centered at $\frac{2k+1}{2^{n+1}}$ with width $\frac{1}{2^n}$. Hence, we may say that $H_{n,k}$ has the *Haar frequency* of 2^n. The functions in the same Haar subspace W_n have the same Haar frequency 2^n. Unlike $\sin x$ or $\cos x$, the Haar function $H_{n,k}$ is a "small" local wave. Thus, the width of the wave provides the frequency information, and the center of the wave provides the spatial (or time) information. In other words, the first index n of $H_{n,k}$ indicates the frequency of the function while the second index k indicates its spatial (or time) location. Thus, the Haar system becomes a useful tool for local time frequency analysis.

Orthonormal bases having structure like the Haar basis are extremely useful in many applications. Hence we give the following definition.

Definition 7.1.1 A function $\psi \in L^2$ is called an *orthonormal wavelet* if $\{\psi_{nm}\}_{n,m \in \mathbb{Z}}$ forms an orthonormal basis of L^2. The basis $\{\psi_{nm}\}_{n,m \in \mathbb{Z}}$ generated by ψ is called an *orthonormal wavelet basis* of L^2.

The term "wavelet" refers to the intuitive idea that the function ψ represents a "small wave." For example, the Haar function $H(x)$ consists of only a small (single) square wave on $[0, 1]$, which can be partially reflected by the property

$$\int_{\mathbb{R}} H(x)\, dx = 0. \tag{1.7}$$

Hence, (1.7) can be used to define wavelet functions in a general sense, as is done in the following.

Definition 7.1.2 In general, a function $\psi \in L^1$ is called a *wavelet* if

$$\int_{\mathbb{R}} \psi(x)\, dx = 0. \tag{1.8}$$

We now return to analyze the relationship between the box function and the Haar wavelet. Note that the box function satisfies the following equation:

$$B(x) = B(2x) + B(2x - 1), \tag{1.9}$$

and the Haar wavelet has the following relationship with $B(x)$:

$$H(x) = B(2x) - B(2x - 1). \tag{1.10}$$

Equation (1.9) reveals the relationship of two box functions with different scales. Hence, we call (1.9) the *two-scale equation* (or the *refinement equation*) of B. Correspondingly, (1.10) is called the *two-scale relation* of H and B. The general definitions of these concepts will be given in the next section. They play a central role in the construction of wavelet bases and in the fast wavelet transform algorithm as well.

1.3 The Decomposition of Functions into Haar Wavelet Series

Since $\{H_{nk}\}_{n,k\in\mathbb{Z}}$ is an orthonormal basis of L^2, any function $f \in L^2$ can be expanded as a Haar series

$$f = \sum d_{nk}H_{nk}, \tag{1.11}$$

where the coefficients d_{nk}, $n, k \in \mathbb{Z}$, can be computed from the inner product

$$d_{nk} = \langle f, H_{nk}\rangle.$$

However, since

$$V_j = \bigoplus_{k<j} W_k$$

and V_j will be reduced to the trivial space as $j \to -\infty$, the hybrid series

$$f = \sum_{k\in\mathbb{Z}} c_{jk}B_{jk} + \sum_{n\geq j}\sum_{k\in\mathbb{Z}} d_{nk}H_{nk} \tag{1.12}$$

sometimes is more useful than the biinfinite Haar series (1.11). Without loss of generality, we can always assume $j = 0$ in (1.12). Thus we have

$$f = \sum_{k\in\mathbb{Z}} c_k B_k + \sum_{n=0}^{\infty}\sum_{k\in\mathbb{Z}} d_{nk}H_{nk}.$$

Besides, in many applications, a function is often represented by a truncated series, along with an error remainder. For example, we use Taylor polynomials (which are truncated Taylor series) to approximate analytic functions, and use trigonometric polynomials (which are truncated Fourier series) to approximate periodic functions. Similarly, we can approximate a function $f \in L^2$ using a truncated Haar series. Let

$$f_N = \sum_{k\in\mathbb{Z}} c_k B_k + \sum_{j=0}^{N-1}\sum_{k\in\mathbb{Z}} d_{jk}H_{jk} \tag{1.13}$$

be a truncated Haar series. Then

$$f_N = \text{Proj}_{V_n} f$$

and

$$\|f_N - f\| \to 0 \quad \text{as } N \to \infty.$$

Using the basis of V_N, we can decompose f_N into another series

$$f_N = \sum_{k \in \mathbb{Z}} c_{Nk} B_{Nk}. \tag{1.14}$$

The algorithm to compute $(c_k), (d_k), (d_{1k}), \cdots, (d_{N-1,k})$ in (1.13) from (c_{Nk}) in (1.14) is called the *Fast (Haar) Wavelet Transform*, and the algorithm doing the reverse is called the *Fast Inverse (Haar) Wavelet Transform* . Using formulas (1.9) and (1.10), we can perform the Fast (Haar) Wavelet Transform as follows:

  (1) Take the sum of two successive boxes numbered $2k, 2k + 1$ on level n for f, and then multiply it by $1/\sqrt{2}$. This gives the kth box function component of f on level $n - 1$.
  (2) Take the difference of two successive boxes numbered $2k, 2k+1$ on level n for f, and then multiply it by $1/\sqrt{2}$. This gives the kth Haar function component of f on level $n - 1$.

EXAMPLE 7.1.1:   Let

$$f(x) = \sum_{k=0}^{4} c_{4k} B_{4k}(x)$$

where $c_{40} = 1, c_{41} = 2, c_{42} = 4, c_{43} = -2$, and $c_{44} = -3$. Then $\mathbf{c}^4 = (c_{4,k})$ is

$$\mathbf{c}^4 = (\cdots, 0, 1, 2, 4, -2, -3, 0, \cdots).$$

By (1.9) and (1.10), we have

$$\mathbf{c}^3 = \sqrt{2}\left(\cdots, 0, \frac{3}{2}, 1, -\frac{3}{2}, 0, \cdots\right),$$

$$\mathbf{b}^3 = \sqrt{2}\left(\cdots, 0, -\frac{1}{2}, 3, -\frac{3}{2}, 0, \cdots\right),$$

$$\mathbf{a}^2 = 2\left(\cdots, 0, \frac{5}{4}, -\frac{3}{4}, 0, \cdots\right),$$

$$\mathbf{b}^2 = 2\left(\cdots, 0, \frac{1}{4}, -\frac{3}{4}, 0, \cdots\right),$$

$$\mathbf{a}^1 = 2\sqrt{2}\left(\cdots, 0, \frac{1}{4}, 0, \cdots\right),$$

$$\mathbf{b}^1 = 2\sqrt{2}(\cdots, 0, 1, 0, \cdots),$$

$$\mathbf{a}^0 = 4\left(\cdots, 0, \frac{1}{8}, 0, \cdots\right),$$

and

$$\mathbf{b}^0 = 4\left(\cdots, 0, \frac{1}{8}, 0, \cdots\right).$$

Thus,

$$f(x) = \frac{1}{2}B(x) + \frac{1}{2}H(x) + 2\sqrt{2}H_{1,0}(x) + \frac{1}{2}H_{2,0}(x) - \frac{3}{2}H_{2,1}(x)$$
$$- \frac{\sqrt{2}}{2}H_{3,0}(x) + 3\sqrt{3}H_{3,1}(x) - \frac{3\sqrt{2}}{2}H_{3,2}(x).$$

Let

$$L = \sqrt{2}\begin{pmatrix} \vdots & 0 & 0 & \vdots & \vdots & \vdots \\ \cdots & 1/2 & 1/2 & 0 & 0 & \cdots \\ \cdots & 0 & 0 & 1/2 & 1/2 & \cdots \\ \vdots & \vdots & \vdots & 0 & 0 & \vdots \end{pmatrix}$$

and

$$H = \sqrt{2}\begin{pmatrix} \vdots & 0 & 0 & \vdots & \vdots & \vdots \\ \cdots & 1/2 & -1/2 & 0 & 0 & \cdots \\ \cdots & 0 & 0 & 1/2 & -1/2 & \cdots \\ \vdots & \vdots & \vdots & 0 & 0 & \vdots \end{pmatrix}$$

Then the Fast Haar Wavelet Transform can be formulated as

$$\mathbf{c}^{j-1} = L\mathbf{c}^j$$

and

$$\mathbf{d}^{j-1} = H\mathbf{c}^j,$$

for $j = N, N-1, \cdots, 1$, while the Fast Inverse Haar Wavelet Transform recovers $\mathbf{c}^N$ from

$$\mathbf{c}^0, \mathbf{d}^0, \cdots, \mathbf{d}^{N-1}$$

using the following algorithm:

$$\mathbf{c}^{j+1} = L^T\mathbf{c}^j + H^T\mathbf{d}^j, \quad j = 0, 1, \cdots, N-1.$$

Exercises

1. Let $f(x), x \in \mathbb{R}$, be a continuous function such that $\lim_{x \to \infty} f(x) = b$ and $\lim_{x \to -\infty} f(x) = a$, where a and b are two real numbers. Let

$$U_n = \left\{ g_n \ \middle| \ g_n = \sum_{k \in \mathbb{Z}} a_{n,k} \chi_{n,k}, \ \sup_{k \in \mathbb{Z}} |a_{n,k}| < \infty \right\}.$$

   Prove that there are functions $f_n \in U_n, n \in \mathbb{Z}$, such that

$$\lim_{n \to \infty} \sup_{x \in \mathbb{R}} |f(x) - f_n(x)| = 0.$$

2. Let $f(x), x \in \mathbb{R}$, be a continuous function and

$$\hat{U}_n = \left\{ g_n \ \middle| \ g_n = \sum_{k \in \mathbb{Z}} a_{n,k} \chi_{n,k} \right\}.$$

   Prove that there are functions $f_n \in \hat{U}_n, n \in \mathbb{Z}$, such that

$$\lim_{n \to \infty} \sup_{x \in \mathbb{R}} |f(x) - f_n(x)| = 0.$$

3. Let $V_n \subset L^2$ be given by (1.1). The distance from a function f to V_n is defined by

$$d(f, V_n) = \inf_{g \in V_n} \|f - g\|.$$

   Let $f(x) = \chi_{[0,1/3]}(x)$. Find the distance $d(f, V_n)$.
4. Prove $\bigcup_{n \in \mathbb{Z}} V_n = L^2$, where $\{V_n\}$ is generated by the box function.
5. Let function $f(x) \in V_2$ be defined by

$$f(x) = \begin{cases} 5, & 0 \le x < \frac{1}{4}, \\ -1, & \frac{1}{4} \le x < \frac{1}{2}, \\ 2, & \frac{1}{2} \le x < 1, \\ 1, & 1 \le x < \frac{3}{2}, \\ 0, & \text{otherwise.} \end{cases}$$

   (a) Expand $f(x)$ as a linear combination of $\{B_{2,k}\}$.
   (b) Decompose $f(x)$, step-by-step, into a wavelet series as

$$f(x) = \sum a_{0,m} B_{0,m} + \sum b_{0,m} H_{0,m} + \sum b_{1,m} H_{1,m}.$$

6. Let $f(x) = \chi_{[0,1/8]}(x)$. Expand it to the series $\sum_{k \in \mathbb{Z}} c_k B_k + \sum_{n \ge 0} \sum_{k \in \mathbb{Z}} d_{nk} H_{nk}$.

7. Let $f(x) = \chi_{[0,a]}(x)$, where $a \in (0, 1)$. Expand it to the series $\sum_{k \in \mathbb{Z}} c_k B_k + \sum_{n \geq 0} \sum_{k \in \mathbb{Z}} d_{nk} H_{nk}$. [Hint: Write $a = \sum_{k=1}^{\infty} a_k 2^{-k}$, where a_k is 0 or 1.]

8. Prove that if a function $g \in L^2(\mathbb{R})$ is in the subspace

$$W = \text{span}\{H_{jk}\}_{k \in \mathbb{Z}, j \geq 0},$$

then

$$\int_k^{k+1} g(x)dx = 0, \quad \text{for all } k \in \mathbb{Z}.$$

9. Let $S_{[a,b]}$ be a subspace of $L^2_{[a,b]}$ defined by

$$S_{[a,b]} = \left\{ f \in L^2_{[a,b]} \,\middle|\, \int_a^b f(x)dx = 0 \right\}.$$

Prove that all Haar wavelets H_{jk} with supp $H_{jk} \subset [a, b]$ form an orthonormal basis of $S_{[a,b]}$.

2 Multiresolution Analysis

In the previous section, we constructed a Haar basis of L^2, which has very simple structure. Unfortunately, the Haar function is not continuous and this limits its application. Hence, we want to construct other wavelet bases. In this section, we establish a general principle for the construction of orthonormal wavelet bases of L^2. As seen in the previous section, the sequence of nested subspaces $\{V_n\}$ plays an important role in the construction. We first study such subspace sequences in L^2.

2.1 Definition of Multiresolution Analysis

Definition 7.2.1 A *multiresolution analysis* (MRA) of L^2 is a nested sequence of subspaces of L^2

$$\cdots \subset V_{-1} \subset V_0 \subset V_1 \subset \cdots$$

that satisfies the following conditions:

(1) $\bigcap_{j \in \mathbb{Z}} V_j = \{0\}$,

(2) $\overline{\bigcup_{j \in \mathbb{Z}} V_j} = L^2$,

(3) $f(\cdot) \in V_j$ if and only if $f(2\cdot) \in V_{j+1}$, and

(4) there exists a function $\phi \in V_0$ such that $\{\phi(x - n)\}_{n \in \mathbb{Z}}$ is an *unconditional basis* of V_0, i.e., $\{\phi(x - n)\}_{n \in \mathbb{Z}}$ is a basis of V_0, and there exist two

constants $A, B > 0$ such that, for all $(c_n) \in l^2$, the following inequality holds:

$$A \sum |c_n|^2 \leq \left\| \sum c_n \phi(\cdot - n) \right\|^2 \leq B \sum |c_n|^2. \tag{2.1}$$

In the literature, an unconditional basis is also called a *Riesz basis* and the constants A and B in (2.1) are called the *lower Riesz bound* and the *upper Riesz bound*, respectively. Condition (2.1) is called a *stable condition*, and a function ϕ satisfying (2.1) is called a *stable function*. The function ϕ described in Definition 7.2.1 is called an *MRA generator*. Furthermore, if $\{\phi(x-n)\}_{n \in \mathbb{Z}}$ is an orthonormal basis of V_0, then ϕ is called an *orthonormal MRA generator*.

From the discussion in the previous section, we know that the box function is an orthonormal MRA generator. When $\{\phi_m\}_{m \in \mathbb{Z}}$ is an unconditional basis of V_0, we can claim that $\{\phi_{n,m}\}_{m \in \mathbb{Z}}$ is an unconditional basis of V_n. In fact, by condition (3) in Definition 7.2.1, $\phi_{n,m} \in V_n$ and

$$\left\| \sum c_m \phi_{n,m} \right\| = \left\| \sum c_m \phi_m \right\|.$$

This implies that $\{\phi_{n,m}\}_{m \in \mathbb{Z}}$ is an unconditional basis of V_n.

Since $\phi \in V_0$ is also in V_1, we can expand ϕ into a linear combination of the basis elements of V_1 :

$$\phi(x) = 2 \sum_{m \in \mathbb{Z}} h(m) \phi(2x - m), \quad (h(m))_{m \in \mathbb{Z}} \in l^2 \tag{2.2}$$

where the coefficient sequence $(h(m))$ is in l^2 because $\{\phi_{1,m}\}$ is an unconditional basis of V_1. In (2.2), we put a factor of 2 on the right hand side to simplify the notation in future discussions.

Equation (2.2) is a generalization of (1.9). We call it a *two-scale equation* (or *refinement equation*). Because of the importance of (2.2), we study it in detail.

Definition 7.2.2 A function $\phi \in L^2$ which satisfies the two-scale equation (2.2) is called a *scaling function* (or *refinable function*). The sequence of coefficients $(h(n))_{n \in \mathbb{Z}}$ in (2.2) is called the *mask* of ϕ, while the series $H(z) := \sum_{m \in \mathbb{Z}} h(m) z^m$ is called the *symbol* of ϕ. If $\{\phi(x - n)\}_{n \in \mathbb{Z}}$ is an orthonormal system, then ϕ is called an *orthonormal scaling function*.

Taking the Fourier transform of (2.2), we obtain

$$\hat{\phi}(\omega) = H(e^{-i\omega/2})\hat{\phi}(\omega/2), \tag{2.3}$$

which represents the two-scale equation in the frequency domain. For convenience, we also call (2.3) a two-scale equation (or refinement equation) of ϕ.

The notation

$$\mathcal{H}(\omega) = H(e^{-i\omega/2})$$

is often used in later discussions. A common technique in wavelet theory is, when dealing with a problem, to go back and forth between the time domain and the frequency domain.

In general, for an arbitrary given sequence $(h(m))$, the L^2-solution of (2.2) may not exist, or the L^2-solution of (2.2) exists, but may not be an MRA generator. A scaling function ϕ which is an MRA generator satisfies two requirements. First, ϕ should be stable. Second, the subspace of L^2 spanned by $\{\phi_{n,m}\}_{m\in\mathbb{Z}}$ must approximate L^2 as $n \to \infty$. We now discuss these two requirements in detail.

2.2 Stability of Scaling Functions

An MRA generator must satisfy the stability condition (2.1). Hence, we give a necessary and sufficient condition for the stability of a function.

Theorem 7.2.1 A function $\phi \in L^2$ satisfies stability condition (2.1) if and only if

$$0 < \operatorname*{ess\,inf}_{\omega\in\mathbb{R}} \sum_{k\in\mathbb{Z}} |\hat{\phi}(\omega + 2k\pi)|^2 \tag{2.4}$$

and

$$\operatorname*{ess\,sup}_{\omega\in\mathbb{R}} \sum_{k\in\mathbb{Z}} |\hat{\phi}(\omega + 2k\pi)|^2 < \infty. \tag{2.5}$$

Proof: Let $f(x)$ be a function in V_0. Then

$$f(x) = \sum_{m\in\mathbb{Z}} c_m \phi(x - m), \quad (c_m) \in l^2.$$

Hence $T(\omega) = \sum_{m\in\mathbb{Z}} c_m e^{-im\omega} \in \tilde{L}^2$ and $||T(\omega)||^2_{\tilde{L}^2} = \sum_{m\in\mathbb{Z}} |c_m|^2$. By the Parseval Formula (Theorem 6.4.2) $||f||^2 = \frac{1}{2\pi}||\hat{f}||^2$ and

$$\hat{f}(\omega) = \int_{\mathbb{R}} \sum_{m\in\mathbb{Z}} c_m \phi(x - m) e^{-ix\omega} \, dx = T(\omega)\hat{\phi}(\omega).$$

Hence,

$$\|f\|^2 = \frac{1}{2\pi} \int_{\mathbb{R}} |T(\omega)\hat{\phi}(\omega)|^2 \, d\omega$$

$$= \frac{1}{2\pi} \sum_{k \in \mathbb{Z}} \int_0^{2\pi} |T(\omega)|^2 |\hat{\phi}(\omega + 2k\pi)|^2 \, d\omega$$

$$= \frac{1}{2\pi} \int_0^{2\pi} |T(\omega)|^2 \left(\sum_{k \in \mathbb{Z}} |\hat{\phi}(\omega + 2k\pi)|^2 \right) d\omega.$$

Let

$$F(\omega) = \sum_{k \in \mathbb{Z}} |\hat{\phi}(\omega + 2k\pi)|^2.$$

$F(\omega)$ is a 2π-periodic, measurable function. Set

$$M_r = \operatorname*{ess\,sup}_{\omega \in \mathbb{R}} F(\omega) \tag{2.6}$$

and

$$M_l = \operatorname*{ess\,inf}_{\omega \in \mathbb{R}} F(\omega). \tag{2.7}$$

If (2.4) and (2.5) both hold, then we have (2.1). On the other hand, if $M_l = 0$, then for an $\epsilon > 0$, the measure of the set

$$E_\epsilon = \{\omega \mid F(\omega) \le \epsilon, \omega \in [0, 2\pi)\}$$

is positive. Let $\delta = M(E_\epsilon)$. We now define a 2π-periodic function $T(\omega)$ by

$$T(\omega) = \begin{cases} \dfrac{\sqrt{2\pi}}{\sqrt{\delta}}, & \omega \in E_\epsilon, \\ 0, & \text{otherwise.} \end{cases}$$

Then $T(\omega) \in \tilde{L}^2$ and $\|T(\omega)\|^2_{\tilde{L}^2} = 1$. Let $T(\omega) = \sum_m c_m e^{im\omega}$ and $f(x) = \sum_m c_m \phi(x - m)$. Then $\sum_m |c_m|^2 = 1$ but

$$\|f\|^2 = \frac{1}{2\pi} \|\hat{f}\|^2 = \frac{1}{2\pi} \int_{E_\epsilon} |T(\omega)|^2 F(\omega) d\omega \le \epsilon,$$

which implies that there does not exist a constant $A > 0$ such that the left part of stability condition (2.1) holds for all $(c_m) \in l^2$. Similarly, if $M_r = \infty$, then there does not exist a constant $B > 0$ such that the right part of stability condition (2.1) holds for all $(c_m) \in l^2$. We leave the details of this part to the reader. ∎

The following corollary is directly derived from Theorem 7.2.1.

Corollary 7.2.1 If the function

$$F(\omega) = \sum_{k \in \mathbb{Z}} |\hat{\phi}(\omega + 2k\pi)|^2$$

is continuous, then $\{\phi(x - k)\}_{k \in \mathbb{Z}}$ is stable if and only if $F(\omega) > 0$ for all $\omega \in \mathbb{R}$.

2.3 Completeness of Scaling Functions

We now discuss the conditions under which a scaling function generates an MRA of L^2. Recall that $|\widehat{\phi_{j,k}}(\omega)| = 2^{-\frac{j}{2}}|\hat{\phi}(2^{-j}\omega)|$, which implies that if $\hat{\phi}(0) = 0$, then for any integer n, the Fourier transform of a function $f \in V_n$ vanishes at $\omega = 0$. However, it is obvious that there exist functions in L^2 whose Fourier transforms do not vanish at 0. This fact shows intuitively that the Fourier transform of an MRA generator does not vanish at 0.

Theorem 7.2.2 Let $\phi \in L^1 \cap L^2$ be an MRA generator. Then $\hat{\phi}(0) \neq 0$.

Proof: Assume ϕ is an orthonormal MRA generator, which generates the MRA $\cdots \subset V_{-1} \subset V_0 \subset V_1 \subset \cdots$. Let $f \in L^2$ be defined by $\hat{f}(\omega) = \chi_{[-\pi,\pi]}(\omega)$. Then $\|f\|_2^2 = \frac{1}{2\pi}\|\hat{f}\|_2^2 = 1$. Since $\overline{\bigcup_{n \in \mathbb{Z}} V_n} = L^2$, there is an integer $n > 0$ such that the function

$$f_n = \sum_{k \in \mathbb{Z}} \langle f, \phi_{n,k} \rangle \phi_{n,k}$$

satisfies $\|f_n - f\|_2^2 < \frac{1}{2}$. Hence, $\|f_n\|_2^2 \geq \frac{1}{2}$. By the orthogonality of $(\phi_{n,k})_{k \in \mathbb{Z}}$, we have

$$\sum_{k \in \mathbb{Z}} |\langle f, \phi_{n,k} \rangle|^2 = \|f_n\|_2^2 \geq \frac{1}{2}.$$

Note that

$$\langle f, \phi_{n,k} \rangle = \frac{1}{2\pi} \langle \hat{f}, \widehat{\phi_{n,k}} \rangle = \frac{1}{2\pi} \int_{-\pi}^{\pi} \overline{\widehat{\phi_{n,k}}(\omega)} d\omega$$

$$= \frac{1}{2\pi} \int_{-\pi}^{\pi} 2^{-\frac{n}{2}} e^{-i2^{-n}k\omega} \overline{\hat{\phi}(2^{-n}\omega)} d\omega$$

$$= \frac{1}{2\pi} \int_{-2^{-n}\pi}^{2^{-n}\pi} 2^{\frac{n}{2}} \overline{\hat{\phi}(\omega)} e^{-ik\omega} d\omega$$

$$= \frac{1}{2\pi} \int_{-\pi}^{\pi} 2^{n/2} \chi_{[-2^{-n}\pi, 2^n\pi]}(\omega) \overline{\hat{\phi}(\omega)} e^{-ik\omega} d\omega,$$

which is the kth Fourier coefficient of the function $2^{n/2}\chi_{[-2^{-n}\pi,2^n\pi]}\overline{\hat{\phi}}$. By Parseval's formula for Fourier series (Theorem 6.4.2), we have

$$\frac{1}{2} \le \sum_{k\in\mathbb{Z}} |\langle f, \phi_{n,k}\rangle|^2 = \|2^{n/2}\chi_{[-2^{-n}\pi,2^n\pi]}\overline{\hat{\phi}}\|_{L^2}^2 = \int_{-2^{-n}\pi}^{2^{-n}\pi} 2^n |\widehat{\phi(\omega)}|^2 d\omega, \quad n > 0.$$

Since $\hat{\phi}(\omega)$ is continuous, we have

$$|\hat{\phi}(0)|^2 = \lim_{n\to\infty} \frac{1}{2\pi} \int_{-2^{-n}\pi}^{2^{-n}\pi} 2^n |\widehat{\phi(\omega)}|^2 d\omega \ge \frac{1}{4\pi} > 0.$$

The theorem is proved for orthonormal MRA generators. In general cases, applying this result to the orthonormalization of ϕ, we can get the conclusion. We leave the proof of this part as an exercise in the next section. ■

We now establish the converse proposition. Assume a nested subspace sequence $\cdots \subset V_{-1} \subset V_0 \subset V_1 \subset \cdots$ is generated by a scaling function ϕ. We prove that $\hat{\phi}(0) \ne 0$ is the sufficient condition for $\overline{\cup_{n\in\mathbb{Z}}V_n} = L^2$. For this, we establish a lemma, which describes the translation invariance of the space $\overline{\cup_{n\in\mathbb{Z}}V_n}$.

Lemma 7.2.1 Let $\phi \in L^2$ be a scaling function. Define

$$V_n = \text{span}_2\{\phi_{n,m}\}_{m\in\mathbb{Z}} \tag{2.8}$$

and

$$U = \overline{\bigcup_{n\in\mathbb{Z}} V_n}. \tag{2.9}$$

Then for any $t \in \mathbb{R}$ and any $f \in U, f(\cdot + t) \in U$.

Proof: It is clear that, if $f \in U$ and

$$t = 2^n m, \quad n, m \in \mathbb{Z},$$

then $f(\cdot + t) \in U$. Assume now $t \in \mathbb{R}$. Then there is a sequence $\{2^{n_k}m_k\}$ such that $2^{n_k}m_k \to t$ as $k \to \infty$. It is known that

$$\lim_{k\to\infty} \|f(\cdot + 2^{n_k}m_k) - f(\cdot + t)\| = 0. \tag{2.10}$$

Since U is a closed subspace of L^2, then $f(\cdot + t) \in U$. ∎

The following theorem gives the condition for $U = L^2$.

Theorem 7.2.3 Let $\phi \in L^1 \cap L^2$ and U be the space defined by (2.9). Then $\hat{\phi}(0) \neq 0$ implies $U = L^2$.

Proof: If $U \neq L^2$, then there is a nonvanishing function $g \in L^2$ such that $g \perp U$. By Lemma 7.2.1, for any $t \in \mathbb{R}$ and $f \in U$, we have $f(\cdot + t) \in U$ and therefore

$$\int_{\mathbb{R}} \overline{g(x)} f(x + t) \, dx = 0, \quad \text{for all } t \in \mathbb{R}, f \in U. \tag{2.11}$$

By Plancherel's Theorem (see Section 6.4 and Theorem 6.4.5),

$$\frac{1}{2\pi} \int_{\mathbb{R}} \overline{\hat{g}(\omega)} \hat{f}(\omega) e^{it\omega} \, d\omega = \int_{\mathbb{R}} \overline{g(x)} f(x + t) \, dx = 0, \quad \text{for all } t \in \mathbb{R}, f \in U, \tag{2.12}$$

which implies

$$\overline{\hat{g}(\omega)} \hat{f}(\omega) = 0 \quad \text{a.e.} \quad \text{for all } f \in U.$$

It is clear that $\phi(2^j x) \in U$. Hence, we have

$$\overline{\hat{g}(\omega)} \hat{\phi}(2^{-j}\omega) = 0, \quad j \in \mathbb{Z}. \tag{2.13}$$

Since $\phi \in L^1 \cap L^2$, $\hat{\phi}$ is continuous on $\mathbb{R}$. By $\hat{\phi}(0) \neq 0$, there is a $\delta > 0$ such that $\hat{\phi}(\omega) \neq 0$ on $(-\delta, \delta)$ and therefore $\hat{\phi}(2^{-j}\omega) \neq 0$ on $(2^{-j}\delta, 2^j \delta)$. Then (2.13) implies

$$\hat{g}(\omega) = 0, \quad \text{a.e.} \quad \text{on } (2^{-j}\delta, 2^j \delta), \quad j \in \mathbb{Z},$$

i.e., $\hat{g}(\omega) = 0$ a.e., which yields a contradiction. Hence, $\hat{\phi}(0) \neq 0$ implies $U = L^2$. ∎

Exercises

1. Prove that if condition (2.5) in Theorem 7.2.1 fails, then there does not exist a constant B such that

$$\left\| \sum c_n \phi(\cdot - n) \right\|^2 \leq B \sum |c_n|^2$$

   holds for all $(c_n) \in l^2$.

2. Prove that the function $\phi = \chi_{[0,3)}$ is a scaling function, but it does not satisfy stability condition (2.1).

3. Let ϕ_1 and ϕ_2 be scaling functions in $L^1 \cap L^2$. Prove that their convolution $\phi_1 * \phi_2$ is a scaling function in L^2.

4. Prove that the function $f(x) = \begin{cases} 1, & 0 \le x \le 3/4, \\ 0, & \text{otherwise,} \end{cases}$ does not satisfy any two-scale equation.

5. Let $\phi(x) \in L^1 \cap L^2$ be a scaling function satisfying

$$\phi(x) = 2 \sum_{m \in \mathbb{Z}} h(m) \phi(2x - m), \quad (h(m))_{m \in \mathbb{Z}} \in l^2.$$

Prove that if $\hat{\phi}(0) \ne 0$ then $\sum_{m \in \mathbb{Z}} h(m) = 1$.

6. Prove Corollary 7.2.1: If the function

$$F(\omega) = \sum_{k \in \mathbb{Z}} |\hat{\phi}(\omega + 2k\pi)|^2$$

is continuous, then $\{\phi(x - k)\}_{k \in \mathbb{Z}}$ is stable if and only if $F(\omega) > 0$ for all $\omega \in \mathbb{R}$.

7. Let $\phi(x) \in L^2$ be a stable scaling function and $B(x)$ be the box function. Prove that the convolution $\phi * B$ is also a stable scaling function.

8. Prove the limit (2.10). That is, $\lim_{k \to \infty} \|f(\cdot + t_k) - f(\cdot + t)\| = 0$ holds for each $f \in L^2$ where $\lim_{k \to \infty} t_k = t$.

9. Let ϕ be a stable scaling function with the lower Riesz bound and upper Riesz bound A and B, respectively. Let $\{V_j\}$ be the MRA generated by ϕ. Assume $f \in V_{-j}$ satisfies $\|f\| = 1$.

   (a) Prove that

   $$\int_{2\pi}^{4\pi} |\hat{f}(\omega)| \, d\omega \le \sqrt{\frac{2\pi}{A}} \left(\int_{2^{2+1}\pi}^{\infty} |\hat{\phi}(\omega)|^2 \, d\omega \right)^{1/2}.$$

   [Hint: Let $f_j(x) = 2^{\frac{j}{2}} f(2^j x)$. Then $f_j \in V_0$. Expand f_j as a series of $\{\phi(x - k)\}$, then apply the stability condition for ϕ in the frequency domain.]

   (b) Use (2.13) to prove that

   $$\bigcap_{j \in \mathbb{Z}} V_j = \{0\}.$$

3 Orthonormal Wavelets from MRA

In the previous section, we learned that a stable two-scaling function ϕ can generate an MRA. In this section we discuss how to construct an orthonormal wavelet basis from an MRA generator. We shall show that an orthonormal wavelet can be easily constructed from an orthonormal MRA generator. Hence, we first discuss how to orthonormalize a stable scaling function.

3.1 Orthonormalization

The stability of $\{\phi(x-k)\}_{k\in\mathbb{Z}}$ does not imply its orthogonality. However, a stable scaling function can be orthonormalized. To show how to orthonormalize it, we give a criterion for orthonormalization of $\{\phi(x-k)\}_{k\in\mathbb{Z}}$.

Theorem 7.3.1 For $\phi \in L^2$, the following statements are equivalent:

(1) $\{\phi(x - n)\}_{n\in\mathbb{Z}}$ is an orthonormal system, i.e.,

$$\int_{\mathbb{R}} \phi(x + k)\overline{\phi(x + j)}\, dx = \delta_{kj}, \quad k, j \in \mathbb{Z}, \tag{3.1}$$

(2)

$$\sum_{k\in\mathbb{Z}} |\hat{\phi}(\omega + 2k\pi)|^2 = 1 \quad \text{a.e.,} \tag{3.2}$$

and
(3)

$$\frac{1}{2\pi} \int_{\mathbb{R}} |\hat{\phi}(\omega)|^2 e^{ik\omega}\, d\omega = \delta_{0k},$$

where $\delta_{0k} = \begin{cases} 1, & \text{if } k = 0, \\ 0, & \text{otherwise.} \end{cases}$

Proof: (1)$\Longleftrightarrow$(2): We have

$$\int_0^{2\pi} \sum_{k\in\mathbb{Z}} |\hat{\phi}(\omega + 2k\pi)|^2\, d\omega = \int_{\mathbb{R}} |\hat{\phi}(\omega)|^2\, d\omega. \tag{3.3}$$

Since $\phi \in L^2$, we have $|\hat{\phi}|^2 \in L^1$ and therefore

$$F(\omega) := \sum_{k\in\mathbb{Z}} |\hat{\phi}(\omega + 2k\pi)|^2 \in \tilde{L}^1.$$

The kth Fourier coefficient of $F(\omega)$ is

$$c_k = \frac{1}{2\pi} \int_0^{2\pi} \sum_{k \in \mathbb{Z}} |\hat{\phi}(\omega + 2k\pi)|^2 e^{-ik\omega} \, d\omega$$

$$= \frac{1}{2\pi} \int_{\mathbb{R}} |\hat{\phi}(\omega)|^2 e^{-ik\omega} \, d\omega$$

$$= \frac{1}{2\pi} \int_{\mathbb{R}} \hat{\phi}(\omega)\overline{\hat{\phi}(\omega)} e^{ik\omega} \, d\omega$$

$$= \int_{\mathbb{R}} \phi(x)\overline{\phi(x + k)} \, dx.$$

By (3.1), $c_k = \delta_{0k}$, i.e., the Fourier series of $\sum_{k \in \mathbb{Z}} |\hat{\phi}(\omega + 2k\pi)|^2$ is 1. Hence,

$$\sum_{k \in \mathbb{Z}} |\hat{\phi}(\omega + 2k\pi)|^2 = 1, \quad \text{a.e.}$$

(1)$\Longrightarrow$(2) is proved. Reversing the preceding proof, we get (2)$\Longrightarrow$(1). Finally, (2)$\Longleftrightarrow$(3) can be proved from (3.3). ∎

EXAMPLE 7.3.1: It is trivial that $B(x)$ is an orthonormal MRA generator. By Theorem 7.3.1,

$$\sum_{k \in \mathbb{Z}} |\hat{B}(\omega + 2k\pi)|^2 = 1 \quad a.e. \tag{3.4}$$

Identity (3.4) can be used to prove the following interesting identity:

$$\sum_{k=-\infty}^{\infty} \frac{1}{(x + k\pi)^2} = \frac{1}{\sin^2 x}, \quad x \in \mathbb{R} \setminus \pi\mathbb{Z}. \tag{3.5}$$

In fact, we have $\hat{B}(\omega) = \frac{1-e^{-i\omega}}{i\omega} = e^{-i\omega/2} \frac{\sin \omega/2}{\omega/2}$, which implies that

$$1 = \sum_{k \in \mathbb{Z}} |\hat{B}(\omega + 2k\pi)|^2 = \sum_{k \in \mathbb{Z}} \frac{\sin^2(\omega/2)}{(\omega/2 + k\pi)^2} \quad a.e.$$

It follows that (3.5) holds a.e. It can be verified that both functions $\sum_{k=-n}^{n} \frac{1}{(x+k\pi)^2}$ and $\frac{1}{\sin^2 x}$ are continuous on each interval $(j\pi, (j+1)\pi), j \in \mathbb{Z}$. Therefore (3.5) holds on $\mathbb{R} \setminus \pi\mathbb{Z}$.

EXAMPLE 7.3.2:

Let $\phi(x) = \chi_{[0,2)}(x)$. Then $\hat{\phi}(\omega) = \frac{1-e^{-2i\omega}}{i\omega}$ and

$$\sum_{k \in \mathbb{Z}} |\hat{\phi}(\omega + 2k\pi)|^2 = \sum_{k \in \mathbb{Z}} \frac{\sin^2 \omega}{(\omega/2 + k\pi)^2} = 4\cos^2 \frac{\omega}{2}.$$

By (3.5), $\sum_{k=-\infty}^{\infty} \frac{1}{(\omega/2+k\pi)^2} = \frac{1}{\sin^2 \omega/2}$. Hence,

$$\sum_{k \in \mathbb{Z}} |\hat{\phi}(\omega + 2k\pi)|^2 = \frac{\sin^2 \omega}{\sin^2 \omega/2} = 4 \cos^2 \frac{\omega}{2}$$

is a continuous function vanishing at $\omega = \pi$. Hence, $\{\phi(x - n)\}_{n \in \mathbb{Z}}$ is not stable.

By Theorem 7.3.1, we can orthonormalize a stable system $\{\phi(x - n)\}_{n \in \mathbb{Z}}$.

Theorem 7.3.2 Assume $\{\phi(x - n)\}_{n \in \mathbb{Z}}$ is an unconditional basis of V_0. Define $\tilde{\phi} \in V_0$ by

$$\hat{\tilde{\phi}}(\omega) = \frac{\hat{\phi}(\omega)}{\left(\sum_{k \in \mathbb{Z}} |\hat{\phi}(\omega + 2k\pi)|^2\right)^{1/2}}. \tag{3.6}$$

Then $\{\tilde{\phi}(x - n)\}_{n \in \mathbb{Z}}$ is an orthonormal basis of V_0.

Proof: We first prove that the function $\tilde{\phi}$ defined by (3.6) is in V_0. Write $F(\omega) = \sum_{k \in \mathbb{Z}} |\hat{\phi}(\omega + 2k\pi)|^2$. Define M_r and M_l as in (2.6) and (2.7), respectively. By Theorem 7.2.1 we have

$$\frac{1}{\sqrt{M_r}} \leq \frac{1}{\sqrt{F(\omega)}} \leq \frac{1}{\sqrt{M_l}}, \quad \text{a.e.} \tag{3.7}$$

Since $\hat{\phi} \in L^2$, by (3.7) we have $\frac{\hat{\phi}(\omega)}{\sqrt{F(\omega)}} \in L^2$. Hence, there is a function $\tilde{\phi} \in L^2$ such that (3.6) holds. Note that (3.7) also implies $\frac{1}{\sqrt{F(\omega)}} \in \tilde{L}^2$. Let $\sum_{n \in \mathbb{Z}} c_n e^{in\omega}$ be the Fourier series of $\frac{1}{\sqrt{F(\omega)}}$. Then $(c_n) \in l^2$, which implies

$$\tilde{\phi}(x) = \sum_{n \in \mathbb{Z}} c_n \phi(x - n) \in V_0.$$

Obviously, the function $\tilde{\phi}$ satisfies condition (2) in Theorem 7.3.1. Hence $\{\tilde{\phi}(x - n)\}_{n \in \mathbb{Z}}$ is an orthonormal basis of V_0.   ∎

If the function ϕ in Theorem 7.3.1 is also a scaling function, then we can derive a property of the symbol of ϕ from identity (3.2).

Corollary 7.3.1 If ϕ satisfies the two-scale equation (2.2), and the system $\{\phi(x - m)\}_{m \in \mathbb{Z}}$ is orthonormal, then

$$|H(e^{-i\omega})|^2 + |H(-e^{-i\omega})|^2 = 1 \quad \text{a.e.,} \tag{3.8}$$

which is equivalent to

$$|\mathbf{H}(\omega)|^2 + |\mathbf{H}(\omega + \pi)|^2 = 1 \quad \text{a.e.,}$$

or

$$2 \sum_{m \in \mathbb{Z}} h(m)h(m - 2k) = \delta_{0k}, \quad \text{for all } k \in \mathbb{Z}. \tag{3.9}$$

Proof: By Theorem 7.3.1 and (2.2), it holds almost everywhere that

$$
\begin{aligned}
1 &= \sum_{k \in \mathbb{Z}} |\hat{\phi}(2\omega + 2k\pi)|^2 \\
&= \sum_{k \in \mathbb{Z}} |H(e^{-i\omega})|^2 |\hat{\phi}(\omega + k\pi)|^2 \\
&= \sum_{l \in \mathbb{Z}} |H(e^{-i\omega})|^2 |\hat{\phi}(\omega + 2l\pi)|^2 + \sum_{l \in \mathbb{Z}} |H(-e^{-i\omega})|^2 |\hat{\phi}(\omega + (2l+1)\pi)|^2 \\
&= |H(e^{-i\omega})|^2 \sum_{l \in \mathbb{Z}} |\hat{\phi}(\omega + 2l\pi)|^2 + |H(-e^{-i\omega})|^2 \sum_{l \in \mathbb{Z}} |\hat{\phi}(\omega + (2l+1)\pi)|^2 \\
&= |H(e^{-i\omega})|^2 + |H(-e^{-i\omega})|^2.
\end{aligned}
$$

The proof of (3.8) is completed. To prove (3.9), we need the Fourier series of $|H(e^{-i\omega})|^2$ and of $|H(-e^{-i\omega})|^2$. Since $H(e^{-i\omega}) = \sum h(n)e^{-in\omega}$, we have

$$H(e^{-i\omega})e^{-ik\omega} = \sum h(n)e^{-i(n+k)\omega}$$

and

$$\overline{H(e^{-i\omega})} = \sum h(n)e^{in\omega}.$$

Therefore

$$
\begin{aligned}
&\frac{1}{2\pi} \int_0^{2\pi} |H(e^{-i\omega})|^2 e^{-ik\omega} \, d\omega \\
&= \frac{1}{2\pi} \int_0^{2\pi} H(e^{-i\omega}) e^{-ik\omega} \overline{H(e^{-i\omega})} \, d\omega \\
&= \sum_{n \in \mathbb{Z}} h(n)h(n - k).
\end{aligned}
$$

Similarly, we have

$$\frac{1}{2\pi} \int_0^{2\pi} |H(-e^{-i\omega})|^2 e^{-ik\omega} \, d\omega = \sum_{n \in \mathbb{Z}} (-1)^n h(n)h(n - k).$$

Since $|H(e^{-i\omega})|^2 + |H(-e^{-i\omega})|^2 = 1$ a.e., we have

$$
\begin{aligned}
\delta_{0k} &= \frac{1}{2\pi} \int_0^{2\pi} \left(|H(e^{-i\omega})|^2 + |H(-e^{-i\omega})|^2 \right) e^{-ik\omega} \, d\omega \\
&= 2 \sum_{n \in \mathbb{Z}} h(n) h(n - k).
\end{aligned}
$$

The proof is completed.   ∎

A sequence $(h(n)) \in l^2$, whose symbol $H(e^{-i\omega}) = \sum_{n\in\mathbb{Z}} h(n) e^{-in\omega}$ satisfies equation (3.8), is called a *conjugate mirror filter*. (Here $h(n)$ is normalized by $\sum_{n\in\mathbb{Z}} h(n) = 1$. In some books, it is normalized by $\sum_{n\in\mathbb{Z}} h(n) = \sqrt{2}$.) Thus, we have already shown that the mask of an orthonormal scaling function is a conjugate mirror filter. However, the converse is not true. For example, the function $\phi = \chi_{[0,3]}$ satisfies the following two-scale equation:

$$
\phi(x) = \phi(2x) + \phi(2x - 3).
$$

The symbol of ϕ is $\frac{1+z^3}{2}$, which satisfies (3.8). But $\{\phi_m\}$ is not an orthonormal system.

3.2 Orthonormal Wavelets

We now can construct an orthonormal wavelet basis from an MRA. We assume that the MRA

$$
\cdots \subset V_{-1} \subset V_0 \subset V_1 \subset \cdots
$$

is generated by an orthonormal MRA generator ϕ. As we did in Section 7.1, let W_0 be the space such that

$$
V_0 \bigoplus W_0 = V_1, \quad W_0 \perp V_0,
$$

and let W_j be the subspace such that $g(\cdot) \in W_j$ if and only if $g(2^{-j}\cdot) \in W_0$. From the multiresolution structure of $\{V_j\}$, we can verify that

$$
V_j \bigoplus W_j = V_{j+1} \quad W_j \perp V_j, \quad j \in \mathbb{Z}.
$$

Thus, we have

$$
L^2 = \bigoplus_{n \in \mathbb{Z}} W_n.
$$

Furthermore, we have the following (which generalizes Lemma 7.1.1):

Lemma 7.3.1 Let $\psi \in L^2$. Then $\{\psi_{n,m}\}_{m\in\mathbb{Z}}$ is an orthonormal basis of W_n if and only if $\{\psi_m\}_{m\in\mathbb{Z}}$ is an orthonormal basis of W_0.

Proof: By definition, $\psi_{n,m} = 2^{\frac{n}{2}}\psi(2^n x - m) \in W_n$. We also have

$$\langle \psi_{n,m}, \psi_{n,k} \rangle = \langle \psi_m, \psi_k \rangle,$$

which implies that $\{\psi_{n,m}\}_{m\in\mathbb{Z}}$ is an orthonormal system in W_n if and only if $\{\psi_m\}_{m\in\mathbb{Z}}$ is an orthonormal system of W_0. All that remains is to prove that $\{\psi_{n,m}\}_{m\in\mathbb{Z}}$ is a basis of W_n if and only if $\{\psi_m\}_{m\in\mathbb{Z}}$ is a basis of W_0. Let g be an arbitrary function in W_n. By the definition of W_n, $g(2^{-n}x) \in W_0$. If $\{\psi_m\}_{m\in\mathbb{Z}}$ is a basis of W_0, then

$$g(2^{-n}x) = \sum_{k\in\mathbb{Z}} d_k \psi(x - k),$$

which yields

$$g(x) = \sum_{k\in\mathbb{Z}} \left(2^{-\frac{n}{2}}d_k\right) 2^{\frac{n}{2}}\psi(2^n x - k).$$

Hence, $\{\psi_{n,m}\}_{m\in\mathbb{Z}}$ is a basis of W_n. If $\{\psi_{n,m}\}_{m\in\mathbb{Z}}$ is a basis of W_n, in the similar way, we can prove that $\{\psi_m\}_{m\in\mathbb{Z}}$ is a basis of W_0. ∎

From Lemma 7.3.1, we see that to construct an orthonormal basis of L^2, we only need to find an orthonormal basis of W_0. Such a basis can be found as follows:

Lemma 7.3.2 Let ϕ be an orthonormal MRA generator, which satisfies the two-scale equation

$$\phi(x) = 2\sum_{k\in\mathbb{Z}} h(k)\phi(2x - k), \quad (h_k) \in l^2. \tag{3.10}$$

Let ψ be defined by

$$\psi(x) = 2\sum_{k\in\mathbb{Z}} g(k)\phi(2x - k), \tag{3.11}$$

where

$$g(k) = (-1)^k h(2l + 1 - k) \quad \text{for some } l \in \mathbb{Z}.$$

Then $\{\psi_m\}_{m\in\mathbb{Z}}$ is an orthonormal basis of W_0.

Proof: We first prove that $\{\psi_m\}_{m \in \mathbb{Z}}$ is an orthonormal system in V_1. Note that

$$\int_{\mathbb{R}} \psi(x - n)\psi(x - m)\,dx = \int_{\mathbb{R}} \psi(x + m - n)\psi(x)\,dx.$$

Hence, we only need to prove

$$\int_{\mathbb{R}} \psi(x - k)\psi(x)\,dx = \delta_{0k}, \quad \text{for all } k \in \mathbb{Z}.$$

By (3.11), it is clear that $\psi_k := \psi(\cdot - k) \in V_1$ for all $k \in \mathbb{Z}$. Note that $\{\phi_{1,m}\}_{m \in \mathbb{Z}}$ is an orthonormal basis of V_1. We have

$$\psi(x - k) = 2 \sum_{n \in \mathbb{Z}} (-1)^n h(2l + 1 - n)\phi(2(x - k) - n)$$

$$= 2^{1/2} \sum_{n \in \mathbb{Z}} (-1)^n h(2l + 1 - n)2^{1/2}\phi(2x - (2k + n))$$

$$= 2^{1/2} \sum_{n \in \mathbb{Z}} (-1)^n h(2l + 1 - n)\phi_{1,2k+n}.$$

Hence,

$$\int_{\mathbb{R}} \psi(x - k)\psi(x)\,dx$$

$$= 2 \int_{\mathbb{R}} \left(\sum_{n \in \mathbb{Z}} (-1)^n h(1 + 2l - n)\phi_{1,n+2k} \right) \left(\sum_{m \in \mathbb{Z}} (-1)^m h(1 + 2l - m)\phi_{1m} \right) dx$$

$$= 2 \int_{\mathbb{R}} \left(\sum_{s \in \mathbb{Z}} (-1)^s h(1 + 2l - s + 2k)\phi_{1,s} \right) \left(\sum_{m \in \mathbb{Z}} (-1)^m h(1 + 2l - m)\phi_{1m} \right) dx$$

$$= 2 \sum_{m \in \mathbb{Z}} \sum_{s \in \mathbb{Z}} (-1)^{s-m} h(1 + 2l + 2k - s)h(1 + 2l - m) \int_{\mathbb{R}} \phi_{1,s}\phi_{1m}\,ds$$

$$= 2 \sum_{m \in \mathbb{Z}} h(1 + 2l + 2k - m)h(1 + 2l - m) = 2 \sum_{s \in \mathbb{Z}} h(s - 2k)h(s) = \delta_{0,k}.$$

The last equality is from (3.9). Therefore $\{\psi_m\}_{m \in \mathbb{Z}}$ is an orthonormal system. Second, we prove that $\{\psi_m\}_{m \in \mathbb{Z}} \subset W_0$. Note that if a function $g \in V_1$ and $g \perp V_0$, then $g(\cdot + m) \perp V_0$ for all $m \in \mathbb{Z}$. Hence, we only need to prove $\psi \perp V_0$.

We have, for any $k \in \mathbb{Z}$,

$$\int_{-\infty}^{\infty} \phi(x - k)\psi(x)\, dx$$

$$= 2 \int_{\mathbb{R}} \left(\sum_{m \in \mathbb{Z}} h(m)\phi_{1m+2k} \right) \left(\sum_{n \in \mathbb{Z}} (-1)^n h(1 + 2l - n)\phi_{1n} \right) dx$$

$$= 2 \sum_{s \in \mathbb{Z}} \sum_{n \in \mathbb{Z}} (-1)^n h(s - 2k)h(1 + 2l - n) \int_{\mathbb{R}} \phi_{1s}\phi_{1n}\, dx$$

$$= 2 \sum_{n \in \mathbb{Z}} (-1)^n h(n - 2k)h(1 + 2l - n)$$

$$= 2 \left(\sum_{n \in \mathbb{Z}} h(2n - 2k)h(1 + 2l - 2n) - \sum_{n \in \mathbb{Z}} h(2n + 1 - 2k)h(2l - 2n) \right)$$

$$= 0,$$

which implies $\{\psi_m\}_{m \in \mathbb{Z}} \subset W_0$. Finally, we prove that $\{\psi_n\}_{n \in \mathbb{Z}}$ is a basis of W_0. It is clear that if $\{\psi_n\}_{n \in \mathbb{Z}} \cup \{\phi_n\}_{n \in \mathbb{Z}}$ is a basis of V_1, then $\{\psi_n\}_{n \in \mathbb{Z}}$ is a basis of W_0. Hence, in order to prove $\{\psi_n\}_{n \in \mathbb{Z}}$ is a basis of W_0, we only need to prove $\{\psi_n\}_{n \in \mathbb{Z}} \cup \{\phi_n\}_{n \in \mathbb{Z}}$ is a basis of V_1. We now show that $\phi_{1,l}, l \in \mathbb{Z}$, can be expanded as a linear combination of $\{\psi_n\}_{n \in \mathbb{Z}}$ and $\{\phi_n\}_{n \in \mathbb{Z}}$. Since spaces V_0 and W_0 both are integer-translate invariant, we only need to prove that $\phi(2x)$ and $\phi(2x - 1)$ can be represented as a linear combination of $\{\psi_n\}_{n \in \mathbb{Z}}$ and $\{\phi_n\}_{n \in \mathbb{Z}}$. Let

$$G(e^{-i\omega}) = -e^{-i(2l+1)\omega}\overline{H(-e^{-i\omega})}. \tag{3.12}$$

Taking the Fourier transform of (3.11), we get

$$\hat{\psi}(\omega) = G(e^{-i\omega/2})\hat{\phi}(\omega/2). \tag{3.13}$$

By (3.12), we have

$$H(e^{-i\omega/2})\overline{G(e^{-i\omega/2})} + H(-e^{-i\omega/2})\overline{G(-e^{-i\omega/2})} = 0, \tag{3.14}$$

which implies

$$\begin{pmatrix} H(e^{-i\omega/2}) & H(-e^{-i\omega/2}) \\ G(e^{-i\omega/2}) & G(-e^{-i\omega/2}) \end{pmatrix} \begin{pmatrix} \overline{H(e^{-i\omega/2})} & \overline{G(e^{-i\omega/2})} \\ \overline{H(-e^{-i\omega/2})} & \overline{G(-e^{-i\omega/2})} \end{pmatrix} = I. \tag{3.15}$$

Write

$$H(z) = H_e(z^2) + zH_o(z^2)$$

and
$$G(z) = G_e(z^2) + zG_o(z^2).$$

From (3.15), we have

$$2 \left(\begin{matrix} H_e(e^{-i\omega}) & H_o(e^{-i\omega}) \\ G_e(e^{-i\omega}) & G_o(e^{-i\omega}) \end{matrix} \right) \left(\begin{matrix} \overline{H_e(e^{-i\omega})} & \overline{G_e(e^{-i\omega})} \\ \overline{H_o(e^{-i\omega})} & \overline{G_o(e^{-i\omega})} \end{matrix} \right) = I. \qquad (3.16)$$

Since

$$\hat{\phi}(\omega) = H(e^{-i\omega/2})\hat{\phi}(\omega/2) = H_e(e^{-i\omega})\hat{\phi}(\omega/2) + e^{-i\omega/2}H_o(e^{-i\omega})\hat{\phi}(\omega/2),$$
$$\hat{\psi}(\omega) = G(e^{-i\omega/2})\hat{\phi}(\omega/2) = G_e(e^{-i\omega})\hat{\phi}(\omega/2) + e^{-i\omega/2}G_o(e^{-i\omega})\hat{\phi}(\omega/2),$$

by (3.16), we have

$$\hat{\phi}(\omega/2) = 2\overline{H_e(e^{-i\omega/2})}\hat{\phi}(\omega) + 2\overline{G_e(e^{-i\omega/2})}\hat{\psi}(\omega) \text{ and}$$
$$e^{-i\omega/2}\hat{\phi}(\omega/2) = 2\overline{H_o(e^{-i\omega/2})}\hat{\phi}(\omega) + 2\overline{G_o(e^{-i\omega/2})}\hat{\psi}(\omega)).$$

In the time domain, we have the representations

$$\phi(2x) = \sum_{n\in\mathbb{Z}} h(2n)\phi(x+n) + \sum_{n\in\mathbb{Z}} g(2n)\psi(x+n) \text{ and}$$
$$\phi(2x-1) = \sum_{n\in\mathbb{Z}} h(2n+1)\phi(x+n) + \sum_{n\in\mathbb{Z}} g(2n+1)\psi(x+n),$$

where $g(n) = (-1)^n h(2l+1-n)$, for all $n \in \mathbb{Z}$. This completes the proof. ∎

We often state Lemma 7.3.2 in the frequency domain as follows: Let ϕ be an orthonormal MRA generator satisfying the two-scale equation:

$$\hat{\phi}(\omega) = H(e^{-i\omega/2})\hat{\phi}(\omega/2). \qquad (3.17)$$

Let $G(e^{-i\omega})$ be defined by (3.12) and ψ defined by (3.13). Then $\{\psi_m\}_{m\in\mathbb{Z}}$ is an orthonormal basis of W_0. Besides, in applications, the integer in (3.12) is often chosen to be 0.

From Lemmas 7.3.1 and 7.3.2, we have the following.

Theorem 7.3.3 Let ϕ be an orthonormal MRA generator satisfying (2.2). Then the function ψ defined by (3.11) is an orthonormal wavelet.

Exercises

1. Use orthonormalization to complete the proof of Theorem 7.2.2: Let $\phi \in L^1 \cap L^2$ be an MRA generator (not necessarily an orthonormal one). Then $\hat{\phi}(0) \neq 0$.

2. Let $\phi \in L^1 \cap L^2$ be an orthonormal MRA generator with the symbol $H(e^{-i\omega})$. Prove the following.
   (a) $H(1) = 1$ and $H(-1) = 0$.
   (b) $\hat{\phi}(2k\pi) = \delta_{0,k}$, for all $k \in \mathbb{Z}$.
   (c) $\sum_{k \in \mathbb{Z}} \phi(x - k) = 1$ almost everywhere.

3. Prove that the results in Exercise 2 also hold for a stable scaling function ϕ.

4. Prove that a stable scaling function $\phi \in L^1 \cap L^2$ is an MRA generator.

5. Prove the equivalence of $\sum_{k \in \mathbb{Z}} |\hat{\phi}(\omega + 2k\pi)|^2 = 1$ a.e., and $\frac{1}{2\pi} \int_{\mathbb{R}} |\hat{\phi}(\omega)|^2 e^{ik\omega} d\omega = \delta_{0k}$. (See Theorem 7.3.1.)

6. Under the conditions of Lemma 7.3.2, prove (3.14), (3.15), and (3.16).

7. Prove that if $\phi \in L^1 \cap L^2$ is an orthonormal MRA generator and ψ is its corresponding wavelet, then $\hat{\psi}(0) = 0$.

8. The function ψ defined by
$$\hat{\psi}(\omega) = \begin{cases} 1, & \omega \in [-2\pi, -\pi) \cup [\pi, 2\pi), \\ 0, & \text{otherwise}, \end{cases}$$
is called the *Shannon wavelet.*
   (a) Prove that $\{\psi_{nm}\}_{n,m \in \mathbb{Z}}$ is an orthonormal basis of L^2.
   (b) Find the corresponding scaling function ϕ for the Shannon wavelet.
   (c) Find the two-scale equation of ϕ and the two-scale relation for ψ.

9. The *Lemarié-Meyer wavelet* ψ is defined as follows: $\hat{\psi}(\omega) = b(\omega)e^{i\omega/2}$, where
$$b(\omega) = \begin{cases} \sin(\frac{3}{4}(|\omega| - \frac{2}{3}\pi)), & \frac{2}{3}\pi < |\omega| \le \frac{3}{4}\pi, \\ \sin(\frac{3}{8}(\frac{8}{3}\pi - |\omega|)), & \frac{4}{3}\pi < |\omega| \le \frac{8}{3}\pi, \\ 0, & \text{otherwise}. \end{cases}$$
   (a) Prove that ψ is an orthonormal wavelet.
   (b) Find the Fourier transform of its corresponding scaling function.

10. Let $\phi = \chi_{[0,3)}$. Derive an explicit expression for $\sum_{k \in \mathbb{Z}} |\hat{\phi}(\omega + 2k\pi)|^2$ and then show that
   (a) $|\hat{\phi}(\omega)|^2 + |\hat{\phi}(\omega + \pi)|^2 = 1$.
   (b) ϕ is not an orthonormal scaling function.

11. Let ϕ be an orthonormal MRA generator, and ψ the corresponding orthonormal wavelet. Prove that
   (a) $|\hat{\phi}(\omega)|^2 = \sum_{j=1}^{\infty} |\hat{\psi}(2^j\omega)|^2$ a.e.
   (b) $\sum_{k \in \mathbb{Z}} \sum_{j=1}^{\infty} |\hat{\psi}(2^j(\omega + 2k\pi))|^2 = 1$ a.e.

12. Assume that $g \in L^2(\mathbb{R})$ and $\{g(\cdot - k) \mid k \in \mathbb{Z}\}$ forms an orthonormal system. Prove that
$$|\text{supp}\,\hat{g}| \ge 2\pi,$$

and equality holds if and only if $|\hat{g}| = \chi_K$, where K is a measurable set with the Lebesgue measure $m(K) = 2\pi$.

4 Orthonormal Spline Wavelets

As an application of the theory of the previous section, we now derive the construction of orthonormal spline wavelets, which were first introduced by Battle (in 1987) and by Lemarié (in 1988) independently. Hence, they are also called Battle–Lemarié wavelets .

4.1 Cardinal B-splines

Splines are piecewise polynomials. Because they are more flexible than polynomials, splines become an important tool in numerical analysis and other fields. Cardinal B-splines are the splines defined on uniform partitions with minimal support.

Definition 7.4.1 The *cardinal B-spline* of order m, denoted by $N_m(x)$, is inductively defined by the multiconvolution of the box function:

$$N_1(x) = B(x), \quad N_m(x) = N_{m-1} * N_1(x) = \int_0^1 N_{m-1}(x - t)\, dt, \qquad (4.1)$$

i.e.,

$$N_m(x) = \overbrace{N_1 * N_1 * \cdots N_1}^{m}(x).$$

It is easy to verify that cardinal B-splines have the following properties.

Theorem 7.4.1 The cardinal B-spline of order m satisfies the following:
   (1) supp $N_m = [0, m]$,
   (2) $N_m \in C^{m-2}(\mathbb{R})$ and N_m is a polynomial of exact degree $m - 1$ on each interval $[k, k + 1], 0 \le k \le m - 1$,
   (3) $N_m(x) > 0$, for all $x \in (0, m)$, and
   (4) $N_m(x)$ is symmetric with respect to $x = m/2$:

$$N_m(x) = N_m(m - x). \qquad (4.2)$$

Proof: We leave the proofs of (1)–(3) to the reader. We now prove (4.2) by mathematical induction. It is trivial that (4.2) is true for $m = 1$. Assume that it is true for $m - 1$ $(m > 1)$. We claim that (4.2) is also true for m since

$$N_m(m - x) = \int_0^1 N_{m-1}(m - x - t)\, dt$$

$$= \int_0^1 N_{m-1}((m - 1) - (x - (1 - t)))\, dt$$

$$= \int_0^1 N_{m-1}(x - (1 - t))\, dt$$

$$= \int_0^1 N_{m-1}(x - t)\, dt = N_m(x).$$

Therefore, (4.2) is true for all $m \in \mathbb{N}$. ∎

We now give the explicit expressions of N_2, N_3, and N_4:

$$N_2(x) = \begin{cases} x, & x \in [0, 1), \\ 2 - x, & x \in [1, 2), \\ 0, & \text{otherwise}, \end{cases} \tag{4.3}$$

$$N_3(x) = \begin{cases} \dfrac{1}{2}x^2, & x \in [0, 1), \\ \dfrac{3}{4} - \left(x - \dfrac{3}{2}\right)^2, & x \in [1, 2), \\ \dfrac{1}{2}(x - 3)^2, & x \in [2, 3), \\ 0, & \text{otherwise}, \end{cases} \tag{4.4}$$

and

$$N_4(x) = \begin{cases} \dfrac{1}{6}x^3, & x \in [0, 1), \\ -\dfrac{1}{2}(x - 2)^3 - (x - 2)^2 + \dfrac{2}{3}, & x \in [1, 2), \\ \dfrac{1}{2}(x - 2)^3 - (x - 2)^2 + \dfrac{2}{3}, & x \in [2, 3), \\ \dfrac{1}{6}(4 - x)^3, & x \in [3, 4), \\ 0, & \text{otherwise}. \end{cases} \tag{4.5}$$

The Fourier transforms of cardinal B-splines are very simple. Since $\hat{N}_1(\omega) = \frac{1 - e^{-i\omega}}{i\omega}$, by the Convolution Theorem (Theorem 6.4.1), the Fourier transform of

N_m is

$$\hat{N}_m(\omega) = \left(\frac{1 - e^{-i\omega}}{i\omega}\right)^m.$$

(4.6)

We can see that N_m is a scaling function. In fact, by (4.6), we have

$$\hat{N}_m(\omega) = \left(\frac{1 + e^{-i\omega/2}}{2}\right)^m \hat{N}_m(\omega/2)$$

(4.7)

and

$$\hat{N}_m(0) = 1.$$

(4.8)

Going back to the time domain, we have the following two-scale equation for $N_m(x)$:

$$N_m(x) = \frac{1}{2^{m-1}} \sum_{k=0}^{m} \binom{m}{k} N_m(2x - k).$$

(4.9)

By (4.2), the center of the cardinal B-spline N_m is at $x = \frac{m}{2}$. In applications, we sometimes prefer the B-splines with center $x = 0$. Hence, we define the *central B-spline* by $N_m^c(x) = N_m(x + \frac{m}{2})$. However, only the central B-splines of even orders are scaling functions. For an even m, we have

$$\widehat{N_m^c}(\omega) = \left(\frac{\sin \omega/2}{\omega/2}\right)^m,$$

$$\widehat{N_m^c}(\omega) = \left(\frac{1 + \cos(\frac{\omega}{2})}{2}\right)^{m/2} \widehat{N_m^c}(\omega/2),$$

(4.10)

and

$$N_m^c(x) = \frac{1}{2^{m-1}} \sum_{k=-m/2}^{m/2} \binom{m}{k + m/2} N_m^c(2x - k).$$

(4.11)

For a cardinal B-spline of an odd order, we shift it to the center $x = 1/2$. We define

$$\widehat{N_m^s}(\omega) = \left(\frac{1 + e^{-i\omega/2}}{2}\right) \left(\frac{1 + \cos(\frac{\omega}{2})}{2}\right)^{[m/2]} \widehat{N_m^s}(\omega/2), \quad m \text{ is odd,}$$

where $[x]$ denotes the integer part of x. Then we have

$$N_m^s(x) = \frac{1}{2^{m-1}} \sum_{k=-[m/2]}^{[m/2]+1} \binom{m}{k + [m/2]} N_m^s(2x - k).$$

In applications, the linear splines (order 2) and the cubic splines (order 4) are more popular than those of other orders.

For splines, we have the following

Theorem 7.4.2 The cardinal B-spline of order m is an MRA generator.

Proof: We prove that N_m satisfies both the stability condition and the completeness condition. In fact, we have

$$\sum_{k\in\mathbb{Z}}|\hat{N}_m(\omega + 2k\pi)|^2 = \sum_{k\in\mathbb{Z}}\left(\frac{\sin(\omega/2)}{\omega/2 + k\pi}\right)^{2m}.$$

Recall that $\sum_{k\in\mathbb{Z}}\left(\frac{\sin(\omega/2)}{\omega/2+k\pi}\right)^{2m}$ is a 2π-periodic function. By the inequality (under the assumption that $\frac{\sin x}{x} = 1$ for $x = 0$),

$$\frac{2}{\pi} \le \frac{\sin x}{x}, \quad x \in \left[0, \frac{\pi}{2}\right].$$

We have

$$\left(\frac{2}{\pi}\right)^{2m} \le \left(\frac{\sin(\omega/2)}{\omega/2}\right)^{2m}, \quad \omega \in [0, \pi],$$

and

$$\left(\frac{2}{\pi}\right)^{2m} \le \left(\frac{\sin(\omega/2)}{\pi - \omega/2}\right)^{2m}, \quad \omega \in [\pi, 2\pi].$$

Thus, when $x \in [0, 2\pi)$,

$$\left(\frac{2}{\pi}\right)^{2m} \le \left(\frac{\sin(\omega/2)}{\omega/2}\right)^{2m} + \left(\frac{\sin(\omega/2)}{\pi - \omega/2}\right)^{2m}$$

$$\le \sum_{k\in\mathbb{Z}}\left(\frac{\sin(\omega/2)}{\omega/2 + k\pi}\right)^{2m}.$$

On the other hand, from the fact that $|\sin x| \le |x|$ and $\frac{1}{\sin^2\omega} = \sum_{k\in\mathbb{Z}}\frac{1}{(\omega+k\pi)^2}$ (see (3.5)), we have

$$\sum_{k\in\mathbb{Z}}\left(\frac{\sin(\omega/2)}{\omega/2 + k\pi}\right)^{2m} = \sum_{k\in\mathbb{Z}}\left(\frac{\sin(\omega/2 + k\pi)}{\omega/2 + k\pi}\right)^{2m}$$

$$\le \sum_{k\in\mathbb{Z}}\left(\frac{\sin(\omega/2 + k\pi)}{\omega/2 + k\pi}\right)^{2}$$

$$= \sum_{k\in\mathbb{Z}}\left(\frac{\sin\omega/2}{\omega/2 + k\pi}\right)^{2} = 1.$$

Hence, $N_m(x)$ is stable. It is clear that $N_m \in L^1 \cap L^2$ and $\hat{N}_m(0) = 1$, which yield the completeness of N_m. The theorem is proved. ∎

4.2 Construction of Orthonormal Spline Wavelets

Since cardinal B-splines are MRA generators, applying the results of the previous section, we can construct orthonormal spline scaling functions and wavelets. Write

$$B_m(\omega) = \sum_{k \in \mathbb{Z}} \left(\frac{\sin(\omega/2)}{\omega/2 + k\pi} \right)^{2m}. \tag{4.12}$$

By Theorem 7.3.2 and Theorem 7.3.3, we have the following:

Theorem 7.4.3 The function $\tilde{N}_m$ defined by

$$\hat{\tilde{N}}_m(\omega) = \frac{\widehat{N_m}(\omega)}{\sqrt{B_m(\omega)}} \tag{4.13}$$

is an orthonormal MRA generator, which satisfies the following refinement equation:

$$\hat{\tilde{N}}_m(\omega) = \tilde{H}(e^{-i\omega/2})\hat{\tilde{N}}_m(\omega/2), \tag{4.14}$$

where

$$\tilde{H}(e^{-i\omega}) = \left(\frac{1 + e^{-i\omega}}{2} \right)^m \frac{\sqrt{B_m(\omega)}}{\sqrt{B_m(2\omega)}},$$

and then the function $\tilde{S}_m$ defined by

$$\hat{\tilde{S}}_m(\omega) = -e^{-i\omega/2}\overline{\tilde{H}(-e^{-i\omega/2})}\hat{\tilde{N}}_m(\omega/2) \tag{4.15}$$

is the corresponding orthonormal wavelet.

Proof: We have $B_m(\omega) = \sum_{k \in \mathbb{Z}} |\hat{N}_m(\omega + 2k\pi)|^2$. By Theorem 7.3.2, the function $\tilde{N}_m$ defined by $\hat{\tilde{N}}_m(\omega) = \frac{\hat{N}_m(\omega)}{\sqrt{B_m(\omega)}}$ is an orthonormal MRA generator. Then the equation $\hat{\tilde{N}}_m(\omega) = (\frac{1+e^{-i\omega/2}}{2})^m \hat{\tilde{N}}_m(\omega/2)$ yields (4.14). Note that $\sqrt{B_m(\omega)}$ is a real valued 2π-periodic function. By Theorem 7.3.3, $\tilde{S}_m(x)$ in (4.15) is the corresponding orthonormal wavelet. ∎

The function $B_m(\omega)$ plays an important rule in the construction of orthonormal spline wavelets. We now give the explicit formula of $B_m(\omega)$.

Theorem 7.4.4 Let $B_m(\omega)$ be the function defined by (4.12). Then

$$B_m(\omega) = \sum_{k=-m+1}^{m-1} N_{2m}(m+k)e^{-ik\omega}, \tag{4.16}$$

or, equivalently,

$$B_m(\omega) = N_{2m}(m) + 2\sum_{k=1}^{m-1} N_{2m}(m-k)\cos k\omega. \tag{4.17}$$

Proof: $B_m(\omega)$ is a 2π-periodic function. When $m > 1$, it is differentiable. Hence, $B_m(\omega) = \sum_k (B_m)^\wedge(k)e^{-ik\omega}$. We have

$$(B_m)^\wedge(k) = \frac{1}{2\pi}\int_{-\pi}^{\pi} B_m(\omega)e^{ik\omega}\,d\omega$$

$$= \frac{1}{2\pi}\int_{-\pi}^{\pi} \sum_{k\in\mathbb{Z}} |\hat{N}_m(\omega+2k\pi)|^2 e^{ik\omega}\,d\omega$$

$$= \frac{1}{2\pi}\int_{\mathbb{R}} |\hat{N}(\omega)|^2 e^{ik\omega}\,d\omega$$

$$= \frac{1}{2\pi}\int_{\mathbb{R}} \overline{\hat{N}(\omega)}\hat{N}(\omega)e^{ik\omega}\,d\omega.$$

By Parseval's formula (see Section 6.4 and Theorem 6.4.2),

$$\frac{1}{2\pi}\int_{\mathbb{R}} \overline{\hat{N}(\omega)}\hat{N}(\omega)e^{ik\omega}\,d\omega = \int_{\mathbb{R}} \overline{N_m(x)}N_m(x+k)\,dy.$$

Since $N_m(x)$ is a real-valued function and $N_m(x) = N_m(m-x)$, we have

$$\int_{\mathbb{R}} \overline{N_m(x)}N_m(x+k)\,dy = \int_{\mathbb{R}} N_m(x)N_m(m-k-x)\,dx$$

$$= (N_m * N_m)(m-k) = N_{2m}(m-k).$$

Thus,

$$B_m(\omega) = \sum_{k \in \mathbb{Z}} N_{2m}(m - k)e^{-ik\omega}$$

$$= N_{2m}(m) + 2\sum_{k=1}^{m-1} N_{2m}(m - k)\cos k\omega.$$

The theorem is proved.   ∎

EXAMPLE 7.4.1: Compute the values of $N_m(k), k \in \mathbb{Z}, m > 1$, numerically and then derive the formula of $B_m(\omega)$.

Note that the support of N_m is $[0, m]$. Hence, $N_m(k) = 0$ if $k \leq 0$ or $k \geq m$. We only need to compute $N_m(k), 1 \leq k \leq m - 1$. From the two-scale equation of N_m we have

$$N_m(1) = \frac{1}{2^{m-1}}(N_m(2) + mN_m(1)),$$

$$N_m(2) = \frac{1}{2^{m-1}}\left(N_m(4) + mN_m(3) + \binom{m}{2}N_m(2) + \binom{m}{3}N_m(1)\right),$$

$$\cdots \qquad \cdots$$

$$N_m(m - 1) = \frac{1}{2^{m-1}}(mN_m(m - 1) + N_m(m - 2)),$$

which is a homogeneous linear system of $[N_m(1), \cdots, N_m(m - 1)]$. The system has a nontrivial solution, which is the 0-eigenvector of the system. Therefore, if $\mathbf{v} = [v_1, \cdots, v_{m-1}]$ is a solution, so is $c\mathbf{v}$. The required solution $[N_m(1), \cdots, N_m(m - 1)]$ should be normalized by the condition $\hat{N}_m(0) = 1$. This condition yields $\sum_{k \in \mathbb{Z}} N_m(k) = 1$ (see Exercise 3).

When $m = 2$, the system is reduced to the equation $N_2(1) = N_2(1)$. Any nonzero real number is a nontrivial solution of this equation. By the normalization condition, we have $N_2(1) = 1$. Hence, $B_2(\omega) = 1$. This also proves that $N_1(x)$ is an orthonormal scaling function.

When $m = 4$, the system is reduced to

$$N_4(1) = \frac{1}{8}(N_4(2) + 4N_4(1)),$$

$$N_4(2) = \frac{1}{8}(4N_4(3) + 6N_4(2) + 4N_4(1)),$$

$$N_4(3) = \frac{1}{8}(4N_4(3) + N_4(2)),$$

which has as a solution $(1, 4, 1)$. Applying the normalization condition, we have $N_4(1) = 1/6, N_4(2) = 2/3, N_4(3) = 1/6$. Therefore, $B_2(\omega) = N_2(2) + 2N_2(1) \cos \omega = \frac{2}{3} + \frac{1}{3} \cos \omega$. We leave the computations of $B_4(\omega)$ as exercises.

The Fourier transform of $\tilde{N}_m$ is already given by (4.13). We now derive the representation of $\tilde{N}_m$ as the linear combination of the translates of $N_m(x)$ in the time domain. Since $\frac{1}{\sqrt{B_m(\omega)}}$ is a differentiable 2π-periodic even function, let $\sum_n c_n e^{-ik\omega}$ be its Fourier series. Then $\frac{1}{\sqrt{B_m(\omega)}} = \sum_n c_n e^{-ik\omega}$ and the coefficient c_n is given by

$$c_n = c_{-n} = \frac{1}{\pi} \int_0^\pi \frac{\cos n\omega}{\sqrt{B_m(\omega)}} d\omega, \quad n \in \mathbb{Z}^+. \tag{4.18}$$

This yields

$$\tilde{N}_m(x) = \sum_n c_n N_m(x - n). \tag{4.19}$$

Similarly, let

$$b_n = b_{-n} = \frac{1}{\pi} \int_0^\pi \cos n\omega \sqrt{B_m(\omega)} \, d\omega, \quad n \in \mathbb{Z}^+.$$

Then

$$N_m(x) = \sum_n b_n \tilde{N}_m(x - n). \tag{4.20}$$

EXAMPLE 7.4.2: Compute the coefficients (c_n) in (4.18) and (b_k) in (4.20) for $m = 2$. (Round them off to six decimal places.)

Recall that

$$c_k = \frac{1}{\pi} \int_0^\pi \frac{\cos k\omega}{\sqrt{\frac{2}{3} + \frac{1}{3} \cos \omega}} d\omega, k \geq 0.$$

By rounding off c_k to six decimal places, we have
$c_0 = 1.291675, c_1 = -.174663, c_2 = 0.035210, c_3 = -.007874, c_4 = 0.001848, c_5 = -0.000446, c_6 = 0.000110, c_7 = -0.000027, c_8 = 0.000007, c_9 = -0.000002, c_k = 0, k \geq 10$.

Similarly, $b_k = \frac{1}{\pi} \int_0^{2\pi} \cos k\omega \sqrt{\frac{2}{3} + \frac{1}{3} \cos \omega} \, d\omega, k \geq 0$. Hence,
$b_0 = 0.802896, b_1 = 0.104705, b_2 = -0.006950, b_3 = 0.000927,$

$b_4 = -0.000155$, $b_5 = 0.000029$, $b_6 = -0.000006$, $b_7 = 0.000001$, $b_k = 0, k \geq 8$.

To obtain the two-scale equation (4.14) in the time domain, we need the Fourier series of $\tilde{H}(e^{-i\omega})$. Recall that $\tilde{H}(e^{-i\omega}) = \left(\frac{1+e^{-i\omega}}{2}\right)^m \frac{\sqrt{B_m(\omega)}}{\sqrt{B_m(2\omega)}}$. Let

$$\frac{\sqrt{B_m(\omega)}}{\sqrt{B_m(2\omega)}} = \sum_k \beta_k e^{-ik\omega}.$$

Then

$$\beta_k = \beta_{-k} = \frac{1}{\pi} \int_0^\pi \sqrt{\frac{B_m(\omega)}{B_m(2\omega)}} \cos n\omega \, d\omega, \quad n \in \mathbb{Z}^+. \tag{4.21}$$

Assume $\tilde{H}(e^{-i\omega}) = \sum_n \alpha_n e^{-ik\omega}$. We have

$$\tilde{H}(e^{-i\omega}) = \left(\frac{1+e^{-i\omega}}{2}\right)^m \sum_k \beta_k e^{-ik\omega}$$

$$= \frac{1}{2^m} \sum_{l=0}^m \binom{m}{l} e^{-il\omega} \sum_k \beta_k e^{-ik\omega},$$

which yields

$$\alpha_n = \alpha_{m-n} = \frac{1}{2^m} \sum_{l=0}^m \binom{m}{l} \beta_{n-l}. \tag{4.22}$$

Thus, we derive the two-scale equation of $\tilde{N}_m(x)$:

$$\tilde{N}_m(x) = 2 \sum_{k \in \mathbb{Z}} \alpha_k \tilde{N}_m(2x - k).$$

By (4.15), we have

$$\tilde{S}_m(x) = 2 \sum_n (-1)^k \alpha_{1-k} \tilde{N}_m(2x - k). \tag{4.23}$$

For the central B-spline N_m^c (with an even m), its orthonormalization is $\tilde{N}_m^c(x) = \tilde{N}_m(x + \frac{m}{2})$. Let $\left(\alpha_k^c\right)$ be the mask of $\tilde{N}_m^c$. Then we define

$$\tilde{S}_m^c(x) = 2 \sum_n (-1)^k \alpha_{1-k}^c \tilde{N}_m^c(2x - k).$$

The function $\tilde{N}_m^c(x)$ (with an even m) is an even function, while $\tilde{S}_m^c(x)$ is symmetric with respect to $x = \frac{1}{2}$. Thus, the function

$$\left(\tilde{N}_m^c\right)_{j,k}(x) = 2^{\frac{j}{2}} \tilde{N}_m^c(2^j x - k)$$

is symmetric about $x = 2^{-j}k$ and

$$\left(\tilde{S}_m^c\right)_{j,k}(x) = 2^{\frac{j}{2}}\tilde{S}_m^c(2^j x - k)$$

is symmetric about $x = 2^{-j}(k + \frac{1}{2})$. Note that both $\tilde{N}_m^c(x)$ (with an even m) and $\tilde{S}_m^c(x)$ have their maxima at their centers and both exponentially decay. Hence, in applications, when m is even, $\tilde{N}_m^c(x)$ and $\tilde{S}_m^c(x)$ are used more often than $\tilde{N}_m(x)$ and $\tilde{S}_m(x)$.

> **EXAMPLE 7.4.3:** Compute the mask of $\tilde{N}_2$, rounding off to six decimal places.
> We first compute β_k in (4.21):
> $\beta_0 = 1.0394978$, $\beta_1 = 0.1168288$, $\beta_2 = -0.1494296$, $\beta_3 = -0.0134167$,
> $\beta_4 = 0.0293394$, $\beta_5 = 0.0027392$, $\beta_6 = -0.0065602$, $\beta_7 = -0.0006023$,
> $\beta_8 = 0.0015367$, $\beta_9 = 0.0001405$, $\beta_{10} = -0.0003706$, $\beta_{11} = -0.0000337$,
> $\beta_{12} = 0.0000910$, $\beta_{13} = 0.0000083$, $\beta_{14} = -0.0000226$, $\beta_{15} = -0.0000020$,
> $\beta_{16} = 0.0000057$, $\beta_{17} = 0.0000006$, $\beta_{18} = -0.0000014$, $\beta_{19} = -0.0000001$,
> $\beta_{20} = 0.0000004$.
> Applying formula (4.22), we obtain
> $\alpha_1 = 0.578163$, $\alpha_2 = 0.280932$, $\alpha_3 = -0.048862$, $\alpha_4 = -0.036731$,
> $\alpha_5 = 0.012000$, $\alpha_6 = 0.007064$, $\alpha_7 = -0.002746$, $\alpha_8 = -0.001557$,
> $\alpha_9 = 0.000653$, $\alpha_{10} = 0.000362$, $\alpha_{11} = -0.000159$, $\alpha_{12} = -0.000087$,
> $\alpha_{13} = 0.000039$, $\alpha_{14} = 0.000021$, $\alpha_{15} = -0.000010$, $\alpha_{16} = -0.000005$,
> $\alpha_{17} = 0.000002$, $\alpha_{18} = 0.000001$, $\alpha_k = 0, k \geq 19$, $\alpha_k = \alpha_{2-k}, k \leq 0$.

Exercises

1. Derive the explicit formulas of $N_2(x)$ in (4.3), $N_3(x)$ in (4.4), and $N_3(x)$ in (4.5), respectively.
2. Apply the Poisson Summation Formula (Theorem 6.5.1) to prove that $\sum_{k \in \mathbb{Z}} N_m(k) = 1$.
3. Prove that $\frac{d}{dx}N_m(x) = N_{m-1}(x) - N_{m-1}(x - 1), m \geq 2$.
4. Use mathematical induction to prove that

$$N_m(x) = \frac{1}{m - 1}(xN_{m-1}(x) + (m - x)N_{m-1}(x - 1)).$$

5. Use the formula in Exercise 4 to derive the explicit expressions of $N_2(x), \cdots, N_6(x)$.
6. Draw the graphs of $N_m(x), 2 \leq m \leq 6$.

7. Prove (see (4.10) and (4.11))

$$\widehat{N^c_m}(\omega) = \left(\frac{1 + \cos(\frac{\omega}{2})}{2}\right)^{m/2} \widehat{N^c_m}(\omega/2)$$

and

$$N^c_m(x) = \frac{1}{2^{m-1}} \sum_{k=-m/2}^{m/2} \binom{m}{k + m/2} N^c_m(2x - k).$$

8. Let $F_m(\omega) = \sum_{k \in \mathbb{Z}} \frac{1}{(\omega + 2k\pi)^{2m}}$. Prove that $\widehat{N^c_m}(\omega) = \frac{1}{\omega^m \sqrt{F_m(\omega)}}$ when m is even, and $\widehat{N^s_m}(\omega) = \frac{e^{-i\omega/2}}{\omega^m \sqrt{F_m(\omega)}}$ when m is odd.

9. Use the method of Example 7.4.1 to compute the value $N_m(k), k \in \mathbb{Z}$, for $m = 4$.

10. Use the formula in Exercise 4 to compute the value $N_m(k), k \in \mathbb{Z}$ for $m = 2, \cdots, 8$.

11. Let m be an even number and $N^c_m(x)$ be the central linear B-spline of order m. Let $\tilde{N}^c_m(x)$ be its orthonormalization. Let $\tilde{N}^c_m(x) = \sum_k c^c_k N^c_m$ $(x - k)$, $\tilde{N}^c_m(x) = 2\sum_k a^c_k \tilde{N}^c_m(2x - k)$, and $\tilde{S}^c_m(x) = 2\sum_k (-1)^k a^c_{1-k} \tilde{N}^c_m$ $(2x - k)$.
   (a) Find the relationship between c^c_k and c_k in (4.19).
   (b) Find the relationship between a^c_k and a_k in (4.22).
   (c) Find the relationship between $\tilde{S}^c_m(x)$ and $\tilde{S}_m(x)$ (in (4.23)).

12. Find the coefficients in the equation

$$S_2(x) = \sum \mu_k N_2(2x - k).$$

13. Draw the graphs of $\tilde{N}^c_2(x)$ and $\tilde{S}^c_2(x)$ and the graphs of $|\widehat{\tilde{N}^c_2}(\omega)|$ and $|\widehat{\tilde{S}^c_2}(\omega)|$.

14. Let $N_4(x)$ be the cubic B-spline and let $\tilde{N}_4(x)$ be its orthonormalization. Use a graphing calculator, or mathematical software such as Maple, Mathematica, or MATLAB, to do the following.
   (a) Find the coefficient c_k, in the form

$$\tilde{N}_4(x) = \sum c_k N_4(x - k).$$

   numerically. (Round off to four decimal places.)
   (b) Find the coefficient b_k in the form (round off to five decimal places)

$$N_4(x) = \sum b_k \tilde{N}_4(x - k).$$

(c) Find the mask (α_k) of $\tilde{N}_4(x)$ numerically. (Round off to five decimal places.)

(d) Let $S_4(x)$ be the orthonormal cubic spline wavelet. Find the coefficients (w_k) and $(\tilde{w}_k)$ in the relationships (round off to five decimal places)

$$S_4(x) = \sum w_k N_4(2x - k)$$

and

$$S_4(x) = \sum \tilde{w}_k \tilde{N}_4(2x - k).$$

15. Draw the graphs of $\tilde{N}_4^c(x)$ and $\tilde{S}_4^c(x)$ and the graphs of $|\widehat{\tilde{N}_4^c}(\omega)|$ and $|\widehat{\tilde{S}_4^c}(\omega)|$.

5 Fast Wavelet Transforms

Let ϕ be an orthonormal MRA generator and ψ be the orthonormal wavelet corresponding to ϕ. Since $\{\psi_{n,m}\}_{n,m\in\mathbb{Z}}$ is an orthonormal basis of L^2, any function $f \in L^2$ can be expanded into a wavelet series

$$f(x) = \sum_{n\in\mathbb{Z}}\sum_{m\in\mathbb{Z}} b_{n,m}\psi_{n,m}, \tag{5.1}$$

where $b_{n,m} = \langle f, \psi_{n,m}\rangle$. The way to compute the coefficients $b_{n,m}$ of the wavelet series of $f(x)$ here is similar to that of computing coefficients in Fourier series. It does not take advantage of the multilevel structure of wavelets. Stépheane Mallat in 1989 discovered a fast way to compute the coefficients in wavelet series. We now introduce Mallat's method.

5.1 Initialization

Mallat's method first finds an approximation of f in a subspace in MRA, say a function $f_n \in V_n$. By the property of MRA, if n is large enough, then the approximation error $\|f - f_n\|$ will be less than a given tolerance. This step is called *initialization* of the fast wavelet transform. We may obtain f_n by orthogonal projection. From the orthogonality of the basis $(\phi_{nm})_{m\in\mathbb{Z}}$, it follows that the orthogonal projection of f on V_n is

$$f_n = \sum_{m\in\mathbb{Z}} \langle f, \phi_{n,m}\rangle \phi_{n,m}.$$

However, sometimes the integral $\langle f, \phi_{n,m} \rangle$ is not easy to compute, or f is obtained experimentally so that only the values of f at sampling points are known. In these cases, interpolation is often used for initialization. A function $f_n \in V_n$ is called an *interpolation* of f if

$$f_n(2^{-n}k) = f(2^{-n}k), \quad k \in \mathbb{Z}. \tag{5.2}$$

Let us write $f_n = \sum_{m \in \mathbb{Z}} a_{nm} \phi_{n,m}$ and $b_k = 2^{-n/2} f(2^{-n}k)$. Then equation (5.2) can be written as $\sum_{m \in \mathbb{Z}} a_{nm} \phi_{n,m}(2^{-n}k) = f(2^{-n}k)$, which yields

$$\sum_{m \in \mathbb{Z}} \phi(k - m) a_{nm} = b_k, \quad k \in \mathbb{Z}. \tag{5.3}$$

This is an infinite linear system. Assume $\sum_{k \in \mathbb{Z}} \phi(k) e^{-ik\omega}$ is convergent everywhere. In Chapter 9, we shall prove the following.

Lemma 7.5.1 The linear system (5.3) is consistent (for all $(b_k) \in l^1$) if and only if $\sum_{k \in \mathbb{Z}} \phi(k) e^{-ik\omega} \neq 0$ for $\omega \in \mathbb{R}$.

Applying the Poisson Summation Formula (Theorem 6.5.1), from this lemma we have the following.

Theorem 7.5.1 If $\phi \in L^1 \cap L^2$ is continuous, the series $\sum_{k \in \mathbb{Z}} \hat{\phi}(\omega + 2k\pi)$ is uniformly convergent, and $(\phi(k))_{k \in \mathbb{Z}} \in l^1$, then the linear system (5.3) is consistent if and only if

$$\sum_{k \in \mathbb{Z}} \hat{\phi}(\omega + 2k\pi) \neq 0, \quad \omega \in \mathbb{R}. \tag{5.4}$$

Proof: We leave the proof as an exercise. ∎

Condition (5.4) is called the *interpolation condition* for ϕ. We say $\phi \in L^1 \cap L^2$ satisfies that interpolation condition if it is continuous, $(\phi(k))_{k \in \mathbb{Z}} \in l^1$, $\sum_{k \in \mathbb{Z}} \hat{\phi}(\omega + 2k\pi)$ is uniformly convergent, and (5.4) holds.

EXAMPLE 7.5.1: We have $N_2^c(k) = \delta_{0,k}$. Hence, $\sum_{k \in \mathbb{Z}} N_2^c(k) e^{-ik\omega} = 1$, which implies that N_2^c satisfies the interpolation condition. In fact, we have $f_n(x) = \sum_{k \in \mathbb{Z}} f(2^{-n}k) N_2^c(2^j x - k) = \sum_{k \in \mathbb{Z}} 2^{-n/2} f(2^{-n}k) (N_2^c)_{j,k}(x)$.

EXAMPLE 7.5.2: We have $N_4^c(-1) = N_4^c(1) = \frac{1}{6}$, $N_4^c(0) = \frac{2}{3}$, and $N_4^c(k) = 0$ for $|k| > 1$. Hence, $\sum_{k \in \mathbb{Z}} N_4^c(k) e^{-ik\omega} = \frac{2}{3} + \frac{1}{3} \cos \omega = \frac{1}{3} + \frac{1}{3} \cos^2 \frac{\omega}{2} \geq \frac{1}{3}$. Therefore, N_4^c satisfies the interpolation condition.

EXAMPLE 7.5.3: We have $N_3(1) = N_3(2) = \frac{1}{2}$ and $N_3^c(k) = 0$ for all integers. Hence, $\sum_{k\in\mathbb{Z}} N_3^c(k)e^{-ik\omega} = \frac{1}{2}e^{-i\omega} + \frac{1}{2}e^{-i2\omega}$, which vanishes at $\omega = \pi$. By Lemma 7.5.1, N_3 does not satisfy the interpolation condition. We can also use Theorem 7.5.1 to verify the conclusion. Recall that $\hat{N}_3(\omega) = \left(\frac{1-e^{-i\omega}}{i\omega}\right)^3$. Hence,

$$\sum_{k\in\mathbb{Z}} \hat{N}_3(\omega + 2k\pi) = \left(\frac{1 - e^{-i\omega}}{i\omega}\right)^3 \sum_{k\in\mathbb{Z}} \left(\frac{1}{\omega + 2k\pi}\right)^3.$$

Let $\omega = \pi$. We have

$$\sum_{k\in\mathbb{Z}} \hat{N}_3(\pi + 2k\pi) = \frac{8i}{\pi^3} \sum_{k\in\mathbb{Z}} \frac{1}{(2k + 1)^3} = 0,$$

which implies the inconsistencey of the interpolation.

In general, when m is even, N_m satisfies the interpolation condition, and when m is odd, it does not. This is why splines of even orders are more often used in applications than splines of odd orders. We leave the proof of the result as an exercise.

Assume now that ϕ satisfies the interpolation condition. For realizing the interpolation, we can construct the *Lagrangian interpolating functions* in V_0.

Definition 7.5.1 If a scaling function $\phi \in L^1 \cap L^2$ satisfies the condition $\phi(k) = \delta_{0,k}$, then it is called an *interpolating scaling function*.

If an MRA generator ϕ satisfies the interpolation condition, then we can construct the interpolating scaling function as follows.

Theorem 7.5.2 Let $\phi \in L^1 \cap L^2$ be a continuous two-scaling function, which satisfies the interpolation condition. Then the function $\phi_{in}(x)$ defined by

$$\hat{\phi}_{in}(\omega) = \frac{\hat{\phi}(\omega)}{\sum_{k\in\mathbb{Z}} \hat{\phi}(\omega + 2k\pi)} \tag{5.5}$$

is an interpolating scaling function. Therefore a function $f \in V_n$ can be represented as

$$f(x) = \sum_{k\in\mathbb{Z}} f(2^{-n}k)\phi_{in}(2^j x - k).$$

Proof: We leave the proof as an exercise.   ∎

Assume now an MRA generator ϕ satisfies the interpolation condition. Let $f \in L^2$ be a continuous function and f_n be the interpolation of f in V_n. It is known that $\|f - f_n\|_2 \to 0$ as $n \to \infty$. (We will not show the proof. Readers can refer to [4].) This result allows us to use the interpolation f_n to approximate f.

In many applications, we are not given the sampling data, say $f(2^{-n}m), m \in \mathbb{Z}$, but given a certain average of f around each sampling point $x = 2^{-n}m$. Write the given data as $(c_m)_{m \in \mathbb{Z}}$. Therefore, we can approximately assume $c_m = \langle f, \phi_{n,m} \rangle$ for a certain function $f \in L^2$. Thus, the function $f_n := \sum c_n \phi_{n,m}$ represents the data (c_m). This is the initialization for the fast wavelet transform.

5.2 Multiscale Decomposition

We now assume the initialization is completed (in a certain way). That is, we have

$$f_n = \sum a_{nm} \phi_{n,m}, \tag{5.6}$$

where $(a_{nm})_{m \in \mathbb{Z}}$ is a known sequence. By (5.1), f_n can also be represented as

$$f_n = \sum_{j<n} \sum_{m \in \mathbb{Z}} b_{j,m} \psi_{j,m}. \tag{5.7}$$

Mallat's algorithm computes $(b_{j,m})$ from $(a_{n,m})$. To explain Mallat's idea, we adopt a concise expression of (5.7):

$$f_n = \sum_{j<n} g_j, \tag{5.8}$$

where $g_j = \sum_{m \in \mathbb{Z}} b_{j,m} \psi_{j,m} \in W_j$.

To avoid the infinity of j-indices in (5.8), we apply the relation

$$V_l = \bigoplus_{j<l} W_j, \quad l \in \mathbb{Z}.$$

Therefore, there is a function $f_l \in V_l$ such that $f_l = \sum_{j<l} g_j$. Thus, for any $l < n$, f_n can be expanded as

$$f_n = f_l + \sum_{j=l}^{n-1} g_j, \quad f_l \in V_l, g_j \in W_j,$$

or equivalently,

$$\sum_{m \in \mathbb{Z}} a_{n,m} \phi_{n,m} = \sum_{m \in \mathbb{Z}} a_{l,m} \phi_m + \sum_{j=l}^{n-1} \sum_{m \in \mathbb{Z}} b_{j,m} \psi_{j,m}. \tag{5.9}$$

Particularly,

$$\sum_{m \in \mathbb{Z}} a_{n,m}\phi_{n,m} = \sum a_{n-1,m}\phi_m + \sum b_{n-1,m}\psi_{n,m},$$

$$\sum_{m \in \mathbb{Z}} a_{n-1,m}\phi_{n,m} = \sum a_{n-2,m}\phi_m + \sum b_{n-2,m}\psi_{n,m},$$

$$\vdots$$

$$\sum_{m \in \mathbb{Z}} a_{l+1,m}\phi_{n,m} = \sum_{m \in \mathbb{Z}} a_{l,m}\phi_m + \sum_{m \in \mathbb{Z}} b_{l,m}\psi_{n,m}.$$

These equations provide a way to compute $(a_{l,m}), (b_{l,m}), \cdots (b_{n-1,m})$ from $(a_{n,m})$. For convenience, without loss of generality, we always assume $n > 0$ and set $l = 0$ in (5.9).

The algorithm used in computing the coefficients $(a_{j-1,m})$ and $(b_{j-1,m})$ from $(a_{j,m})$ is called a *fast wavelet transform* (FWT), and its reverse is called a *fast inverse wavelet transform (FIWT)*. The algorithm computing the coefficients $(a_{0,m})$ and $(b_{j,m}), 0 \leq j \leq n - 1$, from $(a_{n,m})$ is called a *decomposition pyramid algorithm* and its reverse is called a *recovering pyramid algorithm*. They are also called *Mallat's algorithms* because they were first developed by S. Mallat (in 1989).

5.3 The Fast Wavelet Transform

We now develop FWT and FIWT. Let $f_j \in V_j$ have the expansion

$$f_j = \sum a_{j,k}\phi_{j,k}, \tag{5.10}$$

which can be represented as

$$f_j(x) = f_{j-1}(x) + g_{j-1}(x),$$

where

$$f_{j-1}(x) = \sum a_{j-1,k}\phi_{j-1,k}(x) \in V_{j-1} \tag{5.11}$$

and

$$g_{j-1}(x) = \sum b_{j-1,k}\psi_{j-1,k}(x) \in W_{j-1}. \tag{5.12}$$

Theorem 7.5.3 Let ϕ be an orthonormal generator of an MRA $\{V_n\}$, which satisfies the two-scale Equation (3.10) and let ψ be the corresponding wavelet defined by (3.11). Let $f_j \in V_j$, $a_{j,k}$, $a_{j-1,k}$, and $b_{j-1,k}$, be the coefficients in (5.10), (5.11), and (5.12), respectively. Then

$$a_{j-1,k} = \sqrt{2} \sum h(l - 2k)a_{j,l} \tag{5.13}$$

and

$$b_{j-1,k} = \sqrt{2} \sum g(l - 2k)a_{j,l} \tag{5.14}$$

where

$$g(k) = (-1)^l h(2m + 1 - k) \quad \text{for some } m \in \mathbb{Z}.$$

Conversely, the coefficient $a_{j,l}$ can be recovered from $\left(a_{j-1,k}\right)_{k\in\mathbb{Z}}$ and $\left(b_{j-1,k}\right)_{k\in\mathbb{Z}}$ by

$$a_{j,l} = \sqrt{2} \left(\sum h(l - 2k)a_{j-1,k} + \sum g(l - 2k)b_{j-1,k} \right). \tag{5.15}$$

Proof: By (5.10), we have

$$\phi_{j-1,k}(x) = 2^{(j-1)/2}\phi(2^{j-1}x - k)$$

$$= 2^{(j+1)/2} \sum_{i\in\mathbb{Z}} h(i)\phi(2(2^{j-1}x - k) - i)$$

$$= 2^{(j+1)/2} \sum_{i\in\mathbb{Z}} h(i)\phi(2^j x - (2k + i))$$

$$= 2^{(j+1)/2} \sum_{i\in\mathbb{Z}} h(i - 2k)\phi(2^j x - i)$$

$$= \sqrt{2} \sum_{i\in\mathbb{Z}} h(i - 2k)\phi_{j,i}(x)$$

and similarly

$$\psi_{j-1,k}(x) = 2^{(j-1)/2}\psi(2^{j-1}x - k)$$

$$= 2^{(j+1)/2} \sum_{i\in\mathbb{Z}} g(i - 2k)\phi(2^j x - i)$$

$$= \sqrt{2} \sum_{i\in\mathbb{Z}} g(i - 2k)\phi_{j,i}(x),$$

where

$$g(i) = (-1)^i h(1 + 2m - i) \quad \text{for some } m \in \mathbb{Z}.$$

Hence we obtain

$$a_{j-1,k} = \langle f_j, \phi_{j-1,k} \rangle = \left\langle \sum_{l \in \mathbb{Z}} a_{j,l}\phi_{j,l}, \phi_{j-1,k} \right\rangle$$

$$= \sum_{l \in \mathbb{Z}} a_{j,l} \langle \phi_{j,l}, \phi_{j-1,k} \rangle.$$

From the orthogonality of the basis $\{\phi_{j,k} \mid k \in \mathbb{Z}\}$, we have

$$\langle \phi_{j,l}, \phi_{j-1,k} \rangle = \sqrt{2} \sum_{i \in \mathbb{Z}} h(i - 2k)\langle \phi_{j,l}, \phi_{j,i} \rangle = \sqrt{2}h(l - 2k), \qquad (5.16)$$

which yields equation (5.13). Similarly, we can obtain

$$\langle \phi_{j,l}, \psi_{j-1,k} \rangle = \sqrt{2} \sum_{i \in \mathbb{Z}} g(i - 2k)\langle \phi_{j,l}, \phi_{j,i} \rangle = \sqrt{2}g(l - 2k), \qquad (5.17)$$

which yields (5.14).

On the other hand, we have

$$a_{j,l} = \langle f_j, \phi_{j,l} \rangle = \langle f_{j-1} + g_{j-1}, \phi_{j,l} \rangle$$

$$= \sum_{k \in \mathbb{Z}} \left(a_{j-1,k}\langle \phi_{j-1,k}, \phi_{j,l} \rangle + b_{j-1,k}\langle \psi_{j-1,k}, \phi_{j,l} \rangle \right).$$

By (5.16) and (5.17), we obtain (5.15).   ∎

The relationships (5.13) and (5.14) together give a *fast wavelet transform* (*FWT*), while relationship (5.15) gives a *fast inverse wavelet transform* (*FIWT*).

5.4 Pyramid Algorithms

Based on FWT and FIWT, we develop the pyramid algorithms, which perform the multilevel wavelet decomposition and reconstruction. We start with the decomposition

$$f_n = f_0 + \sum_{j=0}^{n-1} g_j, \quad f_0 \in V_0, g_j \in W_j,$$

or, equivalently,

$$\sum_{m \in \mathbb{Z}} a_{n,m}\phi_{n,m} = \sum_{m \in \mathbb{Z}} a_{0,m}\phi_m + \sum_{j=0}^{n-1} \sum_{m \in \mathbb{Z}} b_{j,m}\psi_{n,m}. \qquad (5.18)$$

Let $\mathbf{a}_n = (a_{n,m})$ and $\mathbf{a}_0 = (a_{0,m})$, $\mathbf{b}_0 = (b_{0,m})$, $\cdots$, $\mathbf{b}_{n-1} = (b_{n-1,m})$ be the coefficient sequences in (5.18). Our purpose now is to develop the algorithms for decomposing $\mathbf{a}_n$ into $\mathbf{a}_0$, $\mathbf{b}_0$, $\cdots$, $\mathbf{b}_{n-1}$ and recovering $\mathbf{a}_n$ from $\mathbf{a}_0$, $\mathbf{b}_0$, $\cdots$, $\mathbf{b}_{n-1}$. To represent the algorithms concisely, we introduce the following operators.

Let H and G be two $l^2 \mapsto l^2$ operators defined by

$$Ha(n) = \sqrt{2} \sum_{k \in \mathbb{Z}} a_k h(k - 2n),$$

$$Ga(n) = \sqrt{2} \sum_{k \in \mathbb{Z}} a_k g(k - 2n),$$

where $\mathbf{h} = (h(n))$ and $\mathbf{g} = (g(n))$ are the sequences in (3.10) and (3.11) respectively. For an operator $F : l^2 \mapsto l^2$, the *dual operator* F^* is defined by

$$\langle Fa, b \rangle = \langle a, F^* b \rangle, \quad \text{for all } \mathbf{a}, \mathbf{b} \in l^2.$$

Then

$$(H^* a)(n) = \sqrt{2} \sum_{k \in \mathbb{Z}} a_k h(n - 2k)$$

and

$$(G^* a)(n) = \sqrt{2} \sum_{k \in \mathbb{Z}} a_k g(n - 2k),$$

respectively. Therefore, the FWT algorithm can be written as

$$\mathbf{a}_{j-1} = H\mathbf{a}_j, \quad \mathbf{b}_{j-1} = G\mathbf{a}_j, \tag{5.19}$$

and the FIWT algorithm can be written as

$$\mathbf{a}_j = H^* \mathbf{a}_{j-1} + G^* \mathbf{b}_{j-1}. \tag{5.20}$$

Repeating the FWT, $\mathbf{a}_n$ can be decomposed into $\mathbf{a}_0$, $\mathbf{b}_0$, $\cdots$, $\mathbf{b}_{n-1}$ as follows:

$$\mathbf{a}_n \xrightarrow{H} \mathbf{a}_{n-1} \xrightarrow{H} \mathbf{a}_{n-2} \xrightarrow{H} \cdots \xrightarrow{H} \mathbf{a}_0$$

with G branches going down to $\mathbf{b}_{n-1}$, $\mathbf{b}_{n-2}$, $\cdots$, $\mathbf{b}_0$.

This is called a *decomposition pyramid algorithm*, or Mallat's decomposition algorithm.

Reversing it, we obtain the *recovering pyramid algorithm*, or *Mallat's recovering algorithm*:

$$
\mathbf{a}_0 \xrightarrow[G^*]{H^*} \mathbf{a}_1 \xrightarrow[G^*]{H^*} \cdots \xrightarrow[G^*]{H^*} \mathbf{a}_{n-1} \xrightarrow[G^*]{H^*} \mathbf{a}_n
$$

$$
\mathbf{b}_0 \qquad \mathbf{b}_1 \qquad \cdots \qquad \mathbf{b}_{n-1}
$$

5.5 Approximation of Functions by Wavelet Series

Before we finish this section, we briefly discuss the wavelet series of a function in an intuitive way. For illustration, we use spline scaling functions and wavelets as examples. The analysis of these is applied to other wavelets. Let $\{V_j\}$ be the MRA generated by $\tilde{N}_m$ and $\{W_j\}$ be the corresponding wavelet subspace sequence. It is known that $|\widehat{\tilde{N}}_m(\omega)|$ is essentially concentrated in $[-\pi, \pi]$, whereas $|\widehat{\tilde{S}}_m(\omega)|$ is essentially concentrated in $[-2\pi, -\pi] \cup [\pi, 2\pi]$. (See Section 7.4, Exercises 13 and 15.) Therefore, for any function $f_0 \in V_0$, its frequency is essentially concentrated in $[-\pi, \pi]$ since $|\widehat{\phi(\cdot + k)}(\omega)| = |\widehat{\phi}(\omega)|$. Similarly, for any function $g_0 \in W_0$, its frequency is essentially in $[-2\pi, -\pi] \cup [\pi, 2\pi]$. After dilation, we can see that, for any function in W_j, its frequency is essentially in $[-2^{j+1}\pi, 2^j\pi] \cup [2^{j+1}\pi, 2^j\pi]$. Hence, in the decomposition $f = f_0 + g_0 + g_1 + \cdots$, different functions occupy different frequency channels. Therefore, we can say that f_0 is a coarse version of f while $g_0, g_1, \cdots$ hold the details of f at different levels. Then in some sense the "cut-tail" approximation $f_0 + \sum_{j=0}^{n-1} g_j$ removes high frequency components of f, which contain the details of f which can only be seen in high resolution. This approximation is called *linear approximation of wavelets*. Linear approximation of wavelets behave similar to Fourier series.

However, the "local" behaviors of wavelets are quite different from those of sine and cosine functions. For example, both $\tilde{N}_m^c$ and $\tilde{S}_m^c$ are concentrated about their centers. Hence, although the function $\tilde{S}_m^c$ has propagated waves, the amplitude of the waves decay exponentially so that only the main wave around 0 is significant. For $j > 0$, $(\tilde{S}_m^c)_{j,k}(x)$ shrinks $\tilde{S}_m^c(x)$ horizontally by 2^j, stretches it vertically by $2^{\frac{j}{2}}$, and then shifts it to the center $2^{-j}(k + \frac{1}{2})$. That is, for $(\tilde{S}_m^c)_{j,k}$ the amplitude of the main is enlarged while its length is reduced. We now return to discuss the wavelet series of $f \in V_n$ in the form

$$f(x) = f_0(x) + \sum_{j=0}^{n-1} g_j(x)$$

$$= \sum_{k \in \mathbb{Z}} c_{0,k} \left(\tilde{N}_m^c \right)_{0,k} (x) + \sum_{j=0}^{n-1} \sum_{k \in \mathbb{Z}} d_{j,k} \left(\tilde{S}_m^c \right)_{j,k} (x), \qquad (5.21)$$

where a large $|d_{j,k}|$ indicates that f has a large oscillation around $2^{-j}k$. In other words, a large $|d_{j,k}|$ occurs only if f has a sharp change around $2^{-j}k$. In applications, many functions are often smooth in most parts of their domains and have sharp changes in a relatively small part. Therefore, for these functions, most of their wavelet coefficients $d_{j,k}$ are small and negligible; with only a few of them large and significant. These wavelet coefficients indicate the local sharpness of functions, which is often called the *local singularity* of functions (see [18]). In this sense, wavelet analysis is different from Fourier analysis. As Meyer says [21]: "Wavelet analysis is a way of saying that one is sensitive to change." He also says: "contrary to what happens with Fourier series, the coefficients of the wavelet series translate the properties of the function or distribution simply, precisely, and faithfully, ... the properties that correspond to strong transients: everything that is rupture, discontinuity, the unforeseen." Therefore, wavelets are a mathematical microscope, which detects the details of functions at different resolutions.

Based on the foregoing discussion, we can approximate f by deleting the wavelet terms in (5.21) with small coefficients. To do so, we set a threshold $\epsilon > 0$. By deleting the terms with coefficients less than ϵ, we obtain an approximation of f :

$$\tilde{f}(x) = \sum_{c_{0,k} \geq \epsilon} c_{0,k} \left(\tilde{N}_m^c \right)_{0,k} (x) + \sum_{j=0}^{n-1} \sum_{d_{j,k} \geq \epsilon} d_{j,k} \left(\tilde{S}_m^c \right)_{j,k} (x).$$

An alternate way to obtain an approximation of f is the following. We first sort all the terms in (5.21) by descending coefficients. Then we use the sum of the first N terms as an approximation of f. These approximations are *nonlinear approximation of wavelets* (see [18]).

Exercises

1. Prove Theorem 7.5.1: If $\phi \in L^1 \cap L^2$ is continuous, the series $\sum_{k \in \mathbb{Z}} \hat{\phi}(\omega + 2k\pi)$ is uniformly convergent, and $(\phi(k))_{k \in \mathbb{Z}} \in L^1$, then the linear system (5.3) is consistent if and only if $\sum_{k \in \mathbb{Z}} \hat{\phi}(\omega + 2k\pi) \neq 0$, $\omega \in \mathbb{R}$.

2. Prove that when m is even, N_m satisfies the interpolation condition; when m is odd, it does not.

3. Prove Theorem 7.5.2: Let $\phi \in L^1 \cap L^2$ be a continuous two-scaling function, which satisfies the interpolation condition. Then the function $\phi_{in}(x)$ defined by $\hat{\phi}_{in}(\omega) = \dfrac{\hat{\phi}(\omega)}{\sum_{k \in \mathbb{Z}} \hat{\phi}(\omega + 2k\pi)}$

4. Let ϕ be an orthonormal MRA generator. Prove that $\phi * \phi$ is an interpolating scaling function.

5. Let $\{V_n\}$ be the MRA generated by N_4. Let N_{in} be the interpolating scaling function in V_0. Do the following:
   (a) Get the coefficients (α_k) (rounding off to four decimal places) in the representation of $N_{in}(x) = \sum \alpha_k N_4(x - k)$ and draw the graph of $N_{in}(x)$.
   (b) Find the symbol of N_{in} and then derive the two-scale equation of N_{in}.

6. Derive equation (5.14).

7. Derive equation (5.17).

8. Let $H : l^2 \to l^2$ be an operator defined by
$$Ha(n) = \sqrt{2} \sum_{k \in \mathbb{Z}} a_k h(k - 2n).$$
Prove that its dual operator H^* is
$$(H^* a)(n) = \sqrt{2} \sum_{k \in \mathbb{Z}} a_k h(n - 2k).$$

9. Let ϕ be an orthonormal generator of an MRA $\{V_j\}$, and ψ be the corresponding orthonormal wavelet, which generates the wavelet subspaces $W_j, j \in \mathbb{Z}$.
   (a) Let $f \in V_0$ and write
   $$f = \sum c_k^0 \phi_{0k} = \sum c_k^j \phi_{jk}, \quad j > 0.$$
   Develop a formula to compute (c_k^j) from (c_k^0).
   (b) Let $g \in W_0$ and write
   $$g = \sum d_k^0 \psi_{0k} = \sum c_k^j \phi_{jk}, \quad j > 0.$$
   Develop a formula to compute (c_k^j) from (d_k^0).

10. Use a graphing calculator, or mathematical software, such as Maple, Mathematica, or MATLAB, to calculate the following wavelet decompositions. (Round off to four decimal places for each coefficient.)

(a) Decompose $f(x) = \tilde{N}_2^c(4x)$ into the form

$$f(x) = \sum a_k \tilde{N}_2^c(x - k) + \sum b_k \tilde{S}_2^c(x - k) + \sum c_k \tilde{S}_2^c(2x - k).$$

(b) Decompose $f(x) = \tilde{N}_4^c(4x)$ into the form

$$f = \sum a_k \tilde{N}_4^c(x - k) + \sum b_k \tilde{S}_4^c(x - k) + \sum c_k \tilde{S}_4^c(2x - k).$$

11. Let $f(x) = \tilde{N}_2^c(x) + \sum_{j=0}^{3} \tilde{S}_2^c(2^j x)$. Use a graphing calculator, or mathematical software such as Maple, Mathematica, or MATLAB to find the coefficients (a_k) in $f(x) = \sum a_k \tilde{N}_2^c(2^4 x - k)$. (Round off to four decimal places for each coefficient.)

12. Let $\{V_n\}$ be the MRA generated by N_2. Define $f \in V_4$ by

$$f(x) = \begin{cases} 16x, & x \in [0, 1/16), \\ 1, & x \in [1/16, 1], \\ 16\left(\dfrac{17}{16} - x\right), & x \in \left(1, \dfrac{17}{16}\right], \\ 0, & x > \dfrac{17}{16}, \\ f(-x), & x \le 0. \end{cases}$$

Use a graphing calculator, or mathematical software such as Maple, Mathematica, or MATLAB to do the following.

(a) Find the coefficients $(a_{4,k})$ in $f(x) = \sum a_{4,k} (\tilde{N}_2^c)_{4,k}(x)$. (Round off to five decimal places for each coefficient.)

(b) Use the pyramid algorithm to decompose $f(x)$ into the wavelet series

$$f(x) = \sum_k a_{0,k} \left(\tilde{N}_2^c\right)_{0,k}(x) + \sum_{j=0}^{3} \sum_k b_{j,k} \left(\tilde{S}_2^c\right)_{j,k}(x).$$

(Round off to five decimal places for each coefficient.)

(c) Let $\epsilon = 10^{-3}$. Get a nonlinear approximation $\tilde{f}$ of f by removing the coefficients less than ϵ.

(d) Use the pyramid algorithm to recover $\tilde{f}$ and then draw its graph.

(e) Get a nonlinear approximation f_{ap} of f by finding the 8 terms whose coefficients have the largest absolute values.

(f) Use the pyramid algorithm to recover f_{ap} and then draw its graph.

6 Biorthogonal Wavelet Bases

In the previous section, we discussed how to use FWT and FIWT to decompose and recover functions. As we already have shown, the FWT algorithm needs an initial function $f_n \in V_n$ as a starting point. In some cases, the orthonormal basis of V_n is not consistent with the initialization. For example, let the MRA $\{V_n\}$ be generated by the linear spline N_2^c. Assume we are given the sampling data $(f(2^{-n}k))_{k \in \mathbb{Z}}$ of a function f. If we choose $\{N_2^c(2^n x - k)\}_{k \in \mathbb{Z}}$ as a basis of V_n, then the initialization is simply obtained by setting $f_n(x) = \sum f(2^{-n}k)N_2^c(2^n x - k)$. However, if we choose the orthonormal basis $\{(\tilde{N}_2^c)_{n,k}\}$, then the initialization involves a tedious computation for the coefficients $(a_{n,k})$ in the expansion $f_n(x) = \sum a_{n,k}(\tilde{N}_2^c)_{n,k}(x)$. This motivates us to seek a more flexible wavelet structure than orthonormality.

In this direction, we consider FWT and FIWT in a general framework. Assume now the generator ϕ of the MRA $\{V_n\}$ is not orthonormal. Let the initial function f be in $V_{n+1} : f_{n+1} := \sum c_{n+1,k}\phi_{n+1,k}$. FWT will decompose it into $f_{n+1} = f_n + g_n$, where $f_n = \sum c_{n,k}\phi_{n,k}. \in V_n$ and g_n is in a supplemental subspace, say W_n, of V_n (with respect to V_{n+1}). Since $\{\phi_{n,k}\}$ is no longer an orthonormal basis, $c_{n,k} \neq \langle f, \phi_{n,k} \rangle$. To compute $c_{n,k}$, we need another scaling function $\tilde{\phi}$ such that

$$\langle \phi_{n,k}, \tilde{\phi}_{n,l} \rangle = \delta_{k,l}, \quad \langle g_n, \tilde{\phi}_{n,k} \rangle = 0, \quad k \in \mathbb{Z}. \tag{6.1}$$

Then, we have $c_{n,k} = \langle f, \tilde{\phi}_{n,k} \rangle$.

We now discuss the subspace W_n. Let $\{\tilde{V}_n\}$ be the MRA generated by $\tilde{\phi}$. By the second equation in (6.1), we have $W_n \perp \tilde{V}_n$. Recall that $W_n \oplus V_n = V_{n+1}$ and $\{V_n\}$ is an MRA. Hence, there is a wavelet ψ (it may not be orthonormal) such that $(\psi_{n,k})$ forms a Riesz basis of W_n. Thus, the function g_n can be expanded as $g_n = \sum b_{n,k}\psi_{n,k}$. In order to compute $(b_{n,k})$, we need a wavelet $\tilde{\psi} \in \tilde{V}_1$ such that $\langle \psi_{n,k}, \tilde{\psi}_{n,l} \rangle = \delta_{k,l}$, and $\langle \phi_{n,k}, \tilde{\psi}_{n,l} \rangle = 0$. Thus, the wavelet subspace $\tilde{W}_n$ generated by $\tilde{\phi}$ satisfies $\tilde{W}_n \perp V_n$ and $\tilde{V}_n \oplus \tilde{W}_n = \tilde{V}_{n+1}$.

In this new structure, we call ϕ and $\tilde{\phi}$ biorthogonal scaling functions or dual scaling functions, since $\langle \phi_{n,k}, \tilde{\phi}_{n,l} \rangle = \delta_{l,k}$. The MRA $\{V_n\}$ and $\{\tilde{V}_n\}$, which are generated by ϕ and $\tilde{\phi}$, respectively, are called biorthogonal MRA, or dual MRA. Similarly, $\tilde{\psi}$ and ψ are called biorthogonal wavelets, or dual wavelets. In this section, we shall introduce the principle of the construction of biorthogonal scaling functions and wavelets.

6.1 Construction of Biorthogonal Wavelet Bases

We first give the formal definition of biorthogonal scaling functions.

Definition 7.6.1 Assume the scaling function

$$\phi(t) = 2 \sum_{k \in \mathbb{Z}} h(k)\phi(2t - k) \tag{6.2}$$

generates the MRA $\{V_j\}_{j \in \mathbb{Z}}$ and the scaling function

$$\tilde{\phi}(t) = 2 \sum_{k \in \mathbb{Z}} \tilde{h}(k)\tilde{\phi}(2t - k) \tag{6.3}$$

generates the MRA $\{\tilde{V}_j\}_{j \in \mathbb{Z}}$. If

$$\int_{-\infty}^{\infty} \phi(t - n)\tilde{\phi}(t - m)\,dt = \delta_{nm}, \tag{6.4}$$

then ϕ and $\tilde{\phi}$ are called *biorthogonal scaling functions*, or *dual scaling functions*; $\{V_j\}_{j \in \mathbb{Z}}$ and $\{\tilde{V}_j\}_{j \in \mathbb{Z}}$ are called *biorthogonal MRA*.

From (6.4), we can derive

$$\langle \phi_{jn}, \tilde{\phi}_{jm} \rangle = \delta_{nm}, \text{ for all } j \in \mathbb{Z}.$$

Similarly, we give the following.

Definition 7.6.2 Let ψ and $\tilde{\psi}$ be two functions in L^2. If $\{\psi_{jk}\}_{j,k \in \mathbb{Z}}$ and $\{\tilde{\psi}_{jk}\}_{j,k \in \mathbb{Z}}$ are both bases of L^2 and they satisfy

$$\langle \psi_{i,k}, \tilde{\psi}_{j,l} \rangle = \delta_{ij}\delta_{kl}, \quad \text{for all } i, j, k, l \in \mathbb{Z}, \tag{6.5}$$

then ψ and $\tilde{\psi}$ are called *biorthogonal wavelets* and $\{\psi_{jk}\}_{j,k \in \mathbb{Z}}$ and $\{\tilde{\psi}_{jk}\}_{j,k \in \mathbb{Z}}$ are called *biorthogonal wavelet bases*.

Our purpose is to construct biorthogonal wavelets via biorthogonal scaling functions. To this end, we first establish the following.

Theorem 7.6.1 If ϕ and $\tilde{\phi}$ are biorthogonal scaling functions with masks $\mathbf{h} = (h_k)$ and $\tilde{\mathbf{h}} = (\tilde{h}_k)$, respectively, then

$$2 \sum h(k)\tilde{h}(k - 2l) = \delta_{0l},$$

which is equivalent to

$$\overline{H(z)}\tilde{H}(z) + \overline{H(-z)}\tilde{H}(-z) = 1, \quad z \in \Gamma, \tag{6.6}$$

where $H(z) := \sum h(k)z^k$ and $\tilde{H}(z) := \sum \tilde{h}(k)z^k$ are symbols of ϕ and $\tilde{\phi}$, respectively.

Proof: We leave the proof as an exercise. ∎

In a way similar to the construction of orthonormal wavelets via orthonormal scaling functions, we can construct biorthogonal wavelets via biorthogonal scaling functions.

Theorem 7.6.2 Let ϕ and $\tilde{\phi}$ be biorthogonal scaling functions which satisfy the refinement equations (6.2) and (6.3), respectively. Write

$$\begin{aligned} g(k) &= (-1)^k \tilde{h}(1-k), \\ \tilde{g}(k) &= (-1)^k h(1-k). \end{aligned} \tag{6.7}$$

Define

$$\psi(t) = 2\sum_{k\in\mathbb{Z}} g(k)\phi(2x-k), \tag{6.8}$$

and

$$\tilde{\psi}(t) = 2\sum_{k\in\mathbb{Z}} \tilde{g}(k)\tilde{\phi}(2x-k). \tag{6.9}$$

Then ψ and $\tilde{\psi}$ are biorthogonal wavelets.

Proof: Let $\{V_j\}_{j\in\mathbb{Z}}$ and $\{\tilde{V}_j\}_{j\in\mathbb{Z}}$ be the MRA generated by ϕ and $\tilde{\phi}$, respectively. As usual, for a function f we write $f_{j,k} = 2^{j/2}f(2^j x - k)$. For any $j, k, k' \in \mathbb{Z}$, writing $l = k - k'$, we have

$$\int_{-\infty}^{\infty} \phi_{j,k}(t)\tilde{\psi}_{j,k'}(t)\,dt$$

$$= \int_{-\infty}^{\infty} \phi_{0,k-k'}(t)\tilde{\psi}_{0,0}(t)\,dt$$

$$= \int_{-\infty}^{\infty} 4\sum_{m\in\mathbb{Z}} h(m)\phi(2t-2m-l)\sum_{n\in\mathbb{Z}}\tilde{g}(n)\tilde{\phi}(2t-n)\,dt$$

$$= 2 \int_{\mathbb{R}} \left(\sum_{m \in \mathbb{Z}} h(m) \phi_{1,m+2l}(t) \right) \left(\sum_{n \in \mathbb{Z}} (-1)^n h(1-n) \tilde{\phi}_{1,n}(t) \right) dt$$

$$= 2 \sum_{s \in \mathbb{Z}} \sum_{n \in \mathbb{Z}} (-1)^n h(s - 2l) h(1 - n) \int_{\mathbb{R}} \tilde{\phi}_{1n} \phi_{1s} \, dt$$

$$= 2 \sum_{n \in \mathbb{Z}} (-1)^n h(n - 2l) h(1 - n)$$

$$= 2 \sum_{n \in \mathbb{Z}} h(2n - 2l) h(1 - 2n) - \sum_{n \in \mathbb{Z}} h(2n + 1 - 2l) h(-2n). \qquad (6.10)$$

Setting $n = l - m$ in the second sum of (6.10), we get

$$\int_{-\infty}^{\infty} \phi_{j,k}(t) \tilde{\psi}_{j,k'}(t) \, dt = 0, \text{ for all } j, k, k' \in \mathbb{Z}, \qquad (6.11)$$

which implies $\tilde{\psi}_{j,k'} \perp V_j$. In the same way we can prove $\psi_{j,k} \perp \tilde{V}_j$. It is clear that $\psi_{j,k} \in V_{j+1}$ and $\tilde{\psi}_{j,k} \in \tilde{V}_{j+1}$. Hence we have

$$\langle \tilde{\psi}_{j,k}, \psi_{i,k'} \rangle = 0, \quad i \neq j.$$

The proof that $\langle \tilde{\psi}_{j,k}, \psi_{j,k'} \rangle = \delta_{kk'}$ follows the method of the proof of (6.11). Finally, $\tilde{\psi}_{j,k'} \perp V_j$ and $\psi_{j,k} \perp \tilde{V}_j$ (for all $j \in \mathbb{Z}$) imply that $\{\tilde{\psi}_{j,k}\}_{j,k \in \mathbb{Z}}$ and $\{\psi_{j,k}\}_{j,k \in \mathbb{Z}}$ are bases of L^2. We leave the details as exercises.   ∎

The wavelet subspaces generated by ψ and $\tilde{\psi}$ are constructed in the normal way. Let

$$W_j = \text{ span } L_2\{\psi_{j,k} \mid k \in \mathbb{Z}\}, \quad \tilde{W}_j = \text{ span } L_2\{\tilde{\psi}_{j,k} \mid k \in \mathbb{Z}\}. \qquad (6.12)$$

Then they are wavelet subspaces generated by ψ and $\tilde{\psi}$, respectively.

Corollary 7.6.1 Let ψ and $\tilde{\psi}$ be defined as in Theorem 7.6.2. Then

$$V_{j+1} = V_j \oplus W_j, \quad \tilde{V}_{j+1} = \tilde{V}_j \oplus \tilde{W}_j, \quad \text{for all } j \in \mathbb{Z}, \qquad (6.13)$$

where

$$\tilde{W}_j \perp V_j, \quad W_j \perp \tilde{V}_j \quad \text{for all } j \in \mathbb{Z}. \qquad (6.14)$$

Proof: We leave the proof as an exercise.   ∎

Theorem 7.6.2 provides a way to construct biorthogonal wavelets via the biorthogonal scaling functions ϕ and $\tilde{\phi}$. The construction of other biorthogonal wavelets via ϕ and $\tilde{\phi}$ is possible. We say that ψ and $\tilde{\psi}$ are biorthogonal wavelets corresponding to ϕ and $\tilde{\phi}$ if the wavelet subspaces W_j and $\tilde{W}_j$ satisfy (6.13) and (6.14). It is trivial that if ψ and $\tilde{\psi}$ are biorthogonal wavelets corresponding to ϕ and $\tilde{\phi}$, then for any $k \in \mathbb{Z}$, $\psi(x - k)$ and $\tilde{\psi}(x - k)$ are also biorthogonal wavelets corresponding to ϕ and $\tilde{\phi}$. A more general result is the following.

Corollary 7.6.2 Let ψ and $\tilde{\psi}$ be biorthogonal wavelets corresponding to ϕ and $\tilde{\phi}$. Let $A(\omega)$ be a continuous 2π-periodic function such that $A(\omega)e^{-ik\omega} > 0$ for $k \in \mathbb{Z}$ and for all $\omega \in \mathbb{R}$. Let

$$A(\omega) = \sum_{k \in \mathbb{Z}} a_k e^{-ik\omega}, \quad \frac{1}{A(\omega)} = \sum_{k \in \mathbb{Z}} b_k e^{-ik\omega}.$$

Then $\mu := \sum_{k \in \mathbb{Z}} a_k \psi_{0,k}$ and $\tilde{\mu} := \sum_{k \in \mathbb{Z}} b_k \tilde{\psi}_{0,k}$ are also biorthogonal wavelets corresponding to ϕ and $\tilde{\phi}$.

Proof: We leave the proof as an exercise.   ∎

We often need the Fourier transforms of (6.8) and (6.9). Since

$$\hat{\phi}(\omega) = H(e^{-i\omega})\hat{\phi}(\omega/2), \quad \hat{\tilde{\phi}}(\omega) = \tilde{H}(e^{-i\omega})\hat{\tilde{\phi}}(\omega/2),$$

we have

$$\hat{\psi}(\omega) = -e^{-i\omega/2}\overline{\tilde{H}(-e^{-i\omega/2})}\hat{\phi}(\omega/2), \quad \hat{\tilde{\psi}}(\omega) = -e^{-i\omega/2}\overline{H(-e^{-i\omega/2})}\,\hat{\tilde{\phi}}(\omega/2).$$

As we have done for orthonormal wavelets, we write $G(\omega) = -e^{-i\omega}\overline{\tilde{H}(-e^{-i\omega/2})}$ and $\tilde{G}(\omega) = -e^{-i\omega}\overline{H(-e^{-i\omega/2})}$. From (6.13) and (6.14), we have the following.

Theorem 7.6.3 Let ψ and $\tilde{\psi}$ be biorthogonal wavelets defined by (6.8) and (6.9), respectively. Then their masks (and symbols) satisfy the following:

$$2\sum_k g(k)\tilde{g}(k - 2l) = \delta_{0l},$$
$$\sum_k h(k)\tilde{g}(k - 2l) = 0, \qquad k, l \in \mathbb{Z},$$
$$\sum_k \tilde{h}(k)g(k - 2l) = 0,$$

or equivalently,

$$\overline{G(e^{-i\omega})}\tilde{G}(e^{-i\omega}) + \overline{G(-e^{-i\omega})}\tilde{G}(-e^{-i\omega}) = 1,$$
$$\tilde{G}(e^{-i\omega})\overline{H(e^{-i\omega})} + \tilde{G}(-e^{-i\omega})\overline{H(-e^{-i\omega})} = 0,$$
$$\overline{G(e^{-i\omega})}\tilde{H}(e^{-i\omega}) + \overline{G(-e^{-i\omega})}\tilde{H}(-e^{-i\omega}) = 0.$$

Proof: We leave the proof as an exercise. ∎

We now discuss a special case of biorthogonal scaling functions and wavelets. Assume $V_j = \tilde{V}_j$, i.e., ϕ and $\tilde{\phi}$ generate the same MRA. In this case, $W_j = \tilde{W}_j$, $W_j \perp V_j$, and $W_j \oplus V_j = V_{j+1}$. Therefore, the structure of MRA in this case is the same as in the orthonormal case, but ϕ and ψ may no longer be orthonormal. Hence, we give the following.

Definition 7.6.3 Let ϕ and $\tilde{\phi}$ be biorthogonal scaling functions and ψ and $\tilde{\psi}$ be their corresponding biorthogonal wavelets. If ϕ and $\tilde{\phi}$ generate the same MRA, then ϕ and $\tilde{\phi}$ are called *semiorthogonal scaling functions*; ψ and $\tilde{\psi}$ are called *semiorthogonal wavelets*.

EXAMPLE 7.6.1: Let $\phi = N_m(x), m \geq 2$. Then its dual scaling function $\tilde{\phi}$ is determined by $\widehat{\tilde{\phi}}(\omega) = \widehat{\phi}(\omega)/B_m(\omega)$. Hence, the symbol of $\tilde{\phi}$ is $\tilde{H}(e^{-i\omega}) = \frac{B_m(\omega)}{B_m(2\omega)}(\frac{1+e^{i\omega}}{2})^m$. Thus, we have

$$\widehat{\phi}(\omega) = \left(\frac{1+e^{-i\omega/2}}{2}\right)^m \widehat{\phi}(\omega/2), \quad \widehat{\tilde{\phi}}(\omega) = \frac{B_m\omega/2)}{B_m(\omega)}\left(\frac{1+e^{i\omega/2}}{2}\right)^m \widehat{\tilde{\phi}}(\omega/2).$$

By Theorem 7.6.2, their corresponding biorthogonal wavelets ψ and $\tilde{\psi}$ are determined by

$$\widehat{\psi}(\omega) = -\frac{e^{-i\omega/2}B_m(\omega/2+\pi)}{B_m(\omega)}\left(\frac{1-e^{-i\omega/2}}{2}\right)^m \widehat{\phi}(\omega/2) \qquad (6.15)$$

and

$$\widehat{\tilde{\psi}}(\omega) = -e^{-i\omega/2}\left(\frac{1-e^{i\omega/2}}{2}\right)^m \widehat{\tilde{\phi}}(\omega/2). \qquad (6.16)$$

By Corollary 7.6.2, an alternate choice of the biorthogonal wavelets is the pair of ψ and $\tilde{\psi}$ defined by

$$\widehat{\psi}(\omega) = -e^{i\omega/2}B_m(\omega/2+\pi)\left(\frac{1-e^{-i\omega/2}}{2}\right)^m \widehat{\phi}(\omega/2), \qquad (6.17)$$

and

$$\widehat{\tilde{\psi}}(\omega) = -e^{-i\omega/2}\frac{1}{B_m(\omega)}\left(\frac{1-e^{i\omega/2}}{2}\right)^m \widehat{\tilde{\phi}}(\omega/2). \qquad (6.18)$$

Note that ψ in (7.6.1) is compactly supported with supp $\psi = [0, 2m-1]$. We leave the proof of the results in this example as exercises.

6.2 Decomposition and Recovering of Functions

Biorthogonal wavelets can be used to decompose and recover functions in much the same way as orthonormal wavelets. In this subsection, we derive fast wavelet transform (FWT) and the fast inverse wavelet transform (FIWT) algorithms based on biorthogonal scaling functions and wavelets. Let ψ and $\tilde{\psi}$ be the biorthogonal wavelets. Then a function $f \in L^2$ can be expanded as a wavelet series:

$$f = \sum_{j,k\in} \tilde{b}_{j,k}\psi_{j,k} \quad (\text{or } f = \sum_{j,k\in} b_{j,k}\tilde{\psi}_{j,k})$$

where

$$\tilde{b}_{j,k} = \int_{-\infty}^{\infty} f(t)\tilde{\psi}_{j,k}(t)\,dt \quad (b_{j,k} = \int_{-\infty}^{\infty} f(t)\psi_{j,k}(t)\,dt).$$

As mentioned before, to perform fast wavelet transform, we need to initialize functions. Let ϕ and $\tilde{\phi}$ be biorthogonal scaling functions, which generate biorthogonal MRA $\{V_n\}$ and $\{\tilde{V}\}_n$, respectively. Assume ψ and $\tilde{\psi}$ are biorthogonal wavelets obtained from ϕ and $\tilde{\phi}$ via (6.8) and (6.9). Let $f_n \in V_n$ be the initial function for the fast wavelet transform. Following the strategy in Chapter 7 Section 5, we decompose f_n into

$$f_n = f_0 + g_0 + \cdots + g_{n-1} \tag{6.19}$$

where

$$f_j = \sum_k c_{j,k}\phi_{j,k} \in V_j$$

and

$$g_j = \sum_k d_{j,k}\psi_{j,k} \in W_j.$$

We first develop FWT and FIWT for one-level decomposition and recovering. Write $\mathbf{c}_j = (c_{j,k})$ and $\mathbf{d}_j = (d_{j,k})$. Then we can derive $\mathbf{c}_j$ and $\mathbf{d}_j$ from $\mathbf{c}_{j+1}$ using the following:

   FWT algorithm :

$$c_{j,k} = \sqrt{2}\sum_l \tilde{h}(l - 2k)c_{j+1,l},$$

$$d_{j,k} = \sqrt{2}\sum_l \tilde{g}(l - 2k)c_{j+1,l}. \tag{6.20}$$

To recover $\mathbf{c}_{j+1}$ from $\mathbf{c}_j$ and $\mathbf{d}_j$, we use the following FIWT algorithm:

$$c_{j+1,l} = \sqrt{2}\sum_k h(l - 2k)c_{j,k} + \sqrt{2}\sum_k h(l - 2k)d_{j,k}. \tag{6.21}$$

We now generalize Mallat's algorithm for the multilevel decomposition (6.19). This discussion is parallel to that of the previous section. Let $\mathbf{c}_n = (c_{nm})$ be the coefficient sequence of f_n and $\mathbf{c}_0, \mathbf{d}_0, \cdots, \mathbf{d}_{n-1}$ be the coefficient sequences of $f_0, g_0, \cdots, g_{n-1}$, respectively.

Let $H, G, \tilde{H}, \tilde{G}$ be operators on l^2 defined by

$$Ha(n) = \sqrt{2} \sum_{k \in \mathbb{Z}} a_k h(k - 2n),$$

$$Ga(n) = \sqrt{2} \sum_{k \in \mathbb{Z}} a_k g(k - 2n),$$

$$\tilde{H}a(n) = \sqrt{2} \sum_{k \in \mathbb{Z}} a_k \tilde{h}(k - 2n),$$

$$\tilde{G}a(n) = \sqrt{2} \sum_{k \in \mathbb{Z}} a_k \tilde{g}(k - 2n),$$

respectively. Therefore, the FWT algorithm (6.20) can be represented as

$$\mathbf{c}_{j-1} = \tilde{H}\mathbf{c}_j, \quad \mathbf{d}_{j-1} = \tilde{G}\mathbf{d}_j,$$

and the FIWT algorithm (6.21) can be written as

$$\mathbf{c}_j = H^*\mathbf{c}_{j-1} + G^*\mathbf{d}_{j-1}.$$

Finally, the multilevel decomposition can be completed using the following decomposition pyramid algorithm.

$$
\begin{array}{ccccccc}
\mathbf{c}_n & \xrightarrow{\tilde{H}} & \mathbf{c}_{n-1} & \xrightarrow{\tilde{H}} & \mathbf{c}_{n-2} & \xrightarrow{\tilde{H}} \cdots \xrightarrow{\tilde{H}} & \mathbf{c}_0 \\
& \searrow^{\tilde{G}} & & \searrow^{\tilde{G}} & & \searrow^{\tilde{G}} \quad\quad \searrow^{\tilde{G}} & \\
& & \mathbf{d}_{n-1} & & \mathbf{d}_{n-2} & \cdots & \mathbf{d}_0
\end{array}
\tag{6.22}
$$

The corresponding recovering pyramid algorithm is

$$
\begin{array}{ccccccc}
\mathbf{c}_0 & \xrightarrow{H^*} & \mathbf{c}_1 & \xrightarrow{H^*} & \mathbf{c}_2 & \xrightarrow{H^*} \cdots \xrightarrow{H^*} & \mathbf{c}_n \\
& \nearrow^{G^*} & & \nearrow^{G^*} & & \nearrow^{G^*} \quad\quad \nearrow^{G^*} & \\
\mathbf{d}_0 & & \mathbf{d}_1 & & \mathbf{d}_2 & \cdots &
\end{array}
\tag{6.23}
$$

where H^* and G^* are the dual operators of H and G, respectively.

Definition 7.6.4 Let ϕ and $\tilde{\phi}$ be biorthogonal scaling functions with the symbols $H(z)$ and $\tilde{H}(z)$, respectively. Let ψ and $\tilde{\psi}$ be the corresponding biorthogonal wavelets. In algorithms (6.22) and (6.23), ϕ and ψ are called the *analysis scaling function* and *wavelet*, respectively, while $\tilde{\phi}$ and $\tilde{\psi}$ are called the *synthesis scaling function* and *wavelet*, respectively.

By definition, a scaling function is an analysis scaling function or a synthesis scaling function depending on its function in the algorithm. It is obvious that we can interchange the analysis pair and synthesis pair, which results in a new pair of decomposition and recovering algorithms.

Exercises

1. Prove Theorem 7.6.1: If $\phi \in L^1 \cap L^2$ is continuous, the series $\sum_{k \in \mathbb{Z}} \hat{\phi}(\omega + 2k\pi)$ is uniformly convergent, and $(\phi(k))_{k \in \mathbb{Z}} \in l^1$, then the linear system (5.3) is consistent if and only if $\sum_{k \in \mathbb{Z}} \hat{\phi}(\omega + 2k\pi) \neq 0, \quad \omega \in \mathbb{R}$.

2. Prove that two scaling functions $\phi \in L^1 \cap L^2$ and $\tilde{\phi} \in L^1 \cap L^2$ are biorthogonal if and only if $\sum \hat{\phi}(\omega + 2k\pi)\overline{\hat{\tilde{\phi}}(\omega + 2k\pi)} = 1, \omega \in \mathbb{R}$.

3. Let ψ and $\tilde{\psi}$ be defined as in Theorem 7.6.2. Prove
   (a) $\langle \psi_{j,k}, \tilde{\psi}_{j,k'} \rangle = \delta_{k,k'}$.
   (b) $\{\psi_{j,k}\}$ and $\{\tilde{\psi}_{j,k}\}$ both are bases of L^2.

4. Prove Corollary 7.6.1: Let ψ and $\tilde{\psi}$ be defined as in Theorem 7.6.2. Then

$$V_{j+1} = V_j \oplus W_j, \quad \tilde{V}_{j+1} = \tilde{V}_j \oplus \tilde{W}_j, \quad \text{for all } j \in \mathbb{Z},$$

   where $\tilde{W}_j \perp V_j$ and $W_j \perp \tilde{V}_j$, for all $j \in \mathbb{Z}$.

5. Prove Theorem 7.6.3: Let ψ and $\tilde{\psi}$ be biorthogonal wavelets defined by (6.8) and (6.9), respectively. Then their masks (and symbols) satisfy the following:

$$2 \sum_k g(k)\tilde{g}(k - 2l) = \delta_{0l},$$
$$\sum_k h(k)\tilde{g}(k - 2l) = 0, \quad k, l \in \mathbb{Z},$$
$$\sum_k \tilde{h}(k)g(k - 2l) = 0,$$

or equivalently,

$$\overline{G(e^{-i\omega})}\tilde{G}(e^{-i\omega}) + \overline{G(-e^{-i\omega})}\tilde{G}(-e^{-i\omega}) = 1,$$
$$\tilde{G}(e^{-i\omega})\overline{H(e^{-i\omega})} + \tilde{G}(-e^{-i\omega})\overline{H(-e^{-i\omega})} = 0,$$
$$\overline{G(e^{-i\omega})}\tilde{H}(e^{-i\omega}) + \overline{G(-e^{-i\omega})}\tilde{H}(-e^{-i\omega}) = 0.$$

6. Let $\phi = N_m(x)$, $m \geq 2$ and $\widehat{\tilde{\phi}}(\omega) = \overline{\hat{\phi}(\omega)}/B_m(\omega)$. (See Example 7.6.1.)
   (a) Prove the scaling functions ϕ and $\tilde{\phi}$ are biorthogonal scaling functions.
   (b) Obtain the mask of $\tilde{\phi}$.
7. Let ψ and $\tilde{\psi}$ be the corresponding biorthogonal wavelets of ϕ and $\tilde{\phi}$. (See Example 7.6.1.)
   (a) Prove

   $$\hat{\psi}(\omega) = -\frac{e^{-i\omega/2}B_m(\omega/2 + \pi)}{B_m(\omega)}\left(\frac{1 - e^{-i\omega/2}}{2}\right)^m \hat{\phi}(\omega/2)$$

   and $\widehat{\tilde{\psi}}(\omega) = -e^{-i\omega/2}(\frac{1-e^{i\omega/2}}{2})^m\widehat{\tilde{\phi}}(\omega/2)$.
   (b) Find the sequences (α_k) and (β_k) in

   $$\psi(x) = 2\sum \alpha_k \phi(2k - x)$$

   and in

   $$\tilde{\psi}(x) = 2\sum \beta_k \tilde{\phi}(2k - x).$$

8. Let L_{2m} be the interpolating cardinal spline of order $2m$. Let $\phi = N_m$ be the cardinal B-spline of order m. Let $\{V_n\}$ be the MRA generated by ϕ and $\{W_n\}$ be the corresponding wavelet subspaces.
   (a) Prove that the mth derivative of L_{2m}, $\psi(x) := L_{2m}^{(m)}(x) \in W_0$.
   (b) Find $\tilde{\phi} \in V_0$ and $\tilde{\psi} \in W_0$ such that ϕ and $\tilde{\phi}$ are semiorthogonal scaling functions and ψ and $\tilde{\psi}$ are the corresponding semiorthogonal wavelets.

9. Let ϕ and $\tilde{\phi}$ be biorthogonal generators of biorthogonal MRAs $\{V_j\}$ and $\{\tilde{V}_j\}$, respectively. Assume they satisfy (6.2) and (6.3), respectively. Let ψ and $\tilde{\psi}$ be the corresponding biorthogonal wavelets defined by (6.8) and (6.9), respectively.
   (a) Let $f \in V_0$ and write

   $$f = \sum c_k^0 \phi_{0k} = \sum c_k^j \phi_{jk}, \quad j > 0.$$

   Develop a formula to compute (c_k^j) from (c_k^0).
   (b) Let $g \in W_0$ and write

   $$g = \sum d_k^0 \psi_{0k} = \sum c_k^j \phi_{jk}, \quad j > 0.$$

   Develop a formula to compute (c_k^j) from (d_k^0).

Chapter 8

Compactly Supported Wavelets

In the previous chapter, we introduced multiresolution analysis and derived a method to construct orthonormal wavelets via orthonormal scaling functions. The orthonormal scaling functions were obtained by orthonormalizing existing scaling functions. This approach leads to the construction of the orthonormal spline scaling functions. Note that the masks of an orthonormal spline scaling function form an infinite sequence. Hence, the corresponding fast wavelet transforms (FWT) and fast inverse wavelet transforms (FIWT) require the computation of infinite sums, which will cause truncation errors. From the point of view of numerical computation, the shorter the mask, the faster the FWT and FIWT algorithms. In this chapter, we introduce compactly supported orthonormal scaling functions and wavelets which have finite masks. We will also briefly introduce orthonormal wavelet packets and compactly supported biorthogonal scaling functions and wavelets. All of them provide effective FWT and FIWT algorithms.

I Symbols of Orthonormal Scaling Functions

In the previous chapter, we introduced the orthonormalization approach to orthonormal scaling functions. That method usually leads to the orthonormal scaling functions with infinite masks. While seeking orthonormal scaling

functions with finite masks, Ingrid Daubechies introduced the mask (or symbol) approach in 1989 (see [6]). In this section, we first discuss the structure of
the symbol of an orthonormal scaling function.

Let $\phi \in L^1(\mathbb{R})$ be an MRA generator with a finite mask, i.e., ϕ almost everywhere satisfies a two-scaling equation

$$\phi(x) = 2 \sum_{k=M}^{N} h_k \phi(2x - k), \quad M < N.$$

Later we will assume $M = 0$. Otherwise, writing $\tilde{\phi}(x) = \phi(x + M)$, we have

$$\tilde{\phi}(x) = \phi(x + M) = 2 \sum_{k=M}^{N} h_k \, \phi(2x - k + 2M)$$

$$= 2 \sum_{k=M}^{N} h_k \, \tilde{\phi}(2x - (k - M)) = 2 \sum_{k=0}^{N-M} h_k \tilde{\phi}(2x - k).$$

Then the mask of $\tilde{\phi}$ starts from 0. Since $\tilde{\phi}$ is an integer translate of ϕ, $\tilde{\phi}$, and ϕ
generate the same MRA. Therefore, ϕ can be replaced by $\tilde{\phi}$. We now study the
scaling function ϕ given by

$$\phi(x) = 2 \sum_{k=0}^{N} h_k \phi(2x - k), \quad N > 0, \tag{1.1}$$

where we always assume $h_0 h_N \neq 0$. If ϕ is continuous, then Equation (1.1)
holds everywhere. For simplicity, we agree that (1.1) holds everywhere for
continuous functions and holds almost everywhere for others. Recall that if
ϕ generates an MRA, then $\hat{\phi}(0) \neq 0$. Hence, we always assume the function ϕ
satisfies the normalization condition $\hat{\phi}(0) = 1$.

The Fourier transform of (1.1) is

$$\hat{\phi}(\omega) = H(e^{-i\omega/2})\hat{\phi}(\omega/2), \tag{1.2}$$

where $H(e^{-i\omega}) = \sum_{k=0}^{N} h_k e^{-ik\omega}$ is the symbol of ϕ.

1.1 Basic Properties of the Mask

The first important relation between the mask **h** and the scaling function ϕ is
the following.

Theorem 8.1.1 Assume a scaling function $\phi \in L^2$ satisfies (1.1). Then supp $f \subset$
$[0, N]$.

We skip the proof of the theorem, for it requires the Paley–Wiener–Schwarz Theorem (see [29]), which is beyond the contents of this text. Readers can find the proof in [6]. By Theorem 8.1.1 and Theorem 6.5.4, $\hat{\phi}$ is an entire function. Hence, $\hat{\phi}$ is continuous. From (1.1), we have

$$\hat{\phi}(0) = \int_{\mathbb{R}} \phi(x)\,dx = 2\sum_{k=0}^{N} h_k \int_{\mathbb{R}} \phi(2x - k)\,dx = \left(\sum_{k=0}^{N} h_k\right)\hat{\phi}(0),$$

which yields the following.

Lemma 8.1.1 If the scaling function ϕ in (1.1) is in L^2 and $\hat{\phi}(0) \neq 0$, then $H(1) = 1$, i.e.,

$$\sum_{k=0}^{N} h_k = 1. \tag{1.3}$$

Lemma 8.1.2 If ϕ in (1.1) is an MRA generator, then

$$\sum_{k} h_{2k} = \sum_{k} h_{2k+1} = \frac{1}{2}. \tag{1.4}$$

Proof: We leave the proof as an exercise. ∎

Definition 8.1.1 In (1.1), if (1.4) holds, then we say that ϕ (or the mask h) satisfies the *sum rule*.

The sum rule ensures a unit partition of the integer translates of ϕ.

Theorem 8.1.2 If a scaling function $\phi \in L^2$ in (1.1) satisfies the sum rule, then
  (1) $\hat{\phi}(2k\pi) = \delta_{0,k}$ $k \in \mathbb{Z}$;
  (2) the shifts of ϕ form a unit partition, i.e.,
$$\sum_{k \in \mathbb{Z}} \phi(x - k) = 1 \quad \text{a.e.}$$
In particular, if ϕ is continuous, then $\sum_{k \in \mathbb{Z}} \phi(k) = 1$.

Proof: We first prove (1). $\hat{\phi}(0) = 1$ is the assumption for ϕ. We use mathematical induction to prove $\hat{\phi}(2k\pi) = 0$, for $k \in \mathbb{N}$. When $k = 1$, we have
$$\hat{\phi}(2\pi) = H(e^{-i\pi})\hat{\phi}(\pi) = H(-1)\hat{\phi}(\pi) = 0.$$

Hence, $\hat{\phi}(2k\pi) = 0$ is true for $k = 1$. Assume now $\hat{\phi}(2k\pi) = 0$ for all $1 \leq k \leq n$. We will prove it is true for $k = n + 1$. We have

$$\hat{\phi}(2(n + 1)\pi) = H(e^{-i(n+1)\pi})\hat{\phi}((n + 1)\pi).$$

If n is even, then $H(e^{-i(n+1)\pi}) = H(-1) = 0$, which yields $\hat{\phi}(2(n + 1)\pi) = 0$. If n is odd, then there is a positive integer m such that $m = \frac{n+1}{2}(\leq n)$. By the induction assumption, $\hat{\phi}(2m\pi) = 0$, which yields $\hat{\phi}(2(n + 1)\pi) = 0$. We have already seen that $\hat{\phi}(2k\pi) = 0$, for all $k \in \mathbb{N}$. The proof that $\hat{\phi}(-2k\pi) = 0$, for all $k \in \mathbb{N}$, is similar.

We now prove (2). Note that $f(x) = \sum_{k \in \mathbb{Z}} \phi(x - k)$ is a 1-periodic function in $L^2_{[0,1]}$. We have

$$\int_0^1 f(x)e^{-i2\pi mx}\, dx = \int_0^1 \sum_{k \in \mathbb{Z}} \phi(x - k)e^{-i2\pi mx}\, dx$$

$$= \int_{-\infty}^{\infty} \phi(x)e^{-i2\pi mx}\, dx = \hat{\phi}(2\pi m) = \delta_{0,m}.$$

Hence $f(x) = 1$ a.e. The proof is complete.   ∎

1.2 The Symbol of an Orthonormal Scaling Function

We denote the symbol of an orthonormal scaling function by $m(z)$. As seen in Section 7.3, $m(z)$ is a conjugate mirror filter. Since $m(z)$ is a polynomial, we have (see Exercise 4)

$$|m(z)|^2 + |m(-z)|^2 = 1, \quad z \in \mathbb{C} \setminus \{0\}. \tag{1.5}$$

By Corollary 8.1.1, $m(z)$ must have $(\frac{1+z}{2})$ as a factor. Hence, we assume $m(z)$ has the form

$$m(z) = \left(\frac{1 + z}{2}\right)^L q(z), \quad L \geq 1, \tag{1.6}$$

where $q(z)$ is a polynomial with $q(1) = 1$ and $q(-1) \neq 0$. Since the mask (h_k) is real,

$$|m(z)|^2 = m(z)m(1/z), \quad z \in \mathbb{C} \setminus \{0\}.$$

Write $P(z) = |m(z)|^2$. By (1.5), we have

$$P(z) + P(-z) = 1, \quad z \in \mathbb{C} \setminus \{0\}. \tag{1.7}$$

Applying (1.6), we have

$$P(e^{-i\omega}) = \left(\frac{1 + e^{-i\omega}}{2}\right)^L \left(\frac{1 + e^{i\omega}}{2}\right)^L q(e^{-i\omega})q(e^{i\omega})$$
$$= \left(\cos^2\left(\frac{\omega}{2}\right)\right)^L q(e^{-i\omega})q(e^{i\omega}).$$

Note that $q(e^{-i\omega})q(e^{i\omega})$ is a polynomial of $\cos(\omega)$. Setting $y = \sin^2(\frac{\omega}{2})$, we have

$$\cos^2\left(\frac{\omega}{2}\right) = 1 - y, \quad \cos(\omega) = 1 - 2y.$$

Hence, there is a polynomial $B(y)$ such that $B(y) = q(e^{-i\omega})q(e^{i\omega})$. We now have

$$P(e^{-i\omega}) = (1 - y)^L B(y) \tag{1.8}$$

and

$$P(-e^{-i\omega}) = y^L B(1 - y).$$

Then equality (1.7) becomes

$$(1 - y)^L B(y) + y^L B(1 - y) = 1. \tag{1.9}$$

The polynomial $B(y)$ is given in the following lemma.

Lemma 8.1.3 The polynomial $B(y)$ in (1.9) has the form

$$B(y) = B_L(y) + y^L R\left(\frac{1}{2} - y\right),$$

where

$$B_L(y) = \sum_{k=0}^{L-1} \binom{L+k-1}{k} y^k \tag{1.10}$$

and $R(z)$ is an odd polynomial, chosen such that $B(y) \geq 0$ for $y \in [0, 1]$.

Proof: It is clear that

$$B(y) = q(e^{-i\omega})q(e^{i\omega}) = |q(e^{-i\omega})|^2 \geq 0.$$

Since $(1 - y)^L$ and y^L have no common roots, by Bézout's Theorem (see Exercise 5), there exist two unique polynomials r_1 and r_2 of degree $L - 1$ such that

$$(1 - y)^L r_1(y) + y^L r_2(y) = 1.$$

Substituting $1 - y$ for y, we have

$$(1 - y)^L r_2(1 - y) + y^L r_1(1 - y) = 1.$$

The uniqueness of r_1 and r_2 implies $r_1(y) = r_2(1 - y)$ and therefore

$$(1 - y)^L r_1(y) + y^L r_1(1 - y) = 1.$$

Hence

$$r_1(y) = (1 - y)^{-L}[1 - y^L r_1(1 - y)].$$

Applying the Taylor expansion of $(1 - y)^{-L}$ (see Exercise 6), we have

$$(1 - y)^{-L} = B_L(y) + O(y^L),$$

where $O(y^L)$ stands for a general power series of the form $y^L \sum_{k=0}^{\infty} a_k y^k$. It follows that the Taylor series of $r_1(y)$ has the form

$$\begin{aligned}
r_1(y) &= (B_L(y) + O(y^L)) \left[1 - y^L r_1(1 - y)\right] \\
&= (B_L(y) + O(y^L)) (1 - O(y^L)) \\
&= B_L(y) + O(y^L).
\end{aligned}$$

Since $r_1(z)$ is a polynomial of degree $L - 1$, we obtain $r_1(y) = B_L(y)$. Let $B(y)$ be any polynomial of degree $\geq L$ which satisfies (1.9). Then

$$(1 - y)^L[B(y) - B_L(y)] + y^L[B(1 - y) - B_L(1 - y)] = 0,$$

which implies that y^L is a factor of $B(y) - B_L(y)$. Hence, there is a polynomial $A(y)$ such that

$$B(y) - B_L(y) = y^L A(y).$$

Then

$$\begin{aligned}
&(1 - y)^L[B(y) - B_L(y)] + y^L[B(1 - y) - B_L(1 - y)] \\
&= (1 - y)^L y^L (A(y) + A(1 - y)) = 0,
\end{aligned}$$

which leads to

$$A(y) + A(1 - y) = 0,$$

i.e., $A(y)$ is antisymmetric with respect to $\frac{1}{2}$. The lemma is proved. ∎

We now return to a discussion of the symbol $m(z)$. By (1.8), we have

$$P(e^{-i\omega}) = (1 - y)^L B_L(y) + (1 - y)^L y^L A(y),$$

where $A(y) = -A(1 - y)$. If we choose $A(y) = 0$, we get $P(e^{-i\omega}) = (1 - y)^L B_L(y)$. Later we only discuss $m_L(z)$ such that

$$|m_L(z)|^2 = (1 - y)^L B_L(y).$$

For convenience, we write $z = e^{-i\omega}$ and $P_L(z) = (1 - y)^L B_L(y)$.

Since $y = \sin^2(\frac{\omega}{2})$, we have $1 - y = \left(\frac{1+z}{2}\right)\left(\frac{1+z^{-1}}{2}\right)$ and $y = \left(\frac{1-z}{2}\right)\left(\frac{1-z^{-1}}{2}\right)$. Hence,

$$P_L(z) = \left(\frac{1+z}{2}\right)^L \left(\frac{1+z^{-1}}{2}\right)^L M_L(z), \tag{1.11}$$

where

$$M_L(z) = \sum_{k=0}^{L-1} \binom{L+k-1}{k} \left(\frac{1-z}{2}\right)^k \left(\frac{1-z^{-1}}{2}\right)^k. \tag{1.12}$$

Then

$$m_L(z) = \left(\frac{1+z}{2}\right)^L q_L(z), \tag{1.13}$$

where $|q_L(z)|^2 = M_L(z)$.

Lemma 8.1.4 $M_L(z)$ satisfies the following condition:

$$1 \le M_L(e^{-i\omega}) < 2^{2L-2}, \quad \omega \in \mathbb{R}. \tag{1.14}$$

Proof: We have $\min_{|z|=1} M_L(z) = M_L(1) = 1$ and

$$\max_{|z|=1} |M_L(z)| = M_L(-1) = \sum_{k=0}^{L-1} \binom{L+k-1}{k}$$

$$= \binom{2L-1}{L-1} = \frac{1}{2}\left\{ \binom{2L-1}{L-1} + \binom{2L-1}{L} \right\}$$

$$< \frac{1}{2} \sum_{k=0}^{2L-1} \binom{2L-1}{k} = 2^{2L-2}.$$

The lemma is proved. ■

All $M_L(z)$ are *Laurent polynomials*. $M_L(z), 1 \le L \le 6$, are listed in the following.

$$M_1(z) = 1,$$

$$M_2(z) = -\frac{1}{2}z^{-1} + 2 - \frac{1}{2}z,$$

$$M_3(z) = \frac{3}{8}z^{-2} - \frac{9}{4}z^{-1} + \frac{19}{4} - \frac{9}{4}z + \frac{3}{8}z^2,$$

$$M_4(z) = -\frac{5}{16}z^{-3} + \frac{5}{2}z^{-2} - \frac{131}{16}z^{-1} + 13 - \frac{131}{16}z + \frac{5}{2}z^2 - \frac{5}{16}z^3,$$

$$M_5(z) = \frac{35}{128}z^{-4} - \frac{175}{64}z^{-3} + \frac{95}{8}z^{-2} - \frac{1825}{64}z^{-1} + \frac{2509}{64}$$
$$- \frac{1825}{64}z^1 + \frac{95}{8}z^2 - \frac{175}{64}z^3 + \frac{35}{128}z^4,$$

$$M_6(z) = -\frac{63}{256}z^{-5} + \frac{189}{64}z^{-4} - \frac{4123}{256}z^{-3} + \frac{833}{16}z^{-2} - \frac{13555}{128}z^{-1}$$
$$+ \frac{4335}{32} - \frac{13555}{128}z + \frac{833}{16}z^2 - \frac{4123}{256}z^3 + \frac{189}{64}z^4 - \frac{63}{256}z^5.$$

To find $q_L(z)$ by $q_L(z)q_L(1/z) = M_L(z)$, we need the following Riesz Lemma.

Lemma 8.1.5 [Riesz Lemma] Assume a real Laurent polynomial $R(z)$ satisfies $R(z) = R(1/z)$ and $R(z) \geq 0$, for all $z \in \Gamma$. Then there is a real polynomial $c(z)$ such that

$$R(z) = c(z)c(1/z), \quad z \in \mathbb{C} \setminus \{0\}.$$

Proof: We write $R(z) = \sum_{k=-N}^{N} a_k z^k$, where all $a_k \in \mathbb{R}$. $R(z)$ has $2N$ complex zeros. Since $R(z)$ is a real Laurent polynomial, if $z_0 \notin \mathbb{R}$ is a zero of $R(z)$, so is $\overline{z_0}$. Since $R(z) = R(1/z)$, if $z_0 \neq \pm 1$ is a zero of $R(z)$, then so is $1/z_0$. Because $R(z) \geq 0$ on Γ, any zero of $R(z)$ on Γ must have even multiplicity (Exercise 7). Hence,

$$R(z) = a_N z^{-N} P_1(z) P_2(z) \tag{1.15}$$

where

$$P_1(z) = \prod_{i=1}^{M} (z - z_i)(z - \overline{z_i})\left(z - z_i^{-1}\right)\left(z - \overline{z_i}^{-1}\right), \quad z_i \notin \Gamma,$$

and

$$P_2(z) = \prod_{j=1}^{J} (z - e^{i\omega_j})^2 (z - e^{-i\omega_j})^2 \prod_{k=1}^{K} (z - r_k)(z - r_k^{-1}), \quad \omega_j, r_k \in \mathbb{R}.$$

Recall that, for $z \in \Gamma$,

$$\left| (z - z_i)(z - \overline{z_i}^{-1}) \right| = |z_i|^{-1} |z - z_i|^2$$

and

$$\left| (z - r_k)(z - r_k^{-1}) \right| = |r_k|^{-1} |z - r_k|^2.$$

Consequently, for $z \in \Gamma$, we have

$$R(z) = |R(z)|$$

$$= C^2 \left| \prod_{i=1}^{M} (z - z_i)(z - \overline{z_i}) \right|^2 \left| \prod_{j=1}^{J} (z - e^{i\omega_j})(z - e^{-i\omega_j}) \right|^2 \left| \prod_{k=1}^{K} (z - r_k) \right|^2,$$

where $C = \sqrt{|a_N| \prod_{i=1}^{M} |z_i|^{-2} \prod_{k=1}^{K} |r_k|^{-1}}$. Define

$$c(z) = C \prod_{i=1}^{M} (z - z_i)(z - \overline{z_i}) \prod_{j=1}^{J} (z - e^{i\omega_j})(z - e^{-i\omega_j}) \prod_{k=1}^{K} (z - r_k). \qquad (1.16)$$

Then $c(z)$ is a real polynomial such that

$$R(z) = |c(z)|^2, \quad \text{for all } z \in \Gamma.$$

Hence,

$$R(z) = c(z)c(1/z), \quad \text{for all } z \in \mathbb{C} \setminus \{0\}.$$

The proof is completed. ∎

The following corollary is a direct consequence of Lemma 8.1.5.

Corollary 8.1.1 There exists a real polynomial $q_L(z)$ with $q_L(1) = 1$ and $q_L(e^{-i\omega}) \neq 0$, for all $\omega \in \mathbb{R}$, such that

$$q_L(z)q_L(1/z) = M_L(z).$$

Proof: By Lemma 8.1.5, there is a polynomial $q_L(z)$ such that $q_L(z)q_L(1/z) = M_L(z)$. Recall $q_L^2(1) = M_L(1) = 1$ and $M_L(e^{-i\omega}) > 0, \omega \in \mathbb{R}$. Thus we can select $q_L(z)$ so that $q_L(1) = 1$ and $q_L(e^{-i\omega}) \neq 0, \omega \in \mathbb{R}$. ∎

We have derived the structure of the symbol of an orthonormal scaling function.

Theorem 8.1.3 Let $m_L(z) = (\frac{1+z}{2})^L q_L(z)$. Then $m_L(z)$ is a conjugate mirror filter.

Proof: The proof is straightforward. We leave it as an exercise. ∎

Note that $\deg M_L = 2(L-1)$, which has $L-1$ different roots inside the unit circle Γ, say $z_1, \cdots, z_{L-1}$, and has another $L-1$ roots, $1/\bar{z}_1, \cdots, 1/\bar{z}_{L-1}$, outside the circle. As in the proof of Lemma 8.1.5, we can freely select one of the pairs $(z_i, 1/\bar{z}_i)$, $1 \le i \le L-1$, to construct $q_L(z)$ using formula (1.16). Thus, there are 2^{L-1} different choices for $q_L(z)$. Hence, there are 2^{L-1} different conjugate mirror filters (of degree $2L-1$) in the form $m_L(z) = (\frac{1+z}{2})^L q_L(z)$.

Definition 8.1.2 Let $m_L(z) = (\frac{1+z}{2})^L q_L(z)$ be a conjugate mirror filter of degree $2L-1$. If the polynomial $q_L(z)$ is selected such that the magnitudes of all of its roots are ≥ 1, then it is called the *Daubechies filter* of order L.

EXAMPLE 8.1.1: Let $L = 2$. Then

$$M_L(z) = -\frac{1}{2}z^{-1} + 2 - \frac{1}{2}z = -\frac{z^{-1}}{2}(z - (2 - \sqrt{3}))(z - (2 + \sqrt{3})).$$

If we choose the zero of $q_2(z)$ with the larger magnitude, then we have

$$q_2(z) = -\frac{\sqrt{3}-1}{2}z + \frac{\sqrt{3}+1}{2},$$

and consequently,

$$m_L(z) = \left(\frac{1+z}{2}\right)^2 q_2(z)$$

$$= \frac{1+\sqrt{3}}{8} + \frac{3+\sqrt{3}}{8}z + \frac{3-\sqrt{3}}{8}z^2 + \frac{1-\sqrt{3}}{8}z^3,$$

i.e.,

$$h_2(0) = \frac{1+\sqrt{3}}{8}, h_2(1) = \frac{3+\sqrt{3}}{8}, h_2(2) = \frac{3-\sqrt{3}}{8}, h_3(z) = \frac{1-\sqrt{3}}{8}, \qquad (1.17)$$

which is the Daubechies filter of order 2.

Tables 8.1 and 8.2 list Daubechies filters of order 2 to 10. There the values $\sqrt{2}h_L(k)$ are given because they are used in FWT. (See Section 7.6.)

Table 8.1 Daubechies filters of order 2 to 8.

	k	$\sqrt{2}h_L(k)$		k	$\sqrt{2}h_L(k)$
$L = 2$	0	.482962912145	$L = 7$	0	.077852054085
	1	.836516303738		1	.396539319482
	2	.224143868042		2	.729132090846
	3	$-$.129409522551		3	.469782287405
$L = 3$	0	.332670552950		4	$-$.143906003929
	1	.806891509311		5	$-$.224036184994
	2	.459877502118		6	.071309219267
	3	$-$.135011020010		7	.080612609151
	4	$-$.085441273882		8	$-$.038029936935
	5	.035226291882		9	$-$.016574541631
$L = 4$	0	.230377813309		10	.012550998556
	1	.714846570553		11	.000429577973
	2	.630880767930		12	$-$.001801640704
	3	$-$.027983769417		13	.000353713800
	4	$-$.187034811719	$L = 8$	0	.054415842243
	5	.030841381836		1	.312871590914
	6	.032883011667		2	.675630736297
	7	$-$.010597401785		3	.585354683654
$L = 5$	0	.160102397974		4	$-$.015829105256
	1	.603829269797		5	$-$.284015542962
	2	.724308528438		6	.000472484574
	3	.138428145901		7	.128747426620
	4	$-$.242294887066		8	$-$.017369301002
	5	$-$.032244869585		9	$-$.044088253931
	6	.077571493840		10	.013981027917
	7	$-$.006241490213		11	.008746094047
	8	$-$.012580751999		12	$-$.004870352993
	9	.003335725285		13	$-$.000391740373
$L = 6$	0	.111540743350		14	.000675449406
	1	.464623890398		15	$-$.000117476784
	2	.751133908021			
	3	.315250351709			
	4	$-$.226264693965			
	5	$-$.129766867567			
	6	.097501605587			
	7	.027522865530			
	8	$-$.031562039317			
	9	.000553842201			
	10	.004777257511			
	11	$-$.001077301085			

Table 8.2 Daubechies filters of order 9 and 10.

	k	$\sqrt{2}h_L(k)$		k	$\sqrt{2}h_L(k)$
$L = 9$	0	.038077947363	$L = 10$	0	.026670057901
	1	.243834674637		1	.188176800078
	2	.604823123676		2	.527201188931
	3	.657288078036		3	.688459039453
	4	.133197385822		4	.281172343660
	5	−.293273783272		5	−.249846424326
	6	−.096840783220		6	−.195946274377
	7	.148540749334		7	.127369340336
	8	.030725681478		8	.093057364604
	9	−.067632829060		9	−.071394147166
	10	.000250947115		10	−.029457536822
	11	.022361662123		11	.033212674059
	12	−.004723204758		12	.003606553567
	13	−.004281503682		13	−.010733175483
	14	.001847646883		14	.001395351747
	15	.000230385764		15	.001992405295
	16	−.000251963189		16	−.000685856695
	17	.000039347320		17	−.000116466855
				18	.000093588670
				19	−.000013264203

Exercises

1. Prove the following: Let $\phi \in L^2$ be a scaling function satisfying (1.1). Let $B(e^{-i\omega}) = \sum_{k \in \mathbb{Z}} |\hat{\phi}(\omega + 2k\pi)|^2$. Then

$$|H(e^{-i\omega})|^2 B(e^{-i\omega}) + |H(-e^{-i\omega})|^2 B(-e^{i\omega}) = B(e^{-i2\omega}), \quad \omega \in \mathbb{R},$$

which is equivalent to

$$|H(z)|^2 B(z) + |H(-z)|^2 B(-z) = B(z^2), \quad |z| = 1.$$

2. Use the result in Exercise 1 to prove Lemma 8.1.2: If ϕ in (1.1) is an MRA generator, then $\sum h_{2k} = \sum h_{2k+1} = \frac{1}{2}$.
3. Let $q(z)$ be the polynomial in (1.6). Prove that $q(e^{i\omega})q(e^{-i\omega})$ is a polynomial of $\cos \omega$.
4. Let $P(e^{-i\omega})$ be a trigonometric polynomial. Assume $P(e^{-i\omega}) + P(-e^{-i\omega})$ = 1 for all $\omega \in \mathbb{R}$. Prove

$$P(z) + P(-z) = 1, \quad \text{for all } z \in \mathbb{C} \setminus \{0\}.$$

5. Prove Bézout's Theorem: Let P and Q be two polynomials such that $\deg P = n$, $\deg Q = m$, and let them have no roots in common. Then there exist two polynomials A (of degree $m - 1$) and B (of degree $n - 1$) such that $AP + BQ = 1$.

6. Prove that $(1 - y)^{-L} = \sum_{k=0}^{L-1} \binom{L+k-1}{k} y^k + R(y^L)$, where $R(y)$ is a power series of the form $y^L \sum_{k=0}^{\infty} a_k y^k$.

7. Assume the Laurent polynomial $R(z) \geq 0$ on Γ. Prove that any zero of $R(z)$ on Γ must have even multiplicity.

8. Find a polynomial $c(z)$ such that $|c(z)|^2 = \frac{1}{6} \left(z^{-1} + 4 + z \right)$.

9. Prove Theorem 8.1.3: $m_L(z) = \left(\frac{1+z}{2} \right)^L q_L(z)$ is a conjugate mirror filter.

10. Compute (numerically) the mask of the Daubechies filters of order 3 and 4.

11. Let $m_L(z) := \sum_{k=0}^{2L-1} h_k z^k$ be the Daubechies filter of order L. Let $\tilde{m}_L(z) := \left(\frac{1+z}{2} \right)^L q_L(z)$ be the conjugate mirror filter such that all roots of $q_L(z)$ satisfy $|z| \leq 1$. Prove that $h_k = \tilde{h}_{2L-k}$.

12. Write $P_L(z) = |m_L(z)|^2$. Prove that
    (a) $P_1(e^{-i\omega}) = 1 - \frac{1}{2} \int_0^\omega \sin t \, dt$.
    (b) $P_2(e^{-i\omega}) = 1 - \frac{3}{4} \int_0^\omega \sin^3 t \, dt$.
    (c) $P_3(e^{-i\omega}) = 1 - \frac{15}{8} \int_0^\omega \sin^5 t \, dt$.

2 The Daubechies Scaling Functions

Consider the solution of the refinement equation

$$\phi_L(x) = 2 \sum_{k=0}^{2L-1} h_L(k) \phi_L(2x - k), \qquad (2.1)$$

where $\sum_{k=1}^{2L-1} h_L(k) z^k = m_L(z)$ is the Daubechies filter. We have the following.

Theorem 8.2.1 [Daubechies] Let $m_L(z)$ be the Daubechies filter of order $L(\geq 2)$. Then the refinement equation (2.1) has a unique solution $\phi_L \in L^2$ which is an orthonormal scaling function.

The proof of the theorem is tedious. We will complete it through several steps. Since these orthonormal scaling functions were first discovered by Daubechies, we give the following.

Definition 8.2.1 For $L \geq 2$, the orthonormal scaling function ϕ_L satisfying (2.1) is called the *Daubechies scaling function* of order L. Correspondingly, the wavelet defined by

$$\psi_L(x) = 2 \sum_{k=0}^{2L-1} g_k \phi_L(2x - k), \quad g_k = (-1)^k h_k (2L - 1 - k), \tag{2.2}$$

is called the *Daubechies wavelet* of order L.

We now start to prove Theorem 8.2.1.

2.1 The Infinite Product Form

To prove Theorem 8.2.1, we first derive the infinite product form of $\hat{\phi}_L$. We will discuss it in a more general setting where ϕ satisfies refinement equation (1.1), whose mask only satisfies the sum rule. Note that $\hat{\phi}$ satisfies the following equation:

$$\hat{\phi}(\omega) = H(z)\hat{\phi}(\omega/2), \quad z = e^{-i\omega/2}, \tag{2.3}$$

where $H(z) = \sum_{k=0}^{N} h_k z^k$. Iterating (2.3) n times, we have

$$\hat{\phi}(\omega) = H(e^{-i\omega/2}) \cdots H(e^{-i\omega/2^n})\hat{\phi}(\omega/2^n)$$

$$= \prod_{k=1}^{n} H(e^{-i\omega/2^k})\hat{\phi}(\omega/2^n).$$

If the infinite product $\prod_{k=1}^{\infty} H(e^{-i\omega/2^k})$ converges, then we have

$$\hat{\phi}(\omega) = \lim_{n\to\infty} \prod_{k=1}^{n} H(e^{-i\omega/2^k}) \lim_{n\to\infty} \hat{\phi}(\omega/2^n)$$

$$= \prod_{k=1}^{\infty} H(e^{-i\omega/2^k})\hat{\phi}(0).$$

By $\hat{\phi}(0) = 1$, we obtain

$$\hat{\phi}(\omega) = \prod_{k=1}^{\infty} H(e^{-i\omega/2^k}), \tag{2.4}$$

which is an *infinite product form* of $\hat{\phi}$.

EXAMPLE 8.2.1: For the mth-order cardinal B-spline N_m, we have

$$\hat{N}_m(\omega) = \left(\frac{1 + e^{-i\omega/2}}{2}\right)^m \hat{N}_m(\omega/2).$$

Therefore

$$\hat{N}_m(\omega) = \prod_{k=1}^{\infty} \left(\frac{1 + e^{-i\omega/2^k}}{2} \right)^m \hat{N}_m(0).$$

Since $\int_{\mathbb{R}} N_m(x)\, dx = 1$, we have

$$\hat{N}_m(\omega) = \prod_{k=1}^{\infty} \left(\frac{1 + e^{-i\omega/2^k}}{2} \right)^m = \left(\frac{1 - e^{-i\omega}}{i\omega} \right)^m.$$

It follows that

$$\hat{N}_m(\omega) = e^{-im\omega/2} \left(\frac{\sin \frac{\omega}{2}}{\frac{\omega}{2}} \right)^m. \tag{2.5}$$

We now prove that $\prod_{k=1}^{\infty} H(e^{-i\omega/2^k})$ is always (pointwise) convergent. To this end, we introduce the following lemma.

Lemma 8.2.1 Let (a_k) be a positive sequence. Then $\prod_{k=1}^{\infty} a_k$ is convergent if and only if $\sum_{k=1}^{\infty}(a_k - 1)$ is convergent.

Proof: Let (a_{kj}) be the subsequence of (a_k) obtained by eliminating all terms of (a_k) which are 1. If (a_{kj}) is a finite sequence or is empty, the lemma is trivial. When (a_{kj}) is an infinite sequence, then the convergence of $\sum_k (a_k - 1)$ is equivalent to the convergence of $\sum_j (a_{kj} - 1)$, and the convergence of $\Pi_k a_k$ is equivalent to the convergence of $\Pi_j a_{kj}$. Hence, we can assume $a_k \neq 1$ for each k in (a_k). If $\sum_{k=1}^{\infty}(a_k - 1)$ is convergent, then $\lim_{k\to\infty}(a_k - 1) = 0$. Without loss of generality, we can assume $|a_k - 1| < 1$, $k = 1, 2, \cdots$. Since $\lim_{x\to 0} \frac{\ln(1+x)}{x} = 1$, we have

$$\lim_{a_k \to 1} \frac{\ln a_k}{a_k - 1} = \lim_{(a_k - 1)\to 0} \frac{\ln(1 + (a_k - 1))}{(a_k - 1)} = 1.$$

Then, by the limit form of the comparison test the series $\sum_{k=1}^{\infty} \ln a_k$ is convergent if and only if $\sum_{k=1}^{\infty}(a_k - 1)$ is convergent. Note that

$$\prod_{k=1}^{n} a_k = e^{\sum_{k=1}^{n} \ln a_k}.$$

Hence, $\prod_{k=1}^{\infty} a_k$ is convergent if and only if $\sum_{k=1}^{\infty}(a_k - 1)$ is convergent. ∎

Theorem 8.2.2 Let $H(z)$ be a polynomial with $H(1) = 1$. Then for a given $R > 0$, the infinite product $\prod_{k=1}^{\infty} H(e^{-iw/2^k})$ is uniformly convergent on $|w| \leq R$. Consequently, the function $f(\omega) = \prod_{k=1}^{\infty} H(e^{-iw/2^k})$ is continuous on $\mathbb{R}$.

Proof: By the inequality $|e^{-iw} - 1| \leq |w|$, $\omega \in \mathbb{R}$, we have

$$\left| |H(e^{-iw})| - 1 \right| \leq \left| H(e^{-iw}) - 1 \right|$$

$$\leq \left| \sum_{k=0}^{N} h_k(e^{-ikw} - 1) \right| \leq \left| \sum_{k=0}^{N} kh_k \right| |w|.$$

Write $c = |\sum_{k=0}^{N} kh_k|$. Then

$$\left| |H(e^{-i2^{-j}w})| - 1 \right| \leq c2^{-j}|w| \leq c2^{-j}R, \quad |w| \leq R,$$

which yields

$$\lim_{j \to \infty} \left(|H(e^{-i2^{-j}w})| - 1 \right) = 0$$

and the limit holds uniformly for $|w| \leq R$. By Lemma 8.2.1, $\prod_{k=1}^{\infty}|H(e^{-iw/2^k})|$ is uniformly convergent for $|w| \leq R$. Hence, $\prod_{k=1}^{\infty}|H(e^{-iw/2^k})|$ is a continuous function on $[-R, R]$, which implies that $\prod_{k=1}^{\infty}|H(e^{-iw/2^k})|$ is bounded on $[-R, R]$. We write $\Pi_n(\omega) = \prod_{k=1}^{n} H(e^{-iw/2^k})$. By the uniform convergence of $\prod_{k=1}^{\infty}|H(e^{-iw/2^k})|$, there is an $M > 0$ such that $|\Pi_n(\omega)| \leq M$, $n = 1, 2, \cdots$.

We now prove that $\prod_{k=1}^{\infty} H(e^{-iw/2^k})$ is uniformly convergent for $|\omega| \leq R$. We have

$$|\Pi_n(\omega) - \Pi_{n+1}(\omega)| = |\Pi_n(\omega)||1 - H(e^{-i\omega/2^{n+1}})|$$

$$\leq cMR2^{-n+1}, \quad |\omega| \leq R, n \in \mathbb{N}.$$

Therefore,

$$|\Pi_n(\omega) - \Pi_{n+m}(\omega)| \leq \sum_{k=0}^{m-1} |\Pi_{n+k}(\omega) - \Pi_{n+k+1}(\omega)|$$

$$\leq cMR \sum_{k=1}^{m} 2^{-n+k} \leq cMR2^{-n},$$

which indicates that the sequence $(\Pi_n(\omega))_{n=1}^{\infty}$ is a Cauchy sequence and the infinite product $\prod_{k=1}^{\infty} H(e^{-i\omega/2^k})$ is uniformly convergent on $|w| \leq R$. ∎

2.2 Proof That $\phi_L \in L^2$

We now discuss the conditions under which the solution ϕ of (1.1) is in L^2. From Fourier transform theory, $\phi \in L^2$ if and only if $\hat{\phi} \in L^2$. It is known that if $|f(\omega)| \leq \frac{1}{1+|\omega|^{1/2+\epsilon}}$ for an $\epsilon > 0$, then $f \in L^2$. Based on this fact and $\prod_{k=1}^{\infty} \left(\frac{1+e^{-i\omega/2^k}}{2}\right)^m = \left(\frac{1-e^{-i\omega}}{i\omega}\right)^m$, which decays as fast as $|\omega|^{-m}$, we have the following.

Lemma 8.2.2 If the symbol of ϕ

$$H(z) = \left(\frac{1+z}{2}\right)^m q(z) \tag{2.6}$$

satisfies

$$\max_{\omega \in \mathbb{R}} |q(e^{-i\omega})| < 2^{m-\frac{1}{2}}, \tag{2.7}$$

then $\phi \in L^2$.

   Proof: It is clear that

$$\hat{\phi}(\omega) = \prod_{k=1}^{\infty} H(e^{-i\omega/2^k}) = \left(\frac{1-e^{-i\omega}}{i\omega}\right)^m Q(\omega),$$

where $Q(\omega) = \prod_{k=1}^{\infty} q\left(e^{-i\omega/2^k}\right) \in C$. We now estimate the growth rate of $Q(\omega)$ as $|\omega| \to \infty$. Since

$$\max_{\omega \in \mathbb{R}} |q(e^{-i\omega})| < 2^{m-\frac{1}{2}},$$

there is an $\epsilon > 0$ such that

$$\max_{\omega \in \mathbb{R}} |q(e^{-i\omega})| = 2^{m-\frac{1}{2}-\epsilon}.$$

Write

$$Q(\omega) = \prod_{k=1}^{N} q\left(e^{-i\omega/2^k}\right) Q(2^{-(N+1)}\omega), \quad N \in \mathbb{N}.$$

For an ω such that $2^N \leq |\omega| < 2^{N+1}$,

$$\left|Q(2^{-(N+1)}\omega)\right| \leq C := \max_{|\omega| \leq 1} |Q(\omega)|$$

and by (2.7),

$$\left|\prod_{k=1}^{N} q\left(e^{-i\omega/2^k}\right)\right| \leq 2^{N(m-\frac{1}{2}-\epsilon)}.$$

Since $2^N \leq |\omega|$, we have

$$|Q(\omega)| \leq C2^{N(m-1/2-\epsilon)} \leq C|\omega|^{m-\frac{1}{2}-\epsilon}.$$

Note that

$$\left|\frac{1 - e^{-i\omega}}{i\omega}\right|^m \leq 2^m \min(1, |\omega|^{-m}).$$

Therefore, we have for any $\omega \in \mathbb{R}$,

$$|\hat{\phi}(\omega)| \leq C|\omega|^{m-\frac{1}{2}-\epsilon}2^m \min(1, |\omega|^{-m})$$
$$\leq \frac{C2^m}{1 + |\omega|^{\frac{1}{2}+\epsilon}}.$$

Hence $\hat{\phi} \in L^2$.   ∎

Lemma 8.2.3 The scaling function ϕ_L in (2.1) is in L^2.

  Proof: We have $m_L(z) = \left(\frac{1+z}{2}\right)^L q_L(z)$. By Lemma 8.1.4, $\max_{\omega \in \mathbb{R}} |q_L(e^{-i\omega})| < 2^{L-1}$. By Lemma 8.2.2, $\phi_L \in L^2$.   ∎

2.3 Orthogonality of ϕ_L

We now finish the proof of Theorem 8.2.1 by proving the orthogonality of ϕ_L. For this, we define

$$f_k(\omega) = \left[\prod_{j=1}^{k} m_L(e^{-i\omega/2^j})\right] \chi_{[-2^k\pi, 2^k\pi]}(\omega), \quad k = 0, 1, \cdots.$$

Then it is clear that

$$\lim_{k \to \infty} f_k(\omega) = \hat{\phi}_L(\omega), \quad \omega \in \mathbb{R},$$

where $\hat{\phi}_L \in L^2$. We first prove that there is a $C > 0$ such that

$$|f_k(\omega)| \leq C|\hat{\phi}_L(\omega)|, \quad \omega \in \mathbb{R}. \tag{2.8}$$

Since $f_k(\omega) = 0$, $|\omega| \geq 2^k\pi$, we only need to prove (2.8) for $|\omega| \leq 2^k\pi$. By Lemma 8.1.4,

$$\min_{\omega \in \mathbb{R}} |q_L(e^{-i\omega})| \geq 1,$$

which yields

$$\left| \prod_{j=1}^{\infty} q_L(e^{-i\omega/2^j}) \right| \geq 1.$$

We also have

$$\min_{|\omega| \leq \pi} \left| \left(\frac{1 - e^{-i\omega}}{i\omega} \right) \right|^L = \min_{|\omega| \leq \pi} \left| \frac{\sin \frac{\omega}{2}}{\frac{\omega}{2}} \right|^L \geq \left(\frac{2}{\pi} \right)^L.$$

Thus,

$$\min_{|\omega| \leq \pi} |\hat{\phi}_L(\omega)| \geq \left(\frac{2}{\pi} \right)^L,$$

which implies

$$|f_k(\omega)| = \frac{|\hat{\phi}_L(\omega)|}{|\hat{\phi}_L(2^{-k}\omega)|} \leq \left(\frac{\pi}{2} \right)^L |\hat{\phi}_L(\omega)|, \quad |\omega| \leq 2^k \pi.$$

Inequality (2.8) is proved. By the Lebesgue Dominated Convergence Theorem (Theorem 3.3.1), we have

$$\lim_{k \to \infty} ||f_k - \hat{\phi}_L|| = 0. \tag{2.9}$$

Hence,

$$\lim_{k \to \infty} \int_{\mathbb{R}} |f_k(\omega)|^2 e^{in\omega} d\omega = \int_{\mathbb{R}} |\hat{\phi}(\omega)|^2 e^{in\omega} d\omega, \quad k \in \mathbb{Z}. \tag{2.10}$$

Let

$$I_{k,n} := \frac{1}{2\pi} \int_{\mathbb{R}} |\hat{f}_k(\omega)|^2 e^{in\omega} d\omega, \quad n \in \mathbb{Z}, k \in \mathbb{N}.$$

We have, for $k \in \mathbb{N}$,

$$I_{k,n} = \frac{1}{2\pi} \int_{-2^k\pi}^{2^k\pi} \prod_{j=1}^{k} \left| m_L(e^{-i\omega/2^j}) \right|^2 e^{in\omega} d\omega = \frac{1}{2\pi} \int_{0}^{2^{k+1}\pi} \prod_{j=1}^{k} \left| m_L(e^{-i\omega/2^j}) \right|^2 e^{in\omega} d\omega$$

$$= \frac{1}{2\pi} \int_{0}^{2^k\pi} \prod_{j=1}^{k} \left| m_L(e^{-i\omega/2^j}) \right|^2 e^{in\omega} d\omega + \frac{1}{2\pi} \int_{2^k\pi}^{2^{k+1}\pi} \prod_{j=1}^{k} \left| m_L(e^{-i\omega/2^j}) \right|^2 e^{in\omega} d\omega$$

$$= I_{k,n}^1 + I_{k,n}^2,$$

where

$$I_{k,n}^1 = \frac{1}{2\pi} \int_{0}^{2^k\pi} \prod_{j=1}^{k} \left| m_L(e^{-i\omega/2^j}) \right|^2 e^{in\omega} d\omega$$

$$= \frac{1}{2\pi} \int_{0}^{2^k\pi} \prod_{j=1}^{k-1} \left| m_L(e^{-i\omega/2^j}) \right|^2 \left| m_L(e^{-i\omega/2^k}) \right|^2 e^{in\omega} d\omega$$

and, by changing ω to $2^k\pi + \omega$,

$$I_{k,n}^2 = \frac{1}{2\pi} \int_{2^k\pi}^{2^{k+1}\pi} \prod_{j=1}^{k} \left| m_L(e^{-i\omega/2^j}) \right|^2 e^{in\omega} \, d\omega$$

$$= \int_0^{2^k\pi} \prod_{j=1}^{k-1} \left| m_L(e^{-i\omega/2^j}) \right|^2 \left| m_L(-e^{-i\omega/2^k}) \right|^2 e^{in\omega} \, d\omega.$$

Recall that $\left| m_L(e^{-i\omega/2^k}) \right|^2 + \left| m_L(-e^{-i\omega/2^k}) \right|^2 = 1$. We have

$$I_{k,n} = \frac{1}{2\pi} \int_0^{2^k\pi} \prod_{j=1}^{k-1} \left| m_L(e^{-i\omega/2^j}) \right|^2 e^{in\omega} \, d\omega$$

$$= \frac{1}{2\pi} \int_{-2^{k-1}\pi}^{2^{k-1}\pi} \prod_{j=1}^{k-1} \left| m_L(e^{-i\omega/2^j}) \right|^2 e^{in\omega} \, d\omega = I_{k-1,n}.$$

It follows that $I_{k,n} = I_{k-1,n} = \cdots = I_{0,n}$. Since

$$I_{0,n} = \int_{-\pi}^{\pi} e^{in\omega} \, d\omega = \delta_{0n},$$

we have $I_{k,n} = \delta_{0n}, k \in \mathbb{N}, n \in \mathbb{Z}$. By (2.10), we have

$$\int_{\mathbb{R}} \left| \hat{\phi}_L(\omega) \right|^2 e^{in\omega} \, d\omega = \delta_{0n}, \quad n \in \mathbb{Z}.$$

That is, ϕ_L is an orthonormal scaling function. Theorem 8.2.1 is proved.

Exercises

1. Prove $\frac{1}{2} = \prod_{n=2}^{\infty}(1 - \frac{1}{n^2})$.
2. Prove

$$\frac{1}{1-x} = \prod_{n=1}^{\infty} \left(1 + x^{2^n} \right), \quad |x| < 1.$$

3. It is known that

$$\sin x = x \prod_{n=1}^{\infty} \left(1 - \frac{x^2}{n^2\pi^2} \right), \quad x \in \mathbb{R}.$$

Use this result to prove the infinite product for π :

$$\pi = 2\frac{2 \cdot 2 \cdot 4 \cdot 4 \cdot 6 \cdot 6 \cdots}{1 \cdot 3 \cdot 3 \cdot 5 \cdot 5 \cdot 7 \cdots}$$

$$= 2\prod_{n=1}^{\infty} \frac{4n^2}{(2n-1)(2n+1)}.$$

4. Assume the scaling function ϕ has the symbol $H(z) = \left(\frac{1+z}{2}\right)^2\left(\frac{5}{3} - \frac{2z}{3}\right)$. Prove that $\phi \in L^2$.

5. Let $h(\omega)$ be a 2π-periodic function satisfying

$$|h(\omega)|^2 + |h(\omega + \pi)|^2 = 1$$

and suppose the infinite product $\prod_{n=1}^{\infty} h(2^{-n}\omega)$ is convergent a.e. to a function $\hat{f}$. Write

$$\hat{f}_k(\omega) = \prod_{n=1}^{k} h(2^{-n}\omega)\chi_{[-2^k\pi, 2^k\pi]}(\omega).$$

Prove the following.

(a) $\int_{-2\pi}^{0} \left|\hat{f}_1(\omega)\right|^2 d\omega = \int_{0}^{2\pi} \left|\hat{f}_1(\omega + \pi)\right|^2 d\omega$, and therefore, $\int_{-\infty}^{\infty} \left|\hat{f}_1(\omega)\right|^2 d\omega = 2\pi$.

(b) $\int_{-\infty}^{\infty} \left|\hat{f}_{k+1}(\omega)\right|^2 d\omega = \int_{-\infty}^{\infty} \left|\hat{f}_k(\omega)\right|^2 d\omega$ for all $k \in \mathbb{N}$.

(c) $\hat{f} \in L^2$ and $||\hat{f}||^2 \leq 2\pi$.

6. Suppose the 2π-periodic function $h(\omega)$ in Exercise 5 is continuously differentiable in $[-\frac{\pi}{2}, \frac{\pi}{2}]$ and

$$\min_{|\omega| \leq \pi/2} |h(\omega)| > 0.$$

Let f be the function whose Fourier transform is $\hat{f}(\omega) = \prod_{n=1}^{\infty} h(2^{-n}\omega)$. Use the Lebesgue Dominated Convergence Theorem to prove that f is an orthonormal scaling function. (See the method in the proof of the orthogonality of ϕ_L in Theorem 8.2.1.)

7. Prove the following result: In Theorem 8.2.2 if condition (2.7) is replaced by

$$\max_{\omega \in \mathbb{R}} \left| \prod_{k=1}^{n} q(e^{-i\omega/2^k}) \right| < 2^{n(m-1/2)}, \quad \text{for an } n \in \mathbb{R},$$

then $\phi \in L^2$.

8. Prove the following: Let the symbol (2.6) satisfy

$$\max_{\omega \in \mathbb{R}} |q(e^{-i\omega})| < 2^{m-1}.$$

Then $\hat{\phi} \in L^1$.

9. Prove that the condition in Exercise 8 can be replaced by

$$\max_{\omega \in \mathbb{R}} \left| \prod_{k=1}^{n} q(e^{-i\omega/2^k}) \right| < 2^{n(m-1)}, \quad \text{for an } n \in \mathbb{N}.$$

10. Prove the following: Let ϕ be a scaling function satisfying (1.1) and let its symbol $h(z)$ be (2.6). If

$$\max_{\omega \in \mathbb{R}} |q(e^{-i\omega})| < 2^{(m-s)-1/2},$$

then $\phi \in W_2^s$. The condition can also be replaced by

$$\max_{\omega \in \mathbb{R}} \left| \prod_{k=1}^{n} q(e^{-i\omega/2^k}) \right| < 2^{n(m-s-1/2)}, \quad \text{for an } n \in \mathbb{R}.$$

11. Assume the scaling function ϕ has the symbol $H(z) = (\frac{1+z}{2})^2(\frac{2}{3} + \frac{1}{3}z)$. Prove that $\phi \in W_2^1$.

12. Assume ϕ the scaling function satisfies (1.1) and its symbol $H(z)$ is in the form of (2.6). Prove that, if

$$\max_{\omega \in \mathbb{R}} |q(e^{-i\omega})| < 2^{m-s-1},$$

then ϕ is almost everywhere equal to a function in C^s.

3 Computation of Daubechies Scaling Functions

The Daubechies scaling function ϕ_L is not given in an explicit expression. In this section, we discuss how to compute ϕ_L from its mask $(h_L(k))$.

3.1 The Cascade Algorithm

We introduce an iterative method to compute $\phi_L(x)$. We first illustrate the method in a general setting without consideration for convergence. Let ϕ be the scaling function determined by the refinement equation (1.1). The equation yields the algorithm

$$\phi^j(x) = 2 \sum_{k=0}^{N} h_k \phi^{j-1}(2x - 1), \quad j = 1, 2, \cdots, \tag{3.1}$$

which produces a sequence $\{\phi^j\}_{j=0}^\infty$. If it converges to a function $\phi \in L^2$ (or $\in C$), then ϕ is a solution of the refinement equation (1.1). Algorithm (3.1) is called the *cascade algorithm*. The algorithm needs an initial function $\phi^0 \in L^2$. We always assume the initial function of a cascade algorithm satisfies the following.

Condition 1. ϕ^0 is compactly supported with $\hat{\phi}^0(0) = 1$, and $\sum_{k \in \mathbb{Z}} \phi^0(x - k) = 1$ a.e.

The box function $B(x)$ and the hat function $N_2(x)$ are often chosen as the initial functions. We now denote the cascade algorithm (3.1) by

$$\phi^j = T_{\mathbf{h}} \phi^{j-1} \tag{3.2}$$

and call $T_{\mathbf{h}}$ the *transaction operator* (with respect to $\mathbf{h}$).
The cascade algorithm has the following properties.

Lemma 8.3.1 Let (ϕ^j) be the sequence obtained in (3.1). Then
(1) $\int_{\mathbb{R}} \phi^j(x - k)\, dx = 1$,
(2) $\sum_{k \in \mathbb{Z}} \phi^j(x - k) = 1$, a.e. Particularly, if $\phi^0 \in C$, then $\sum_{k \in \mathbb{Z}} \phi^j(x - k) = 1$ holds everywhere.

Proof: We have $\int_{\mathbb{R}} \phi^j(x - k)\, dx = \int_{\mathbb{R}} \phi^j(x)\, dx$ for $k \in \mathbb{Z}$, and

$$\int_{\mathbb{R}} \phi^j(x)\, dx = 2 \sum_k h_k \int_{\mathbb{R}} \phi^{j-1}(2x - k)\, dx$$

$$= \sum_k h_k \int_{\mathbb{R}} \phi^{j-1}(x)\, dx = \int_{\mathbb{R}} \phi^{j-1}(x)\, dx,$$

which inductively yields $\int_{\mathbb{R}} \phi^j(x - k)\, dx = \int_{\mathbb{R}} \phi^0(x)\, dx = 1$ for all $j \geq 0$. We now prove (2). Assume

$$\sum_{k \in \mathbb{Z}} \phi^0(x - k) = 1.$$

Recall that in (3.1), $\mathbf{h}$ satisfies the sum rule. Thus,

$$\sum_k \phi^j(x - k)$$

$$= 2 \sum_k \sum_l h_l \phi^{j-1}(2x - 2k - l) = 2 \sum_k \sum_s h_{s-2k} \phi^{j-1}(2x - s)$$

$$= \sum_s \left(2 \sum_k h_{s-2k} \right) \phi^{j-1}(2x - s) = \sum_s \phi^{j-1}(2x - s),$$

which inductively yields

$$\sum_{k\in\mathbb{Z}} \phi^j(x-k) = \sum_{k\in\mathbb{Z}} \phi^0(2^j x - k) = 1, \quad \text{for all } j \geq 0.$$

The lemma is proved.   ∎

In practice, to compute the values of $\phi^j(x)$, we write the cascade algorithm (3.1) in a vector form. For ϕ, we define the vector-valued function $\Phi(x)$ on $[0,1]$ by

$$\Phi(x) = (\phi(x), \phi(x+1), \cdots, \phi(x+N-1))^T, \quad x \in [0,1]. \tag{3.3}$$

Thus, the ith component of $\Phi(x)$ is $\phi|_{[i-1,i]}$. Let T_0 and T_1 be two $N \times N$ matrices defined by

$$T_0 = 2(h_{2m-n-1})_{m,n=1}^N = 2 \begin{pmatrix} h_0 & & & \\ h_2 & h_1 & h_0 & \\ \cdots & \cdots & \cdots & \cdots \\ & & h_N & h_{N-1} \end{pmatrix} \tag{3.4}$$

and

$$T_1 = 2(h_{2m-n})_{m,n=1}^N = 2 \begin{pmatrix} h_1 & h_0 & & & \\ \cdots & \cdots & \cdots & \cdots \\ \cdots & h_N & h_{N-1} & h_{N-2} \\ & & & h_N \end{pmatrix}, \tag{3.5}$$

respectively. For convenience, we call (T_0, T_1) the CA-(matrix) pair for **h**. By (1.1), we have

$$\Phi(x) = \begin{cases} T_0 \Phi(2x), & x \in [0, 1/2], \\ T_1 \Phi(2x-1), & x \in [1/2, 1]. \end{cases} \tag{3.6}$$

Let $\tau : [0,1] \to [0,1]$ be the mapping

$$\tau x = \begin{cases} 2x, & x \in [0, 1/2), \\ 2x-1, & x \in [1/2, 1). \end{cases}$$

For $0 \leq x < 1$, define the binary sequence $d(x) := (d_n(x))_{n=1}^\infty$ by

$$x = \sum_{n=1}^\infty d_n(x) 2^{-n}.$$

Thus,

$$d_1(x) = \begin{cases} 0, & x \in [0, 1/2), \\ 1, & x \in [1/2, 1). \end{cases}$$

Now (3.6) becomes

$$\Phi(x) = T_{d_1(x)}\Phi(\tau x), \quad x \in [0,1). \tag{3.7}$$

In general,

$$\tau^j x = \begin{cases} 2^j x, & x \in [0, 2^{-j}), \\ 2^j x - 1, & x \in [2^{-j}, 2(2^{-j})), \\ \cdots & \cdots \\ 2^j x - (2^j - 1), & x \in [(2^{-j} - 1)2^{-j}, 1), \end{cases}$$

and

$$d_l(x) = d_1(\tau^{l-1}x).$$

Therefore,

$$\Phi(x) = T_{d_1(x)} T_{d_2(x)} \cdots T_{d_j(x)} \Phi(\tau^j x), \quad j \geq 1,$$

which yields the vector form of the cascade algorithm (3.1):

$$\Phi^j(x) = T_{d_1(x)} T_{d_2(x)} \cdots T_{d_j(x)} \Phi^0(\tau^j x), \quad j \geq 1.$$

Thus, $\Phi^j(x)$ converges to $\Phi(x)$ if and only if $\phi^j(x)$ converges to $\phi(x)$.

EXAMPLE 8.3.1: Let us consider the CA-pair for (1.17). We have

$$T_0 = \frac{1}{4} \begin{pmatrix} 1+\sqrt{3} & 0 & 0 \\ 3-\sqrt{3} & 3+\sqrt{3} & 1+\sqrt{3} \\ 0 & 1-\sqrt{3} & 3-\sqrt{3} \end{pmatrix},$$

$$T_1 = \frac{1}{4} \begin{pmatrix} 3+\sqrt{3} & 1+\sqrt{3} & 0 \\ 1-\sqrt{3} & 3-\sqrt{3} & 3+\sqrt{3} \\ 0 & 0 & 1-\sqrt{3} \end{pmatrix}.$$

Let $\phi^0(x) = N_2(x) = \begin{cases} x, & x \in [0,1] \\ 2-x, & x \in [1,2] \end{cases}$. Then $\Phi^0(x) = (x, 1-x, 0)^T$, $x \in [0,1]$, and

$$\Phi^1(x) = \begin{cases} T_0\Phi^0(2x) = \begin{pmatrix} \left(\frac{1+\sqrt{3}}{2}\right)x \\ \left(\frac{3+\sqrt{3}}{4}\right) - \sqrt{3}x \\ \left(\frac{1-\sqrt{3}}{4}\right) - \left(\frac{1-\sqrt{3}}{2}\right)x \end{pmatrix}, & x \in [0, \frac{1}{2}], \\ \\ T_1\Phi^0(2x-1) = \begin{pmatrix} \left(\frac{-1+\sqrt{3}}{4}\right) + x \\ \left(\frac{5-\sqrt{3}}{4}\right) - x \\ 0 \end{pmatrix}, & x \in [\frac{1}{2}, 1], \end{cases}$$

i.e.,

$$\phi^1(x) = \begin{cases} \left(\frac{1+\sqrt{3}}{2}\right) x, & x \in [0, \frac{1}{2}], \\ \left(\frac{-1+\sqrt{3}}{4}\right) + x, & x \in [\frac{1}{2}, 1], \\ \left(\frac{3+5\sqrt{3}}{4}\right) - \sqrt{3}x, & x \in [1, \frac{3}{2}], \\ \left(\frac{9-\sqrt{3}}{4}\right) - x, & x \in [\frac{3}{2}, 2], \\ \left(\frac{5-5\sqrt{3}}{4}\right) - \left(\frac{1-\sqrt{3}}{2}\right) x, & x \in [2, \frac{5}{2}], \\ 0, & x \in [\frac{5}{2}, 3]. \end{cases}$$

3.2 The Recursion Algorithm

If the scaling function ϕ is continuous, then it is uniquely determined by its values at all dyadic numbers, which can be computed by the *recursion algorithm*.

First we compute the values of $\phi(x)$ at all integers, i.e., get the value of $\Phi(0)$. By (3.7), we have

$$\Phi(0) = T_0\Phi(0).$$

Hence, $\Phi(0)$ is the right 1-eigenvector of the matrix T_0. Once $\Phi(0)$ is obtained, then the values of ϕ at half integers can be computed by

$$\Phi\left(\frac{1}{2}\right) = T_1\Phi(0).$$

Continuing, we have

$$\Phi\left(\frac{1}{4}\right) = T_0\Phi\left(\frac{1}{2}\right),$$

$$\Phi\left(\frac{3}{4}\right) = T_1\Phi\left(\frac{1}{2}\right),$$

and so on. In general,

$$\Phi\left(\frac{2k+1}{2^j}\right) = \begin{cases} T_0\Phi\left(\frac{2k+1}{2^{j-1}}\right), & 2k+1 < 2^{j-1} \\ T_1\Phi\left(\frac{2k+1}{2^{j-1}} - 1\right), & 2^{j-1} < 2k+1 < 2^j \end{cases}. \tag{3.8}$$

Thus, we can compute the values of ϕ at all dyadic numbers recursively.

The recursion algorithm starts from $\Phi(0)$, which is a right 1-eigenvector of T_0. Since $\mathbf{h}$ in (1.1) satisfies the sum rule, then the vector $\mathbf{e} = (1, 1, \cdots, 1)^T$ is a left 1-eigenvector of T_0. The relation between **e** and $\Phi(0)$ can be derived from the following lemma.

Lemma 8.3.2 Let λ be a simple eigenvalue of an $n \times n$ matrix M. Let **u** be a left λ-eigenvector of M and **v** be a right λ-eigenvector of M. Then $\mathbf{u}^T\mathbf{v} \neq 0$.

Proof: Since M has a simple eigenvalue λ, there is an invertible matrix S such that

$$M = S \begin{pmatrix} \lambda & 0 \\ 0 & B \end{pmatrix} S^{-1},$$

where B is an $(N-1) \times (N-1)$ matrix. The first column of S, denoted by **c**, is a right λ-eigenvector of M, and the first row of S^{-1}, denoted by $\mathbf{r}^T$, is a left λ-eigenvector of M. Since $S^{-1}S = I$, we have $\mathbf{r}^T\mathbf{c}=1$, which yields $\mathbf{u}^T\mathbf{v} \neq 0$. ■

From the lemma, we have the following.

Corollary 8.3.1 Assume **h** satisfies the sum rule and (T_0, T_1) is the CA-pair of **h**. Assume also the eigenvalue 1 of T_0 is simple. Let **v** be the right 1-eigenvector of T_0. Then $\sum_{k\in\mathbb{Z}} \mathbf{v}(k) \neq 0$. Besides, for any vector $\mathbf{u} \in \mathbb{R}^N$, $\sum_{k\in\mathbb{Z}}(T_0\mathbf{u})(k) = \sum_{k\in\mathbb{Z}} \mathbf{u}(k)$.

Proof: Under the condition of the corollary, T_0 has the unique left 1-eigenvector **e** (up to a constant). By Lemma 8.3.2, $\mathbf{e}^T\mathbf{v} = \sum_{k\in\mathbb{Z}} \mathbf{v}(k) \neq 0$. We also have

$$\sum_{k\in\mathbb{Z}}(T_0\mathbf{u})(k) = \mathbf{e}^T(T_0\mathbf{u}) = (\mathbf{e}^T T_0)\mathbf{u} = \mathbf{e}^T\mathbf{u} = \sum_{k\in\mathbb{Z}} \mathbf{u}(k).$$

The corollary is proved. ■

Lemma 8.3.3 If $\phi \in L^2$ in (1.1) is a continuous MRA generator, then $\sum_k \phi(k) = 1$.

Proof: We leave the proof as an exercise. ■

Note that $\Phi(0) = (\phi(0), \phi(1), \cdots \phi(N-1))^T$ is the right 1-eigenvector of the matrix T_0. By Lemma 8.3.3, $\mathbf{e}^T \Phi(0) = \sum_k \phi(k) = 1$. Hence, it is uniquely determined. Now $\phi(0) = 0$, $\Phi(0)=(\phi(0), \mathbf{a})^T$, where $\mathbf{a} = (\phi(1), \cdots, \phi(N-1))^T$ is the right 1-eigenvector of $\tilde{T}_0$, i.e.,

$$\mathbf{a} = \tilde{T}_0\mathbf{a} \tag{3.9}$$

where

$$\tilde{T}_0 = 2(h_{2m-n-1})_{m,n=2}^N = 2 \begin{pmatrix} h_1 & h_0 & \\ \cdots & \cdots & \cdots \\ & h_N & h_{N-1} \end{pmatrix}.$$

EXAMPLE 8.3.2: Let us still consider the CA-pair for (1.17) in the previous example. We have

$$\tilde{T}_0 = \frac{1}{4} \begin{pmatrix} 3 + \sqrt{3} & 1 + \sqrt{3} \\ 1 - \sqrt{3} & 3 - \sqrt{3} \end{pmatrix}.$$

Solving the equation $\mathbf{a} = \tilde{T}_0\mathbf{a}$, we get $\mathbf{a} = \left(\frac{1+\sqrt{3}}{2}, \frac{1-\sqrt{3}}{2} \right)^T$, i.e.,

$$\Phi_2(0) = \left(0, \frac{1+\sqrt{3}}{2}, \frac{1-\sqrt{3}}{2} \right)^T.$$

Then

$$\Phi_2 \left(\frac{1}{2} \right) = T_1\Phi_2(0) = \begin{pmatrix} \frac{1}{2} + \frac{1}{4}\sqrt{3} \\ 0 \\ \frac{1}{2} - \frac{1}{4}\sqrt{3} \end{pmatrix},$$

$$\Phi_2 \left(\frac{1}{4} \right) = T_0\Phi_2 \left(\frac{1}{2} \right) = \begin{pmatrix} \frac{5}{16} + \frac{3}{16}\sqrt{3} \\ \frac{1}{8} + \frac{1}{8}\sqrt{3} \\ \frac{9}{16} - \frac{5}{16}\sqrt{3} \end{pmatrix},$$

$$\Phi_2 \left(\frac{3}{4} \right) = T_1\Phi_2 \left(\frac{1}{2} \right) = \begin{pmatrix} \frac{9}{16} + \frac{5}{16}\sqrt{3} \\ \frac{1}{8} - \frac{1}{8}\sqrt{3} \\ \frac{5}{16} - \frac{3}{16}\sqrt{3} \end{pmatrix},$$

i.e., $\phi_2(0) = 0, \phi_2(\frac{1}{4}) = \frac{5}{16} + \frac{3}{16}\sqrt{3}, \phi_2(\frac{1}{2}) = \frac{1}{2} + \frac{1}{4}\sqrt{3}, \phi_2(\frac{3}{4}) = \frac{9}{16} + \frac{5}{16}\sqrt{3}, \phi_2(1) = \frac{1+\sqrt{3}}{2}, \cdots.$

3.3 Convergence of the Cascade Algorithm

We now briefly discuss the convergence of the cascade algorithm (3.1). It is clear that if the algorithm converges, it is convergent to the solution of equation (1.1). However, the following example shows that the converse may not be true.

EXAMPLE 8.3.3: Consider the refinement equation $\phi(x) = \phi(2x) + \phi(2x + 3)$. Under the condition $\hat{\phi}(0) = 1$, it has the unique solution $\phi(x) = \frac{1}{3}\chi_{[0,3)} \in L^2$. We consider the cascade algorithm $\phi^j(x) = \phi^{j-1}(2x) + \phi^{j-1}(2x + 3)$ with the initial function $\phi^0(x) = \chi_{[0,1)}$. Let

$$I_j = \cup_{k=0}^{2^j-1} \left[\frac{3k}{2^j}, \frac{3k+1}{2^j} \right).$$

Then $\phi^j = \chi_{I_j}$, whose range contains only 0 and 1. It is obvious that $\phi^j(x)$ is not convergent to $\phi(x)$.

We now prove a convergence theorem for the cascade algorithm (3.1).

Theorem 8.3.1 Let (ϕ^j) be the sequence generated by the cascade algorithm (3.1) with $\phi^0 \in C$. Assume 1 is a simple eigenvalue of T_0 and T_1, and the modulus of other eigenvalues of T_0 and T_1 are less than 1. Then $\{\phi^j\}$ uniformly converges to a function $\phi \in C$.

Proof: Let

$$\Phi^j(x) = \left(\phi^j(x), \cdots, \phi^j(x + N - 1)\right)^T, \quad x \in [0, 1]$$

and $e = (1, \cdots, 1)^T$. Then $e^T \Phi^j(x) = \sum_k \phi^j(x - k) = 1$. By Lemma 8.3.1, we have

$$e^T (\Phi^n(x) - \Phi^m(x)) = 0, \quad \text{for all } x \in [0, 1], \quad n, m \in \mathbb{N}. \quad (3.10)$$

We denote the orthogonal complement space of e in $\mathbb{R}^N$ by E. By (3.10),

$$\Phi^n(x) - \Phi^m(x) \in E, \quad \text{for all } x \in [0, 1].$$

Write

$$A = \max(\|T_0\|_E, \|T_1\|_E).$$

Since 1 is a simple eigenvalue of T_0 and T_1, and the modulus of other eigenvalues of T_0 and T_1 are less than 1, we have $A < 1$, which yields

$$\|\Phi^{j+1}(x) - \Phi^j(x)\|_E$$
$$\leq \|T_{d_1(x)} \cdots T_{d_j(x)}\|_E \|\Phi^1(\tau^j x) - \Phi^0(\tau^j x)\|_E$$
$$\leq CA^j, \quad \text{for all } x \in [0, 1]$$

where

$$C = \max_{x\in[0,1]} ||\Phi^1(x) - \Phi^0(x)||_E$$

$$\leq \max_{x\in[0,N]} (|\phi^1(x)| + |\phi^0(x)|).$$

Hence $(\Phi^n(x))$ is a Cauchy sequence that uniformly converges to a vector-valued function $\Phi(x)$ on $[0,1]$. Hence, there is a continuous function ϕ such that

$$\lim_{n\to\infty} \max_{x\in[0,N]} |\phi^n(x) - \phi(x)| = 0.$$

Since for any j, $\phi^j \in C$, then $\phi \in C$. The theorem is proved. ∎

EXAMPLE 8.3.4: We apply Theorem 8.3.1 to study the convergence of the cascade algorithm for Daubechies scaling function ϕ_2. Recall that we have

$$T_0 = \frac{1}{4}\begin{pmatrix} 1+\sqrt{3} & 0 & 0 \\ 3-\sqrt{3} & 3+\sqrt{3} & 1+\sqrt{3} \\ 0 & 1-\sqrt{3} & 3-\sqrt{3} \end{pmatrix},$$

$$T_1 = \frac{1}{4}\begin{pmatrix} 3+\sqrt{3} & 1+\sqrt{3} & 0 \\ 1-\sqrt{3} & 3-\sqrt{3} & 3+\sqrt{3} \\ 0 & 0 & 1-\sqrt{3} \end{pmatrix}.$$

T_0 has three eigenvalues $1, 1/2$, and $\frac{1+\sqrt{3}}{4}$ and T_1 has three eigenvalues $1, 1/2$, and $\frac{1-\sqrt{3}}{4}$. By Theorem 8.3.1, the cascade algorithm for Daubechies scaling function ϕ_2 is convergent and $\phi \in L^2$.

Exercises

1. Let ϕ_0 be the hat function $N_2(x)$. Let $\mathbf{T}_m$ be the cascade algorithm created by the binomial filters of order $m = 3$ and 4, respectively, where the binomial filter of order 3 is $\mathbf{h} = \frac{1}{8}(1,3,3,1)$ and of order 4 is $\mathbf{h} = \frac{1}{16}(1,4,6,4,1)$. Run the cascade program 20 times to obtain ϕ_m^{20} (for $m = 3$ and 4, respectively.) Then draw the graphs of ϕ_3^{20} and ϕ_4^{20}.
2. Find the maximum error of $\max_{x\in\mathbb{R}} |N_m(x) - \phi_m^{20}(x)|$ and L^2- error $||N_m - \phi_m^{20}||$, $m = 3, 4$.
3. Let ϕ_0 be the hat function $N_2(x)$. Let $\mathbf{T}_L$ be the cascade algorithm created by Daubechies filter of order $L = 2, 3$, and 4, respectively. (See Table 8.1.) Run the cascade program 20 times (numerically) to obtain ϕ_L^{20} (for $L = 2, 3$, and 4, respectively). Then draw the graph of ϕ_L^{20}.
4. Estimate the error $||\phi_L^{21} - \phi_L^{20}||_C$ and $||\phi_L^{21} - \phi_L^{20}||_2$.

5. Prove the following without using Theorem 8.1.1: If ϕ satisfies (1.1) and the cascade algorithm with respect to h is convergent, then supp $\phi \subset [0, N]$.

6. Let $N_m(x)$ be the cardinal B-spline of order m. Find the CA-pair for $N_m(x)$, $m = 2, 3$, and 4. Then use $\tilde{T}_0$ to compute the values $N_m(k)$, $k \in \mathbb{Z}$.

7. Use the recursion algorithm to compute $N_m(\frac{k}{16})$, $k \in \mathbb{Z}, m = 2, 3, 4$.

8. Let $\phi_L(x)$ be the Daubechies scaling function of order L. Find the CA-pair for $\phi_L(x)$, $L = 3, 4$. Then use $\tilde{T}_0$ to compute the values $\phi_L(k)$, $k \in \mathbb{Z}$.

9. Use the values in Example 8.3.2 and Exercise 7, and apply the recursion algorithm to compute $\phi_L(\frac{k}{32})$, $k \in \mathbb{Z}, L = 2, 3, 4$. Draw the graph of $\phi_L, L = 2, 3, 4$.

10. Prove Lemma 8.3.3: If $\phi \in L^2$ in (1.1) is a continuous MRA generator, then $\sum_k \phi(k) = 1$.

11. Prove that if the cascade algorithm is convergent to a continuous function ϕ, then for any vector $\mathbf{v} \in \mathbb{R}^{N-1}$ with $\sum \mathbf{v}(k) = 1$,

$$\lim_{n \to \infty} \left(\tilde{T}_0 \right)^n \mathbf{v} = (\phi(1), \cdots, \phi(N-1))^T.$$

12. Let ϕ^j be the function obtained in the cascade algorithm (3.1). Assume ϕ_0 is the initial function of the algorithm. Prove

$$\widehat{\phi^j}(\omega) = \prod_{k=1}^{j} H(e^{-i\omega/2^k}) \widehat{\phi^0}(2^{-j}\omega).$$

13. Let $\phi^0 = N_2$. Prove that the cascade algorithms for Daubechies scaling functions ϕ_3 and ϕ_4 are uniformly convergent.

4 Wavelet Packets

4.1 The Construction of Wavelet Packets

Assume ϕ is an orthonormal MRA generator, and its corresponding orthonormal wavelet is ψ. Then

$$\phi(x) = 2 \sum_{k=0}^{N} h(k)\phi(2x - k), \tag{4.1}$$

$$\psi(x) = 2 \sum_{k=0}^{N} g(k)\phi(2x - k), \tag{4.2}$$

where $g(k) = (-1)^k h_{n-k}$ for some odd integer n. Both $\{\psi_{jk}\}_{j,k\in\mathbb{Z}}$ and $\{\phi_{0k}, \psi_{jk} \mid j \geq 0, k \in \mathbb{Z}\}$ are orthonormal bases of L^2. To find other orthonormal bases with structures similar to these wavelet bases, we introduce wavelet packets.

To simplify notation, in what follows, we denote

$$\begin{cases} \mu_0(x) = \phi(x) \\ \mu_1(x) = \psi(x) \end{cases} . \tag{4.3}$$

Correspondingly, we write

$$p_0(z) = \sum_{k\in\mathbb{Z}} h(k)z^k, \quad p_1(z) = \sum_{k\in\mathbb{Z}} g(k)z^k.$$

Then the Fourier transforms of (4.1) and (4.2) are

$$\begin{cases} \hat{\mu}_0(\omega) = p_0(e^{-i\omega/2})\hat{\mu}_0(\omega/2) \\ \hat{\mu}_1(\omega) = p_1(e^{-i\omega/2})\hat{\mu}_0(\omega/2) \end{cases} . \tag{4.4}$$

Definition 8.4.1 Let the set of functions $\{\mu_l\}_{l=0}^\infty$ be defined inductively by

$$\begin{cases} \hat{\mu}_{2n}(\omega) = p_0(e^{-i\omega/2})\hat{\mu}_n(\omega/2) \\ \hat{\mu}_{2n+1}(\omega) = p_1(e^{-i\omega/2})\hat{\mu}_n(\omega/2) \end{cases} \quad n = 0, 1, \cdots . \tag{4.5}$$

Then $\{\mu_l\}_{l=0}^\infty$ is called a *wavelet packet* of L^2 relative to ϕ.

To find the infinite product form of $\hat{\mu}_l$, $0 \leq l < \infty$, we assume the dyadic expansion of a nonnegative integer n is

$$n = \sum_{j=1}^\infty e_j 2^{j-1}, \quad e_j \in \{0, 1\}. \tag{4.6}$$

Then we have the following.

Theorem 8.4.1 Let n be a nonnegative integer and let its dyadic expansion be given by (4.6). Then the infinite product form of $\hat{\mu}_n$ is given by

$$\hat{\mu}_n(\omega) = \prod_{j=1}^\infty p_{e_j}(e^{-i\omega/2^j}).$$

Proof: We leave the proof as an exercise. ∎

We now prove the orthogonality of the integer translates of the wavelet packet.

Theorem 8.4.2 Let μ_n be defined by (4.5). Then

$$\int_{\mathbb{R}} \mu_n(x-j)\mu_n(x-k)\,dx = \delta_{jk} \qquad (4.7)$$

and

$$\int_{\mathbb{R}} \mu_n(x-j)\mu_{n+1}(x-k)\,dx = 0. \qquad (4.8)$$

Proof: We prove (4.7) by mathematical induction. It is clear that (4.7) is true for $n = 0$. Assume that (4.7) is true for all k, $0 \le k \le n-1$. We prove it is also true for n. In fact, since $\{\mu_k(x-l)\}_{l \in \mathbb{Z}}$ is orthonormal,

$$\sum_{l=-\infty}^{\infty} |\hat{\mu}_k(\omega + 2l\pi)|^2 = 1.$$

Let the dyadic expansion of n be given by (4.6). We denote the integer part of x by $[x]$. Then

$$n = 2\left[\frac{n}{2}\right] + e_1.$$

By (4.5), we have

$$\hat{\mu}_n(\omega) = p_{e_1}(e^{-i\omega/2})\hat{\mu}_{[\frac{n}{2}]}\left(\frac{\omega}{2}\right).$$

Note that $[\frac{n}{2}] < n$ for $n \ge 1$. Hence,

$$\int_{\mathbb{R}} \mu_n(x-j)\mu_n(x-k)\,dx = \frac{1}{2\pi}\int_{\mathbb{R}} |\hat{\mu}_n(\omega)|^2 e^{i(k-j)\omega}\,d\omega$$

$$= \frac{1}{2\pi}\int_{\mathbb{R}} |p_{e_1}(e^{-i\omega/2})|^2 \left|\hat{\mu}_{[\frac{n}{2}]}\left(\frac{\omega}{2}\right)\right|^2 e^{i(k-j)\omega}\,d\omega$$

$$= \frac{1}{2\pi}\sum_{k=-\infty}^{\infty}\int_{4k\pi}^{4(k+1)\pi} |p_{e_1}(e^{-i\omega/2})|^2 \left|\hat{\mu}_{[\frac{n}{2}]}\left(\frac{\omega}{2}\right)\right|^2 e^{i(k-j)\omega}\,d\omega.$$

Recall that $\left|p_{e_1}(e^{-i\omega/2})\right|^2$ is a 4π-periodic function, so we have

$$\int_{\mathbb{R}} \mu_n(x-j)\mu_n(x-k)\,dx$$

$$= \frac{1}{2\pi} \int_0^{4\pi} \left|p_{e_1}(e^{-i\omega/2})\right|^2 \sum_{k=-\infty}^{\infty} \left|\hat{\mu}_{[\frac{n}{2}]}\left(\frac{\omega}{2}+2l\pi\right)\right|^2 e^{i(k-j)\omega}\,d\omega$$

$$= \frac{1}{2\pi} \int_0^{4\pi} \left|p_{e_1}(e^{-i\omega/2})\right|^2 e^{i(k-j)\omega}\,d\omega$$

$$= \frac{1}{2\pi} \int_0^{2\pi} \left(\left|p_{e_1}(e^{-i\omega/2})\right|^2 + \left|p_{e_1}(-e^{-i\omega/2})\right|^2\right) e^{i(k-j)\omega}\,d\omega$$

$$= \frac{1}{2\pi} \int_0^{2\pi} e^{i(k-j)\omega}\,d\omega = \delta_{jk},$$

which is (4.7). We leave the proof of (4.8) as an exercise. ∎

4.2 Orthonormal Bases from Wavelet Packets

We now discuss how to use a wavelet packet to construct various orthonormal bases of L^2. Let $\{\mu_n\}_{n=0}^{\infty}$ be the wavelet packet relative to the orthonormal scaling function ϕ. Write

$$\mu_{n,j,k}(x) = 2^{j/2}\mu_n(2^j x - k).$$

We define

$$U_j^n = \text{clos}_2 \text{ span } \{2^{j/2}\mu_n(2^j x - k) \mid k \in \mathbb{Z}\}, \quad j \in \mathbb{Z}, \quad n \in \mathbb{Z}^+. \tag{4.9}$$

Let

$$\cdots \subset V_{-1} \subset V_0 \subset V_1 \subset \cdots$$

be the MRA generated by ϕ and $\{W_j\}_{j\in\mathbb{Z}}$ be the wavelet subspaces of L^2 generated by ψ. Then we have

$$W_j \oplus V_j = V_{j+1}, \quad W_j \perp V_j.$$

Converting to the new notation, we have

$$\begin{cases} U_j^0 = V_j, \ j \in \mathbb{Z}, \\ U_j^1 = W_j, \ j \in \mathbb{Z}, \end{cases}$$

and

$$U_j^0 \oplus U_j^1 = U_{j+1}^0, \quad U_j^0 \perp U_j^1.$$

We can generalize these relationships to the subspaces U_j^n, $n \in \mathbb{Z}^+$, $j \in \mathbb{Z}$.

Theorem 8.4.3 For any $n \in \mathbb{Z}^+$, we have

$$U_{j+1}^n = U_j^{2n} \oplus U_j^{2n+1}, \quad U_j^{2n} \perp U_j^{2n+1}, \quad j \in \mathbb{Z}.$$

Proof: It is obvious that $U_j^{2n} \subset U_{j+1}^n$ and $U_j^{2n+1} \subset U_{j+1}^n$. The relation $U_j^{2n} \perp U_j^{2n+1}$ is a consequence of Theorem 8.4.2. We now prove $U_{j+1}^n = U_j^{2n} \oplus U_j^{2n+1}$. Recall that U_{j+1}^n is spanned by $\{\mu_{n,j+1,k}\}_{k,j \in \mathbb{Z}}$. We only need to prove that each $\mu_{n,j+1,k} \in U_j^{2n} \oplus U_j^{2n+1}$. Recall that

$$|p_0(z)|^2 + |p_1(z)|^2 = 1, \quad z \in \Gamma,$$

which implies

$$\sum_k (h(l - 2k)h(m - 2k) + g(l - 2k)g(m - 2k)) = \delta_{lm}.$$

Thus, for any $m, j \in \mathbb{Z}$,

$$2 \sum_k \left(h(m - 2k)\mu_{2n}(2^j x - k) + g(m - 2k)\mu_{2n+1}(2^j x - k) \right)$$
$$= \sum_k \sum_l (h(m - 2k)h(l) + g(m - 2k)g(l))\mu_n(2^{j+1} x - 2k - l)$$
$$= \sum_k \sum_l (h(m - 2k)h(l - 2k) + g(m - 2k)g(l - 2k))\mu_n(2^{j+1} x - l)$$
$$= \mu_n(2^{j+1} x - m).$$

The theorem is proved. ∎

By Theorem 8.4.3, we can split each wavelet subspace in various ways.

Corollary 8.4.1 For each $j = 1, 2, \cdots$, and each k, $1 \leq k \leq j$, we have

$$W_j = U_{j-k}^{2^k} \oplus U_{j-k}^{2^k+1} \oplus \cdots \oplus U_{j-k}^{2^{k+1}-1}, \quad 1 \leq k \leq j.$$

Proof: We leave the proof as an exercise. ∎

We now may use wavelet packets to construct various orthonormal bases of L^2, by splitting W_j in different ways. We may choose to split some W_j spaces less often, or to split some of its subspaces more than others. To obtain the best bases for an application, we set an objective functional for the application, then choose the bases minimizing the objective functional.

EXAMPLE 8.4.1: Let μ_0 be the box function and μ_1 be the Haar wavelet. Then

$$\mu_2(x) = \chi_{[0,1/4)\cup[1/2,3/4)}(x) - \chi_{[1/4,1/2)\cup[3/4,1)}(x),$$
$$\mu_3(x) = \chi_{[0,1/4)\cup[3/4,1)}(x) - \chi_{[1/4,3/4)}(x),$$
$$\cdots\cdots$$

Consider the space $S = \text{span } \{\mu_{0,3,k} \mid k = 0,\cdots,7\}$. Using the wavelet packet relative to μ_0, we have

$$
\begin{aligned}
S &= \text{span } \{\mu_{i,0,0},\mu_{1,1,k},\mu_{1,2,l} \mid i = 0,1; k = 0,1; l = 0,\cdots,3\}\\
&= \text{span } \{\mu_{i,2,k} \mid i = 0,1; k = 0,\cdots,3\}\\
&= \text{span } \{\mu_{0,2,k},\mu_{2,1,i},\mu_{3,1,i} \mid k = 0,\cdots,3; i = 0,1\}\\
&= \text{span } \{\mu_{0,2,k},\mu_{i,0,0}, \mid k = 0,\cdots,3; i = 4,\cdots,7\}\\
&= \text{span } \{\mu_{i,0,0},\mu_{1,2,k}, \mid k = 0,\cdots,3,; i = 0,\cdots,3\}\\
&= \text{span } \{\mu_{i,1,k} \mid i = 0,\cdots 3; k = 0,1\}\\
&= \text{span } \{\mu_{i,1,k},\mu_{l,0,0} \mid i = 0,1; k = 0,1; l = 4,\cdots,7\}\\
&= \text{span } \{\mu_{i,1,k},\mu_{l,0,0} \mid i = 2,3; k = 0,1; l = 0,\cdots,3\}\\
&= \text{span } \{\mu_{l,0,0} \mid l = 0,\cdots,7\}\\
&= \cdots\cdots
\end{aligned}
$$

We now set $I_1 = [0,1/8) \cup [1/4,3/8) \cup [1/2,5/8) \cup [3/4,7/8), I_2 = [1/8,1/4) \cup 3/8,1/2)$, and define $f(x) = \chi_{I_1}(x) - \chi_{I_2}(x)$. Then

$$f = 2^{-3/2}(\mu_{0,3,0} - \mu_{0,3,1} + \mu_{0,3,2} - \mu_{0,3,3}, + \mu_{0,3,4} + \mu_{0,3,6})$$

$$= \frac{1}{2}(\mu_{1,2,0} + \mu_{1,2,1} + \mu_{1,2,2} + \mu_{1,2,3}) - \frac{\sqrt{2}}{2}\mu_{0,1,1}$$

$$= \mu_{5,0,0} - \frac{\sqrt{2}}{2}\mu_{0,1,1}.$$

If we choose the bases minimizing the number of the nonzero coefficients in the decomposition of f, then one of these bases is $\{\mu_{0,1,k}, \mu_{1,1,k}, \mu_{i,0,0} \mid k = 0,1; i = 4,5,6,7\}$.

Exercises

1. Prove Theorem 8.4.1: Let the set of functions $\{\mu_l\}_{l=0}^{\infty}$ be defined inductively by (4.5). Let n be a nonnegative integer and let its dyadic expansion be given by (4.6). Then the infinite product form of $\hat{\mu}_n$ is given by

$$\hat{\mu}_n(\omega) = \prod_{j=1}^{\infty} p_{e_j}(e^{-i\omega/2^j}).$$

2. Prove (4.8) in Theorem 8.4.2: Let μ_n be defined by (4.5). Then $\int_{\mathbb{R}} \mu_n(x-j)\mu_{n+1}(x-k)\,dx = 0$.

3. Prove Corollary 8.4.1: For each $j = 1, 2, \cdots$, and each $k, 1 \le k \le j$, we have

$$W_j = U_{j-k}^{2^k} \oplus U_{j-k}^{2^k+1} \oplus \cdots \oplus U_{j-k}^{2^{k+1}-1}, \quad 1 \le k \le j,$$

where U_j^n is defined by (4.9).

4. Let $\{\mu_l\}_{l \in \mathbb{Z}}$ be the wavelet packet in Example 8.4.1. Prove that

$$\mu_{2^n} = \sum_{k=0}^{2^n-1} \mu_1(2^n x - k).$$

5. Let $\{\mu_l\}_{l \in \mathbb{Z}}$ be the wavelet packet in Example 8.4.1. Assume

$$f(x) = \sum_{k \in \mathbb{Z}} c_k \mu_{0,3,k}(x).$$

   Develop an algorithm to compute the coefficients in the series.

6. Let ϕ be the *Shannon scaling function* defined by $\hat{\phi}(\omega) = \chi_{[-\pi,\pi)}(\omega)$. Let μ_l be defined by $\hat{\mu}_l(\omega) = \chi_{[-(l+1)\pi,-l\pi)\cup\{l\pi,(l+1)\pi)}(\omega)$. Prove that $\{\mu_l\}_{l \in \mathbb{Z}}$ is a wavelet packet relative to ϕ.

7. Let $\{\mu_l\}_{l \in \mathbb{Z}}$ be the wavelet packet in Exercise 6. Let $I_{j,l} = [2^j l\pi, 2^j(l+1)\pi)$. Let Λ and Ξ be two subsets of $\mathbb{Z}$. Prove that if $\cup_{j \in \Lambda, l \in \Xi} I_{j,l} = [0, \infty)$, and $I_{j,l} \cap I_{i,k} = \emptyset$, then $\{\mu_{j,l,k} \mid j \in \Lambda, l \in \Xi, k \in \mathbb{Z}\}$ is an orthonormal basis of L^2.

8. Let ϕ_2 be the Daubechies scaling function of order 2. Set $\mu_0 = \phi_2$. Assume $\{\mu_l\}_{l \in \mathbb{Z}}$ is the wavelet packet relative to ϕ_2. Draw the graph of $\mu_i, i = 2, 3$.

9. Let $\{\mu_l\}_{l \in \mathbb{Z}}$ be the wavelet packet in Exercise 6. Assume

$$f(x) = \sum_{k \in \mathbb{Z}} c_k \mu_{0,3,k}(x).$$

Develop an algorithm to compute the coefficients in the series

$$f(x) = \sum_{j=0}^{7} \sum_{k\in\mathbb{Z}} c_{j,k}\mu_{j,0,k}$$

from $(c_k)_{k\in\mathbb{Z}}$.

10. Let $\{\mu_l\}_{l\in\mathbb{Z}}$ be the wavelet packet in Exercise 6. Assume $f(x) = \mu_{2,0,0} + \mu_{4,0,0}$. Write

$$f(x) = \sum_{k\in\mathbb{Z}} c_k\mu_{0,0,k} + \sum_{j=0}^{2} \sum_{k\in\mathbb{Z}} d_k^j\mu_{1,j,k}.$$

Find the coefficients $(c_k), (d_k^j)$.

5 Compactly Supported Biorthogonal Wavelet Bases

The compactly supported orthonormal wavelets offer effective algorithms to decompose functions into wavelet series and recover them from their wavelet series. In many applications the symmetry of wavelets plays an important role. But compactly supported wavelets are asymmetric, except the Haar wavelet (see [6]). In Section 7.6, a more flexible structure of wavelet bases was obtained from biorthogonal scaling functions and the corresponding wavelets. In this section we study the construction of compactly supported biorthogonal scaling functions and wavelets, which are symmetric or antisymmetric. In this section, a scaling function is always assumed to be compactly supported.

5.1 Symmetry of the Scaling Function and Its Mask

We first discuss the symmetric biorthogonal scaling functions.

Definition 8.5.1 A real function $f(x) \in L^2$ is said to be *symmetric* (about $x = c$) if and only if there is $c \in \mathbb{R}$ such that

$$f(c - x) = f(c + x), \quad a.e.$$

A sequence $\mathbf{h} = \{h_k\}_{k=-\infty}^{\infty}$ is said to be *symmetric* (about $N/2$) if $h_k = h_{N-k}$ for all $k \in \mathbb{Z}$.

As usual, we write $H(z) = \sum_{k\in\mathbb{Z}} h_k z^k$, which engineers call the *z-transform* of $\mathbf{h}$. We say that $H(z)$ is symmetric if $\mathbf{h} = \{h_k\}_{k=-\infty}^{\infty}$ is symmetric.

Lemma 8.5.1 A real function $f \in L^2$ is symmetric about $x = c$ if and only if $e^{ic\omega}\hat{f}(\omega)$ is a real function. A real sequence $\mathbf{h}$ is symmetric about $N/2$ if and only if $e^{i\omega N/2}H(e^{-i\omega})$ is a real function.

Proof: We have

$$\int_{\mathbb{R}} f(c - x)e^{-i\omega x}\,dx = e^{-ic\omega}\int_{\mathbb{R}} f(c - x)e^{i\omega(c-x)}\,dx$$

$$= e^{-ic\omega}\int_{\mathbb{R}} f(y)e^{-i\omega y}\,dy = \overline{e^{ic\omega}\hat{f}(\omega)}$$

and

$$\int_{\mathbb{R}} f(c + x)e^{-i\omega x}\,dx = e^{ic\omega}\hat{f}(\omega).$$

Note that

$$f(c - x) = f(c + x) \text{ if and only if } e^{ic\omega}\hat{f}(\omega) = \overline{e^{ic\omega}\hat{f}(\omega)},$$

where the second equation means that $e^{ic\omega}\hat{f}(\omega)$ is a real function. We now prove the second statement. $h_k = h_{N-k}$ if and only if

$$\sum_{k\in\mathbb{Z}} h_k e^{-ik\omega} = \sum_{k\in\mathbb{Z}} h_{N-k} e^{-ik\omega}.$$

Note that

$$\sum_{k\in\mathbb{Z}} h_{N-k} e^{-ik\omega} = e^{-iN\omega}\sum_{k\in\mathbb{Z}} h_{N-k} e^{i(N-k)\omega}$$

$$= e^{-iN\omega}\sum_{k\in\mathbb{Z}} \overline{h_k e^{-ik\omega}}.$$

Hence $\mathbf{h}$ is symmetric about $N/2$ if and only if

$$\sum_{k\in\mathbb{Z}} h_k e^{-ik\omega} = e^{-iN\omega}\sum_{k\in\mathbb{Z}} \overline{h_k e^{-ik\omega}},$$

i.e., if and only if $e^{iN\omega/2}h(e^{-i\omega})$ is a real function. ∎

For understanding the importance of symmetry, we introduce the notion of phase. For $f \in L^2 \cap L^1$, we write

$$\hat{f}(\omega) = |\hat{f}(\omega)|\, e^{i\theta(\omega)}.$$

The function $\theta(\omega)$ is called the *phase* of f. Sometimes, in order to keep the continuity of $\theta(\omega)$, we write

$$\hat{f}(\omega) = R(\omega)e^{i\theta(\omega)},$$

where $R(\omega)$ is a real function such that $\theta(\omega)$ is continuous. In this case, $\theta(\omega)$ is called a *generalized phase* of f. If the phase (or generalized phase) function $\theta(\omega)$ is linear, then we say that f has a *linear* (or *generalized linear*) *phase*. Lemma 8.5.1 indicates that a symmetric function has a generalized linear phase.

In applications, when a function is given in the frequency domain, most devices can only measure $|\hat{f}(\omega)|$ or $R(\omega)$. Without the information of the phase $\theta(\omega)$, we cannot recover f from $|\hat{f}|$ or R, unless the phase is linear (or generalized linear). We will discuss phase in detail in Section 9.2.

The following theorem describes the relationship between the symmetry of a scaling function and its mask.

Theorem 8.5.1 Let ϕ be a compactly supported scaling function satisfying (1.1). Then ϕ is symmetric if and only if its mask **h** is symmetric.

Proof: By the lemma, if **h** is symmetric, then $e^{iN\omega/2}H(e^{-i\omega})$ is a real function. We write $H_r(e^{-i\omega}) = e^{iN\omega/2}H(e^{-i\omega})$. Since

$$\hat{\phi}(\omega) = \prod_{k=1}^{\infty} H(e^{-i\omega/2^k}),$$

and $\sum_{k=1}^{\infty} e^{iN\omega/2^{k+1}} = e^{iN\omega/2}$, we have

$$e^{iN\omega/2}\hat{\phi}(\omega) = \prod_{k=1}^{\infty} H_r(e^{-i\omega/2^k}),$$

which is a real function. It follows that $\phi(x)$ is symmetric about $x = N/2$. On the other hand, if ϕ is symmetric about $x = a$ for an $a \in \mathbb{R}$, then $e^{ia\omega}\hat{\phi}(\omega)$ is a real function. Because ϕ is compactly supported, its Fourier transform is an entire function. Then $H(e^{-i\omega/2}) = \dfrac{\hat{\phi}(\omega)}{\hat{\phi}(\omega/2)}$ holds almost everywhere. We now have that

$$e^{ia\omega/2}H(e^{-i\omega/2}) = \frac{e^{ia\omega}\hat{\phi}(\omega)}{e^{ia\omega/2}\hat{\phi}(\omega/2)}$$

is a real function. Hence, a must be an integer and **h** is symmetric about $a/2$. Recall that $h(k) = 0$, for $k > N$ and $k < 0$, and $h(0)h(N) \neq 0$. Hence a must be equal to N. The theorem is proved. ∎

When **h** is symmetric, it is convenient to set its symmetric center at 0 or 1. Thus, if in (1.1) N is even, by setting $M = N/2$, $p(k) = h(M + k)$, and $\varphi(x) =$

$\phi(x + M)$, we have

$$\varphi(t) = 2 \sum_{k=-M}^{M} p(k)\varphi(2t - k), \quad p(k) = p(-k), \tag{5.1}$$

which is an even function. Similarly if, in (1.1), N is odd then by setting $N = 2M + 1$, $p(k) = h(M + k)$, and $\varphi(x) = \phi(x + M)$, we have

$$\varphi(t) = 2 \sum_{k=-M}^{M+1} p(k)\varphi(2t - k), \quad p(k) = p(-k + 1), \tag{5.2}$$

which is symmetric about $x = \frac{1}{2}$.

5.2 The Construction of Symmetric Biorthogonal Scaling Functions and Wavelets

Assume ϕ and $\tilde{\phi}$ are biorthogonal scaling functions. From the preceding discussion, we see that we may always assume ϕ and $\tilde{\phi}$ are even or symmetric about $x = \frac{1}{2}$. Let $\tilde{h}$ and h be the masks of $\tilde{\phi}$ and ϕ, respectively. By Theorem 7.6.1, we have

$$2 \sum h(k)\tilde{h}(k - 2l) = \delta_{0l}, \tag{5.3}$$

which is equivalent to

$$\overline{H(z)}\tilde{H}(z) + \overline{H(-z)}\tilde{H}(-z) = 1, \tag{5.4}$$

which is a necessary condition for ϕ and $\tilde{\phi}$ to be biorthonormal scaling functions.

Similar to the construction of Daubechies scaling functions, we can construct the biorthogonal scaling function starting from (5.4). Set

$$\overline{H(z)}\tilde{H}(z) = P(z).$$

Then $P(z)$ satisfies (1.11). That is,

$$P_L(z) = \left(\frac{1+z}{2}\right)^L \left(\frac{1+z^{-1}}{2}\right)^L M_L(z),$$

where

$$M_L(z) = \sum_{k=0}^{L-1} \binom{L+k-1}{k} \left(\frac{1-z}{2}\right)^k \left(\frac{1-z^{-1}}{2}\right)^k.$$

To construct ϕ and $\tilde{\phi}$, we factor $P(z)$ into $\overline{H(z)}\tilde{H}(z)$. Recall that we want both ϕ and $\tilde{\phi}$ to be even or symmetric about $x = \frac{1}{2}$. By Theorem 8.5.1, we may choose $H(z)$ and $\tilde{H}(z)$ as follows.

Let ϕ be a cardinal B-spline function of order p. Thus, if p is even, we choose

$$H_p(z) := \left(\frac{1+z}{2}\right)^r \left(\frac{1+z^{-1}}{2}\right)^r, \quad r = \frac{p}{2},$$

which yields

$$\tilde{H}_{2L-p}(z) = \left(\frac{1+z}{2}\right)^{L-r} \left(\frac{1+z^{-1}}{2}\right)^{L-r} M_L(z).$$

In this case, both $H_p(z)$ and $\tilde{H}_{2L-p}(z)$ are symmetric about 0. If p is odd, we choose

$$H_p(z) := \left(\frac{1+z}{2}\right)^r \left(\frac{1+z^{-1}}{2}\right)^{r-1}, \quad p = 2r - 1,$$

which yields

$$\tilde{H}_{2L-p}(z) = \left(\frac{1+z}{2}\right)^{L-r} \left(\frac{1+z^{-1}}{2}\right)^{L-r+1} M_L(z).$$

In this case, $H_p(z)$ is symmetric about $1/2$ while $\tilde{H}_{2L-p}(z)$ is symmetric about $-1/2$. We now set $\tilde{p} = 2L - p$ and $\tilde{H}_{\tilde{p},p}(z) = \tilde{H}_{2L-p}(z)$. It is known that $H_p(z)$ is the symbol of ϕ_p^c. Let the scaling function with the symbol $\tilde{H}_{\tilde{p},p}(z)$ be denoted by $\tilde{\phi}_{\tilde{p},p}$, where we assume L is so large that $\tilde{\phi}_{\tilde{p},p} \in L^2$. Then, $\tilde{\phi}_{\tilde{p},p}$ is the dual scaling function of ϕ_p^c. Different L will create a different version of dual scaling functions of ϕ_p^c.

We call ϕ_p^c and $\tilde{\phi}_{\tilde{p},p}$ the *spline biorthogonal scaling functions* of order $(p, \tilde{p})$. Table 8.3 gives the symbols $\tilde{H}_{\tilde{p},p}(z)$ of $\tilde{\phi}_{\tilde{p},p}$.

Using the method in Section 7.6, we can construct the corresponding biorthogonal wavelets for the pair of ϕ_p and $\tilde{\phi}_{\tilde{p},p}$. Write

$$G_{\tilde{p},p}(z) = z\overline{\tilde{H}_{\tilde{p},p}(-z)},$$
$$\tilde{G}_p(z) = z\overline{H_p(-z)}.$$

We define $\psi_{\tilde{p},p}$ by

$$\widehat{\psi_{\tilde{p},p}}(\omega) = G_{\tilde{p},p}(e^{-i\omega/2})\phi_p(\omega/2) \tag{5.5}$$

and define $\tilde{\psi}_{\tilde{p},p}$ by

$$\widehat{\tilde{\psi}_{\tilde{p},p}}(\omega) = \tilde{G}_p(e^{-i\omega/2})\widehat{\tilde{\phi}_{\tilde{p},p}}(\omega/2). \tag{5.6}$$

Table 8.3 The symbols of ϕ_p^c and $\tilde{\phi}_{\tilde{p},p}$.

p	H_p	$\tilde{p}$	$\tilde{H}_{\tilde{p},p}$
1	$\frac{1}{2}(1+z)$	3	$\frac{1}{16}(-z^{-2}+z^{-1}+8+8z+z^2-z^3)$
		5	$\frac{1}{256}(3z^{-4}-3z^{-3}-22z^{-2}+22z^{-1}+128$ $+128z+22z^2-22z^3-3z^4+3z^5)$
2	$\frac{1}{4}(z^{-1}+2+z)$	2	$\frac{1}{8}(-z^{-2}+2z^{-1}+6+2z-z^2)$
		4	$\frac{1}{128}(3z^{-4}-6z^{-3}-16z^{-2}+38z^{-1}+90$ $+38z-16z^2-6z^3+3z^4)$
		6	$\frac{1}{1024}(-5z^{-6}+10z^{-5}+34z^{-4}-78z^{-3}$ $-123z^{-2}+324z^{-1}+700+\cdots)$
		8	$\frac{1}{2^{15}}(35z^{-8}-70z^{-7}-300z^{-6}+670z^{-5}$ $+1228z^{-4}-3126z^{-3}-3796z^{-2}$ $+10718z^{-1}+22050+\cdots)$
3	$\frac{1}{8}(z^{-1}+3+3z+z^2)$	3	$\frac{1}{64}(3z^{-3}-9z^{-2}-7z^{-1}+45+45z+\cdots)$
		5	$\frac{1}{512}(-5z^{-5}+15z^{-4}+19z^{-3}-97z^{-2}$ $-26z^{-1}+350+350z+\cdots)$
		7	$\frac{1}{2^{14}}(35z^{-7}-105z^{-6}-195z^{-5}+865z^{-4}$ $+363z^{-3}-3489z^{-2}-307z^{-1}$ $+11025+11025z+\cdots)$
		9	$\frac{1}{2^{17}}(-63z^{-9}+189z^{-8}+469z^{-7}-1911z^{-6}$ $-1308z^{-5}+9188z^{-4}+1140z^{-3}-29676z^{-2}$ $+190z^{-1}+87318+87318z+\cdots)$

Then $\psi_{\tilde{p},p}$ and $\tilde{\psi}_{\tilde{p},p}$ are biorthogonal wavelets. They are symmetric or anti-symmetric. It is clear that the scaling function ϕ_p^c and the wavelet $\psi_{\tilde{p},p}$ are spline functions (but $\tilde{\phi}_{\tilde{p},p}$ and $\tilde{\psi}_{\tilde{p},p}$ are not).

Definition 8.5.2 Let ϕ_p^c and $\tilde{\phi}_{\tilde{p},p}$ be the spline biorthogonal scaling functions of order $(p, \tilde{p})$. Then their corresponding biorthogonal wavelets $\psi_{\tilde{p},p}$ and $\tilde{\psi}_{\tilde{p},p}$ are called *spline biorthogonal wavelets* of order $(p, \tilde{p})$.

Exercises

1. Assume a real function $f \in L^2 \cap L^1$ is antisymmetric, i.e., here is a $c \in \mathbb{R}$, such that $f(c-x) = -f(c+x)$ a.e. Prove that f has a generalized linear phase and find its phase function.

2. Prove that if $\phi \in L^1 \cap L^2$ is an MRA generator, then it cannot be antisymmetric.

3. A sequence $(a_k)_{k \in \mathbb{Z}} \in l^1$ is said to be *antisymmetric* if there is an integer $m \in \mathbb{Z}$ such that $a_{m-k} = -a_k$ for all $k \in \mathbb{Z}$. Let $f \in L^2$ be a compactly supported symmetric function and $(a_k)_{k \in \mathbb{Z}} \in l^1$ be an antisymmetric sequence. Prove that the function $\sum a_k f(x - k)$ is antisymmetric.

4. A Laurent polynomial $P(z)$ is called *reciprocal* if $P(1/z) = z^k P(z)$ for a $k \in \mathbb{Z}$. Prove that the coefficient of a reciprocal Laurent polynomial is symmetric.

5. Assume $P(z)$ is a reciprocal Laurent polynomial. Under what conditions is $P(-z)$ also reciprocal? Under what conditions are the coefficients of $P(-z)$ antisymmetric?

6. Apply the result in Exercise 7 of Section 8.2 (setting $n = 2$) to prove $\tilde{\phi}_{2,2} \in L^2$.

7. Prove that for even p, $\psi_{\tilde{p},p}$ and $\tilde{\psi}_{\tilde{p},p}$ are both symmetric, and for odd p, they are antisymmetric.

8. Obtain the masks for the biorthogonal scaling functions $\tilde{\phi}_{\tilde{p},p}$ for

    (a) $p = 4, \tilde{p} = 4$.

    (b) $p = 5, \tilde{p} = 5$.

9. Use the cascade algorithm to draw the graphs of the biorthogonal scaling functions $\tilde{\phi}_{1,3}, \tilde{\phi}_{1,5}, \tilde{\phi}_{2,2}, \tilde{\phi}_{2,4}, \tilde{\phi}_{3,7}$, and $\tilde{\phi}_{4,4}$.

10. Use the results in Exercise 9 to draw the graphs of the following wavelets: $\tilde{\psi}_{1,3}, \tilde{\psi}_{1,5}, \tilde{\psi}_{2,2}, \tilde{\psi}_{2,4}, \tilde{\psi}_{3,7}$, and $\tilde{\psi}_{4,4}$.

11. Based on the graphs of the central cardinal B-splines $\phi_1^c, \phi_2^c, \phi_3^c$, and ϕ_4^c, draw the graphs of $\psi_{1,3}, \psi_{1,5}, \psi_{2,2}, \psi_{2,4}, \psi_{3,7}$, and $\psi_{4,4}$.

Chapter 9

Wavelets in Signal Processing

In this chapter, we introduce an application of wavelets to signal processing: how to use wavelet bases to transform signals. Signals exist everywhere in our world. Speeches, images, geological data, even records of stock price fluctuations can be considered as signals. In our contemporary scientific and technological activity, we often need to transmit signals (as in telecommunications), analyze them (for satellite or medical images, or for stock price fluctuations), and compress or synthesize them (as in computer vision and visualization). These tasks are the objectives of signal processing. As Yves Meyer mentions [21]: "The objectives of signal processing are to analyze accurately, code efficiently, transmit rapidly, and then to reconstruct carefully at the receiver the delicate oscillations or fluctuations of this function of time. This is important because all of the information contained in the signal is effectively present and hidden in the complicated arabesques appearing in its graphical representation." To accomplish these tasks, we need to transform signals (also called code/decode) into a particular form for a certain task. Fast wavelet transformation provides effective algorithms for signal coding and decoding.

I Signals

In this section, we introduce the basic concepts of signals. Mathematically, a one-dimensional signal appears as a function of time, say $x(t)$. If the time variable is changed continuously, then the signal $x(t)$ is called an *analog signal* or a *continuous signal*. If the time variable runs through a discrete set, then $x(t)$ is called a *discrete signal*, or a *digital signal*, which is a numerical representation of an analog signal. (In a strict sense, a digital signal means a quantized discrete signal, whose range is a finite number set. However, we often do not distinguish digital signals from discrete signals since the discrete signals are more conducive to a mathematical framework.) We often use the short-term *signals* for discrete digitals because in signal processing, we mainly deal with discrete digitals.

1.1 Analog Signals

Let $x(t)$ be an analog signal. If $x \in L^2$, then x is called an analog signal with *finite energy* and $\|x\|$ is called the *energy* of x. If $x \in \tilde{L}^2_{2\sigma}$, then x is called a *periodic analog signal* and its energy is $\|x\|_{\tilde{L}^2_{2\sigma}}$. In this chapter we mainly discuss the analog signals in L^2. The term "function" is often used to stand for "analog signal." Oscillations or fluctuations are very important feature of analog signals. We use *frequency* to characterize this feature. The term "frequency" originally comes from the study of waves. It is known that the frequency of the cosine wave

$$c(t) = A\cos(2\pi ft - \theta) \tag{1.1}$$

is f (which is the reciprocal of the period). If the unit of time t in (1.1) is seconds, then we say that the frequency of $c(t)$ is f Hz. The *circular frequency* of $c(t)$ is $2\pi f$ rad/s. The frequency of a complex-valued wave is defined in the same way. For example, the complex simple harmonic oscillator $e^{i\omega t}$ has frequency $\frac{\omega}{2\pi}$ Hz and its circular frequency is ω rad/s. If an analog signal x is a combination of several simple harmonic motions with different frequencies, say $x(t) = \sum_{k=0}^{n} A_k \cos(k\omega t - \theta_k)$, then x has a frequency spectrum which covers the frequency set $\{\frac{k\omega}{2\pi} \mid k = 0, 1, \cdots, n\}$, where the highest frequency is $\frac{n\omega}{2\pi}$. In general, if $x \in \tilde{L}^2_{2\sigma}$, then the frequency spectrum of x is exhibited by its Fourier series $\sum_{k=-\infty}^{\infty} c_k e^{ik\frac{\pi}{\sigma}t}$. If a function $x(t) \in L^2$, then its Fourier transform $\hat{x}(\omega)$ provides the frequency information of x. Note that, for a real-valued function $x(t)$, $\hat{x}(-\omega) = \overline{\hat{x}(\omega)}$. Hence, a real-valued function is uniquely determined by $\hat{x}(\omega)$, $\omega \geq 0$.

Frequency information is important in analyzing an analog signal. In Chapter 6, we saw how to use Fourier series to analyze periodic functions and how to use the Fourier transform to analyze the functions on $\mathbb{R}$. An important principle in frequency analysis is: High frequency indicates violent oscillations or fluctuations. For example, we can use this principle to analyze noise.

An analog signal often carries noise. Let $x(t)$ be a received or observed analog signal. It can be written as

$$x(t) = y(t) + n(t),$$

where $y(t)$ is the *proper function* and $n(t)$ is the *noise*. Noise is broadly defined as an additive contamination. It is not predictable. Hence we often use the term "random noise" to describe an unknown contamination added on a function. Intuitively, noise has violent fluctuations, i.e., its frequency spectrum occurs in the high-frequency region. Removing high-frequency components from a function results in a smooth function with less noise.

1.2 Approximation

Changing an analog signal to a discrete signal is called *discretization*. A popular way to discretize an analog signal is sampling. In many applications, the analog signals are continuous functions. Let $x \in L^2$ be continuous. Then an observer records a value of an analog signal $x(t)$ after a time step T to obtain a discrete signal $\{x(nT)\}_{n \in \mathbb{Z}}$. This method is called *sampling*. We call T the *sampling period*, call $\nu_s = \frac{1}{T}$ the *sampling rate* (or sampling frequency), and call $\{x(nT)\}_{n \in \mathbb{Z}}$ the *sampling data* (of the signal x). If $x(t)$ can be completely recovered from its sampling data $\{x(nT)\}_{n \in \mathbb{Z}}$, then the sampling is called a *lossless sampling*, and x is called *well sampled*. Otherwise it is called a *lossy sampling* and x is said to be *undersampled*. If $\{x(nT)\}_{n \in \mathbb{Z}}$ has a proper subsequence which forms lossless sampling data of x, then we say x is *oversampled*.

In general, a function in L^2 cannot be well sampled, only functions in some subspaces of L^2 can. Among these subspaces, shift-invariant spaces play an important role.

Definition 9.1.1 A subspace $U_h \subset L^2$ is called a *(h-)shift-invariant space*, if there is an $h > 0$ such that $f \in U_h$ if and only if $f(\cdot + h) \in U_h$.

In general, we can create shift-invariant spaces as follows. Let $\phi \in L^2$ be a function such that $\sum_{k \in \mathbb{Z}} c_k \phi(t - k)$ is convergent almost everywhere for any $(c_k) \in l^2$. Let

$$U = \text{span } _{L^2}\{\phi(t - k) \mid k \in \mathbb{Z}\}. \tag{1.2}$$

Then U is a 1-shift-invariant space, and

$$U_h = \left\{ f \mid f\left(\frac{t}{h}\right) \in U \right\}$$

is an h-shift-invariant space. For $x \in L^2$, we denote its orthogonal projection on U_h by x_h. We have the following.

Theorem 9.1.1 If ϕ in (1.2) is an MRA generator, then $\lim_{h \to 0} \|x - x_h\| = 0$.

Proof: Let $\{V_n\}_{n \in \mathbb{Z}}$ be the MRA generated by ϕ and let ϕ^* be the orthonormal scaling function in V_0. Then for any $y \in L^2$,

$$y_{2^{-j}}(t) = \sum_{k \in \mathbb{Z}} \langle y, \phi_{j,k}^* \rangle \phi_{j,k}^*(t)$$

is the orthogonal projection of y on V_j. Since $\{V_n\}_{n \in \mathbb{Z}}$ is an MRA, for $\epsilon > 0$, there is a $J > 0$ such that

$$\|y(t) - y_{2^{-j}}(t)\| < \epsilon, \quad j \geq J.$$

Let $h > 0$ satisfy $h \leq 2^{-J}$. Let $j \in \mathbb{N}$ and $c > 0$ be numbers such that $c = \frac{2^{-j}}{h}$ and $1 \leq \frac{2^{-j}}{h} < 2$. Hence $j \geq J$. For $x \in L^2$, we define $y(t) = x(ct)$. Let

$$\tilde{x}_h(t) = \sum_k \langle y, \phi_{j,k}^* \rangle \phi_{j,k}^*(t/c) = y_{2^{-j}}(t/c).$$

Then $\tilde{x}_h \in U_h$, and for $h \leq 2^{-J}$,

$$\|x(t) - x_h(t)\| \leq \|x(t) - \tilde{x}_h(t)\| = \|y(t/c) - y_{2^{-j}}(t/c)\| \leq \sqrt{2}\epsilon.$$

The theorem is proved. ∎

It is clear that, if $\{\phi(x - k)\}_{k \in \mathbb{Z}}$ is an orthonormal basis of U, then $\left\{\frac{1}{\sqrt{h}} \phi(\frac{t}{h} - k)\right\}_{k \in \mathbb{Z}}$ is an orthonormal basis of U_h and

$$x_h(t) = \sum_k \frac{1}{h} \left\langle x, \phi\left(\frac{t}{h} - k\right) \right\rangle \phi\left(\frac{t}{h} - k\right)$$

is an approximation of x as $h \to 0$. Similarly, if $\phi^* \in U$ is a dual function of ϕ, then

$$\left\{\frac{1}{\sqrt{h}} \phi\left(\frac{t}{h} - k\right)\right\}_{k \in \mathbb{Z}}$$

and

$$\left\{\frac{1}{\sqrt{h}} \phi^*\left(\frac{t}{h} - k\right)\right\}_{k \in \mathbb{Z}}$$

are biorthonormal bases in U_h and

$$x_h(t) = \sum_k \frac{1}{h} \left\langle x, \phi^*\left(\frac{t}{h} - k\right) \right\rangle \phi\left(\frac{t}{h} - k\right)$$

is an approximation of x as $h \to 0$.

Speeches and sounds are important signals. They can be considered as frequency-bounded signals, which are defined as follows.

Definition 9.1.2 If the Fourier transform of a signal is compactly supported, then the signal is called a *frequency-bounded signal*. Let $x \in L^2$ be a frequency-bounded signal such that

$$\omega_0 = \max\{|\omega| \mid \omega \in \text{supp } \hat{x}\}.$$

Then $\frac{\omega_0}{2\pi}$ is called the *highest frequency* of x.

It is clear that the space

$$S_\sigma = \{f \in L^2 \mid \hat{f} \in [-\sigma\pi, \sigma\pi]\}$$

is a shift-invariant space (for any $h > 0$). For $x \in L^2$, let x_σ be the orthogonal projection of x on S_σ. Then $\hat{x}_\sigma(\omega) = \hat{x}(\omega)\chi_{[-\sigma\pi, \sigma\pi]}(\omega)$ and

$$\lim_{\sigma \to \infty} \|x - x_\sigma\| = 0.$$

Hence, $x \in L^2$ can be approximated by frequency bounded functions.

EXAMPLE 9.1.1:   Let

$$\text{sinc } (t) := \frac{\sin \pi t}{\pi t},$$

where sinc $(0) = 1$. It is an entire function on $\mathbb{R}$. The Fourier transform of sinc (t) is

$$\widehat{\text{sinc}} \, (\omega) = \begin{cases} 1, & |\omega| \leq \pi, \\ 0, & |\omega| > \pi. \end{cases} \tag{1.3}$$

Hence, sinc (t) is a frequency bounded function with highest frequency $\frac{1}{2}$.

1.3 Sampling Theorems

Sampling theorems state the well-sampling conditions for analog signals. The following theorem is for frequency-bounded functions.

Theorem 9.1.2 [Shannon Sampling Theorem] Let $x \in L^2$ be a frequency bounded analog signal, which satisfies

$$\hat{x}(\omega) = 0, \quad |\omega| > \pi\sigma, \quad \sigma > 0. \tag{1.4}$$

Write $T = \frac{1}{\sigma}$. Then

$$x(t) = \sum_n x(nT)\frac{\sin \pi\sigma(t - nT)}{\pi\sigma(t - nT)} = \sum_n x(nT) \text{ sinc } (\sigma(t - nT)). \qquad (1.5)$$

Proof: Let $\hat{x}_p(\omega)$ be the $2\pi\sigma$-periodization of $\hat{x}(\omega)$. Then $\hat{x}_p \in L^2_{2\pi\sigma}$ and $\hat{x}_p$ can be expanded as a Fourier series

$$\hat{x}_p(\omega) = \sum_n C_n e^{-inT\omega},$$

where

$$\begin{aligned}
C_n &= \frac{1}{2\pi\sigma} \int_{-\pi\sigma}^{\pi\sigma} \hat{x}_p(\omega)e^{inT\omega}\, d\omega \\
&= \frac{1}{2\pi\sigma} \int_{-\infty}^{\infty} \hat{x}(\omega)e^{inT\omega}\, d\omega \\
&= \frac{1}{\sigma}x(nT).
\end{aligned}$$

Therefore, we have

$$\hat{x}(\omega) = \frac{1}{\sigma} \sum x(nT)e^{-inT\omega}, \quad |\omega| \le \pi\sigma. \qquad (1.6)$$

Hence,

$$\begin{aligned}
x(t) &= \frac{1}{2\pi} \int_{\mathbb{R}} \hat{x}(\omega)e^{it\omega}\, d\omega = \sum_n C_n \left(\frac{1}{2\pi} \int_{-\pi\sigma}^{\pi\sigma} e^{-inT\omega}e^{it\omega}\, d\omega\right) \\
&= \sum_n C_n \frac{\sin \pi\sigma(t - nT)}{\pi(t - nT)} = \sum_n x(nT) \text{ sinc } (\sigma(t - nT)).
\end{aligned}$$

The theorem is proved.   ∎

Let $\bar{\sigma}$ be the smallest number such that (1.4) holds. Then the number $\bar{\sigma}$ is called the *Nyquist rate* of x (or the Nyquist frequency of x). Since the highest frequency of x is

$$\nu = \frac{\pi\bar{\sigma}}{2\pi} = \frac{\bar{\sigma}}{2},$$

the Nyquist rate is twice that of the signal's highest frequency. The Shannon Sampling Theorem confirms the following.

Corollary 9.1.1 In order to get a lossless sampling for a frequency-bounded signal, the sampling rate must be equal to or greater than the Nyquist rate.

> **EXAMPLE 9.1.2:** If we want to sample a speech signal with the highest frequency 4 kHz without distortion, then the sampling rate is at least 8 kHz. High-quality music signals are assumed to have highest frequency 22.05 kHz, so the sampling rate for them has to be at least 44.1 kHz.

Note that sinc (t) is a Lagrangian interpolation function in the space S_1, that is, sinc $(k) = \delta_{0,k}$. Then (1.5) can be considered as a Lagrangian interpolation formula for functions in S_σ. From the interpolation point of view, we can generalize the Shannon sampling theorem to other shift-invariant spaces. Recall that (see Section 7.5) a function ϕ is said to satisfy the interpolation condition if $\sum_{k\in\mathbb{Z}} \hat{\phi}(\omega + 2k\pi) \neq 0$ for all $\omega \in \mathbb{R}$. In this case, the function $\tilde{\phi}$ defined by

$$\widehat{\tilde{\phi}}(\omega) = \frac{\hat{\phi}(\omega)}{\sum_{k\in\mathbb{Z}} \hat{\phi}(\omega + 2k\pi)} \tag{1.7}$$

is a Lagrangian interpolation function in $U = \text{span }_{L^2}\{\phi(t-k)\}_{k\in\mathbb{Z}}$. The general sampling theorem is the following.

Theorem 9.1.3 Assume ϕ satisfies the interpolation condition. Let $U = \text{span}$ $_{L^2}\{\phi(t - k)\}_{k\in\mathbb{Z}}$ and $U_\sigma = \{f \mid f(\frac{\cdot}{\sigma}) \in U\}$. Then for any function $f \in U_\sigma$, we have

$$f(t) = \sum_{k\in\mathbb{Z}} f(k/\sigma)\tilde{\phi}(\sigma t - k), \tag{1.8}$$

where $\tilde{\phi}$ is the Lagrangian interpolation function defined by (1.7).

Proof: We leave the proof as an exercise. ∎

Later, we call ϕ (or $\tilde{\phi}$) the underlying function for the sampling.

I.4 Discrete Signals

A *discrete signal* (or briefly, a *signal*) is a number sequence. In this chapter we assume any signal $x = (x(n))_{-\infty}^{\infty} \in l^2$, which is called a *finite energy signal*. Then $\|x\|_2$ is the *energy* of x. If x has finite nonzero terms, then it is a *finite signal*. Most signals are sampling data of analog signals.

To recover an analog signal from its sampling data, we need to know the sampling period and the underlying function. For example, assume the signal x is the sampling data of a function $f(t)$ in the space U_h generated by ϕ. Let $\tilde{\phi}$ be the Lagrangian interpolation function in U. Then x represents the function

$$f(t) = \sum x(n)\tilde{\phi}\left(\frac{t}{h} - n\right) \in U_h. \tag{1.9}$$

In applications, usually the underlying function $\tilde{\phi}$ is unknown. Hence, we have to *guess* $\tilde{\phi}$ according to known properties of $x(t)$. For example, if x is sample data of a speech signal, then $x(t)$ can be considered as a frequency bounded function.

Definition 9.1.3 The *z-transform* of a signal x is the Laurent series

$$X(z) = \sum_{n\in\mathbb{Z}} x(n)z^n.$$

The *math reduced notation* (or *Fourier form*) of the z-transform of x is

$$X(\omega) = \sum_n x(n)e^{-in\omega}, \qquad z = e^{-i\omega},$$

which is also called the *symbol* of x.

By (1.9), when the sampling period h and the underlying function ϕ are given, we can obtain the Fourier transform of $f(t)$ from the formula

$$\hat{f}(\omega) = hX(h\omega)\widehat{\tilde{\phi}}(h\omega). \tag{1.10}$$

EXAMPLE 9.1.3:   If x is the sampling data of a frequency-bounded analog signal x with supp $\hat{x}(\omega) \subset [-\pi, \pi]$, then x represents the function

$$x(t) = \sum_n x(n) \text{ sinc } (x - n). \tag{1.11}$$

It follows that

$$\hat{x}(\omega) = X(\omega), \quad |\omega| \leq \pi. \tag{1.12}$$

By (1.12), the frequency property of the signal $x(t)$ is totally determined by the z-transform of x. It is reasonable to use $X(\omega)$ to characterize the frequency property of the signal x.

Exercises

1. Let $I = [n\sigma\pi, (n+1)\sigma\pi] \cup [-(n+1)\sigma\pi, -n\sigma\pi], \sigma > 0$. Assume the Fourier transform of the signal $w(t)$ is $\hat{w}(\omega)$ with supp $\hat{w} \subset I$. Find the formula that represents $w(t)$ by its sampling data $\{w(n/\sigma)\}_{n\in\mathbb{Z}}$.

2. Let $I_\sigma = [-\sigma\pi, \sigma\pi]$ and $\hat{f}_\sigma$ be defined by $\hat{f}_\sigma(\omega) = \hat{f}(\omega)\chi_{I_\sigma}$. Write $h_\sigma(t) = \sigma$ sinc (σt). Prove that $f_\sigma(t) = (f * h_\sigma)(t)$.

3. Let the Fourier transform of $x(t)$ be $\hat{x}(\omega) = \cos(\omega)\chi_{[-\pi,\pi]}(\omega)$. Derive the formula for $x(n)$.

4. Let the Fourier transform of $x(t)$ be $\hat{x}(\omega) = |\omega|\chi_{[-\pi,\pi]}(\omega)$. Derive the formula for $x(n)$.

5. Prove Theorem 9.1.3: Assume ϕ satisfies the interpolation condition. Let $U = \text{span}_{L^2}\{\phi(t-k)\}_{k\in\mathbb{Z}}$ and $U_\sigma = \{f \mid f(\frac{\cdot}{\sigma}) \in U\}$. Then for any function $f \in U_\sigma$, we have
$$f(t) = \sum_{k\in\mathbb{Z}} f(k/\sigma)\tilde{\phi}(\sigma t - k),$$

6. Assume the digital signal $\mathbf{f} = (f(n))_{n\in\mathbb{Z}}$ is the sampling data of the function $f \in L^2$ which satisfies the condition for the Poisson summation formula. Let $X(z) = \sum_{k\in\mathbb{Z}} f(n)z^n$ be the z-transform of $\mathbf{f}$. Prove that
$$X(e^{-i\omega}) = \sum_{k\in\mathbb{Z}} \hat{f}(\omega + 2k\pi).$$

7. Let the z-transform of a signal $\mathbf{x}$ be $H(z) = \sum x(n)z^n$. Assume the z-transform of the signal $\mathbf{y}$ is $H(z^2)$. Represent $y(n)$ by $(x(j))_{j\in\mathbb{Z}}$.

8. Let $x(n) = f(n/2^j), j \geq 0$, and the z-transform of $\mathbf{x}$ be $X(z)$. Find the z-transform of $\{f(n)\}$.

9. Let $f(t) = \sum x(n)\tilde{\phi}\left(\frac{t}{h} - n\right)$. Prove $\hat{f}(\omega) = h\mathbf{X}(h\omega)\hat{\tilde{\phi}}(h\omega)$ [see (1.10)].

10. Let $\mathbf{x}$ be the sampling data (with sampling period 1) of a frequency-bounded function x with supp $\hat{x} \subset [-\pi, \pi]$. Prove that $\|\mathbf{x}\|_2 = \|x(t)\|$.

11. Let $\mathbf{x}$ be the sampling data (with sampling rate σ) of a frequency-bounded function x with supp $\hat{x} \subset [-\sigma\pi, \sigma\pi]$. What is the relation of $\|\mathbf{x}\|_2$ to $\|x(t)\|$?

12. Assume $\mathbf{x}$ is sampling data of an analog function $x(t)$. If $\mathbf{x}$ is a finite signal, then is $x(t)$ compactly supported?

13. Let the continuous function $x(t)$ be compactly supported. Is there a finite signal $\mathbf{x}$ that can completely represent $x(t)$?

14. List a condition so that a compactly supported function can be lossless sampled by a finite signal $\mathbf{x}$.

2 Filters

To analyze, code, reconstruct signals and so on, special operators on signals are needed. Among these operators, a filter is the most important one. In this section, we introduce linear filters, which are convolution operators on l^2. We first introduce the natural basis in l^2. Let $\delta_k = (\delta_{k,j})_{j \in \mathbb{Z}}$, which vanishes everywhere except at the kth term, where it has value 1. Then $\{\delta_k\}_{k \in \mathbb{Z}}$ is an orthonormal basis of l^2. Any $x \in l^2$ can be represented as

$$x = \sum_{k \in \mathbb{Z}} x(k)\delta_k. \tag{2.1}$$

In signal processing, δ_k is called a *unit impulse* at time k.

Definition 9.2.1 An operator S on l^2 is called a *shift operator* (also called a *time-delay operator*) if

$$(Sx)(n) = x(n-1), \quad x \in l^2,$$

an operator T on l^2 is called *time-invariant* if

$$ST = TS,$$

and an operator T on l^2 is called a *linear operator* if for any $x \in l^2$,

$$Tx = \sum_{k \in \mathbb{Z}} x(k)T\delta_k. \tag{2.2}$$

A linear and time-invariant operator is called a *filter*. If F is a filter, then Fx is called the *response* of x.

EXAMPLE 9.2.1: Let T be the operator defined by

$$(Tx)(n) = [x(n)]^2.$$

Then T is a time-invariant operator, but is not linear. Let L be the operator such that

$$(Lx)(n) = x(2n).$$

Then L is a linear operator, but not time-invariant.

Definition 9.2.2 The *(discrete) convolution* of two sequences h and x is a sequence $h * x$ given by

$$(h * x)(n) = \sum_k h(k)x(n-k), \quad n \in \mathbb{Z}, \tag{2.3}$$

provided the series in (2.3) is convergent for each $n \in \mathbb{Z}$.

The following theorem identifies a filter with a sequence.

Theorem 9.2.1 $\mathbf{H}$ is a filter if and only if there is a sequence $\mathbf{h}$ such that $\mathbf{Hx} = \mathbf{h} * \mathbf{x}$.

Proof: We first prove that the convolution

$$\mathbf{y} = \mathbf{h} * \mathbf{x}, \quad \text{for all } \mathbf{x} \in l^2,$$

defines a filter $\mathbf{H} : \mathbf{y} = \mathbf{Hx}$. It is clear that $\mathbf{H}\delta_k(n) = \sum_k h(k)\delta_k(n - k) = h(n - k)$. Hence,

$$\mathbf{Hx}(n) = \sum_k h(k)x(n - k) = \sum_k h(n - k)x(k)$$

$$= \sum_k x(k)\mathbf{H}\delta_k(n), \textit{ for all } n \in \mathbb{Z}.$$

Hence $\mathbf{H}$ is linear. We also have

$$(\mathbf{HSx})(n) = \sum_k h(k)x(n - k - 1) = \sum_k h(k - 1)x(n - k)$$

$$= (\mathbf{SHx})(n),$$

which implies that $\mathbf{H}$ is time-invariant. Hence $\mathbf{H}$ is a filter and $\mathbf{Hx} = \mathbf{h} * \mathbf{x}$. We now prove the converse. Assume $\mathbf{H}$ is a filter. Let $\mathbf{h} = \mathbf{H}\delta$. We claim that

$$\mathbf{Hx} = \mathbf{h} * \mathbf{x}, \quad \text{for all } \mathbf{x} \in l.$$

In fact, we have

$$\mathbf{x} = \sum_k x(k)\mathbf{S}^k \delta.$$

Since $\mathbf{H}$ is linear and time-invariant,

$$\mathbf{Hx} = \sum_k x(k)\mathbf{HS}^k \delta = \sum_k x(k)\mathbf{S}^k \mathbf{H}\delta = \sum_k x(k)\mathbf{S}^k \mathbf{h}.$$

Therefore

$$(\mathbf{Hx})(n) = \sum_k x(k)\left(\mathbf{S}^k \mathbf{h}\right)(n) = \sum_k x(k)h(n - k),$$

i.e., $\mathbf{Hx} = \mathbf{h} * \mathbf{x}$. ∎

Since a filter can be identified with a sequence, we shall directly call $\mathbf{h}$ a filter. From the computational point of view, finite filters are desirable.

Definition 9.2.3 If **h** is finite, then **h** is called a *finite impulse response filter* (*FIR filter*); otherwise it is called an *infinite impulse response filter* (*IIR filter*). A filter **h** with $h(n) = 0$, for all $n < 0$, is called a *causal filter*.

The following lemma gives a sufficient condition for filters on the space l^2.

Lemma 9.2.1 If $\mathbf{h} \in l^1$, then it is a filter from l^2 to l^2, and for $\mathbf{x} \in l^2$,

$$||\mathbf{h} * \mathbf{x}||_2 \leq ||\mathbf{h}||_1 ||\mathbf{x}||_2.$$

Proof: We leave the proof as an exercise. ∎

EXAMPLE 9.2.2: The *moving average filter*, $\mathbf{h}_0$, is defined by

$$h_0(0) = 1/2, \quad h_0(1) = 1/2,$$
$$h_0(k) = 0, \quad k \neq 0, 1.$$

Let

$$\mathbf{x} = (\cdots, 0, 1, 2, 1, 0, \cdots), \tag{2.4}$$

and $\mathbf{y} = \mathbf{h}_0 * \mathbf{x}$. Then $y(n) = \frac{1}{2}(x(n) + x(n-1))$ and

$$\mathbf{y} = \frac{1}{2}(\cdots, 0, 1, 3, 3, 1, \cdots).$$

EXAMPLE 9.2.3: The *moving difference filter*, $\mathbf{h}_1$, is defined by

$$h_1(0) = 1/2, h_1(1) = -1/2,$$
$$h_1(k) = 0, \quad k \neq 0, 1.$$

Let **x** be the signal in (2.4) and $\mathbf{z} = \mathbf{h}_1 * \mathbf{x}$. Then $z(n) = \frac{1}{2}(x(n) - x(n-1))$ and

$$\mathbf{z} = \frac{1}{2}(\cdots, 0, 1, 1, -1, -1, \cdots).$$

The *identity filter* is the sequence δ since $\delta * \mathbf{x} = \mathbf{x}$. The shift operator **S** is also a filter, called the *shift filter*. We have $\mathbf{Sx} = \delta_1 * \mathbf{x}$.

2.1 Representing Filters in the Time Domain

In the time domain, a filter is represented as a biinfinite Toeplitz matrix.

Definition 9.2.4 A biinfinite matrix T_h is called a *Toeplitz matrix* relative to a sequence $\mathbf{h}$ if

$$
T_h = \begin{pmatrix}
\cdots & \cdots & \cdots & & \cdots & & \cdots & \cdots \\
\cdots & h(0) & h(-1) & \cdots & & \cdots & \cdots & \cdots \\
\cdots & h(1) & h(0) & h(-1) & \cdots & & \cdots & \cdots \\
\cdots & \cdots & h(1) & h(0) & h(-1) & & \cdots & \cdots \\
\cdots & \cdots & & h(1) & h(0) & h(-1) & \cdots & \\
\cdots & \cdots & \cdots & & h(1) & h(0) & \cdots & \\
\cdots & \cdots & \cdots & & \cdots & & \cdots & \cdots
\end{pmatrix}
$$

Let $\mathbf{x} = (\cdots , x(-1), x(0), x(1), \cdots)^T$. Then $\mathbf{h} * \mathbf{x} = T_h \mathbf{x}$.

EXAMPLE 9.2.4: The moving average filter $\mathbf{h}_0$ performs

$$
y(n) = \frac{1}{2}(x(n) + x(n-1)).
$$

It can be represented by

$$
\begin{pmatrix}
\vdots \\
y(-1) \\
y(0) \\
y(1) \\
\vdots
\end{pmatrix}
=
\begin{pmatrix}
\vdots & \vdots & \vdots & \vdots & \vdots & \vdots \\
\cdots & 1/2 & 1/2 & 0 & \cdots & 0 \\
\cdots & 0 & 1/2 & 1/2 & \cdots & 0 \\
\cdots & \cdots & 0 & 1/2 & 1/2 & 0 \\
\vdots & \vdots & \vdots & \vdots & \vdots & \vdots
\end{pmatrix}
\begin{pmatrix}
\vdots \\
x(-1) \\
x(0) \\
x(1) \\
\vdots
\end{pmatrix}
$$

Similarly, the moving differences filter $\mathbf{h}_1$ performs $z(n) = \frac{1}{2}(x(n) - x(n-1))$. It can be represented by

$$
\begin{pmatrix}
\vdots \\
z(-1) \\
z(0) \\
z(1) \\
\vdots
\end{pmatrix}
=
\begin{pmatrix}
\vdots & \vdots & \vdots & \vdots & \vdots & \vdots \\
\cdots & -1/2 & 1/2 & 0 & \cdots & \cdots \\
\cdots & 0 & -1/2 & 1/2 & 0 & \cdots \\
\cdots & \cdots & 0 & -1/2 & 1/2 & \cdots \\
\vdots & \vdots & \vdots & \vdots & \vdots & \vdots
\end{pmatrix}
\begin{pmatrix}
\vdots \\
x(-1) \\
x(0) \\
x(1) \\
\vdots
\end{pmatrix}
$$

2.2 Filters in the Frequency Domain

Using the z-transform of signals in l^2, we can obtain the representation of filters in the frequency domain. The following Discrete Convolution Theorem plays the central role.

Theorem 9.2.2 [Discrete Convolution Theorem] Let **h** be a filter and

$$\mathbf{y} = \mathbf{h} * \mathbf{x}.$$

Then

$$Y(z) = H(z)X(z),$$

or, in the math reduced notation,

$$\mathbf{Y}(\omega) = \mathbf{H}(\omega)\mathbf{X}(\omega). \tag{2.5}$$

Proof: We leave the proof as an exercise.   ∎

By Lemma 9.2.1, we know that $\mathbf{h} \in l^1$ is a filter on l^2. But there exist filters on l^2 which are not sequences in l^1. From the discrete convolution theorem, we can characterize filters on l^2.

Theorem 9.2.3 A filter **h** is a bounded operator from l^2 to l^2 if and only if $\mathbf{H}(\omega) \in \tilde{L}^\infty$.

Proof: Note that $\|\mathbf{x}\|_2 = \|\mathbf{X}(\omega)\|_{\tilde{L}^2}$ and $\|\mathbf{y}\|_2 = \|\mathbf{Y}(\omega)\|_{\tilde{L}^2}$. Hence, to prove the theorem we only need to prove the following: for any $\mathbf{X}(\omega) \in \tilde{L}^2$, $\mathbf{H}(\omega)\mathbf{X}(\omega) \in \tilde{L}^2$ if only if $\mathbf{H}(\omega) \in \tilde{L}^\infty$. We now assume $\mathbf{X}(\omega) \in \tilde{L}^2$ and $\mathbf{H}(\omega) \in \tilde{L}^\infty$. Then

$$\|\mathbf{H}(\omega)\mathbf{X}(\omega)\|_{\tilde{L}^2} \leq \|\mathbf{H}(\omega)\|_{\tilde{L}^\infty}\|\mathbf{X}(\omega)\|_{\tilde{L}^2},$$

which implies that $\mathbf{H}(\omega)\mathbf{X}(\omega) \in \tilde{L}^2$. Conversely, assume $\mathbf{X}(\omega)$ and $\mathbf{H}(\omega)\mathbf{X}(\omega)$ both are in $\tilde{L}^2$. If $\mathbf{H}(\omega) \notin \tilde{L}^\infty$, then there is a strictly increasing sequence $(n_k)_{k=1}^\infty$, $n_k \in \mathbb{N}$, such that the set

$$E_k = \{\omega \in [-\pi, \pi] \mid n_k \leq |\mathbf{H}(\omega)| < n_{k+1}\}$$

has positive measure for each k. Let m_k be the Lebesgue measure of E_k. We define a 2π-periodic function $\mathbf{X}(\omega)$ as

$$\mathbf{X}(\omega) = \begin{cases} \frac{1}{k\sqrt{m_k}}, & \omega \in E_k, k \in \mathbb{N}, \\ 0 & \text{otherwise.} \end{cases}$$

Since

$$\int_{-\pi}^{\pi} |X(\omega)|^2 \, d\omega = \sum_{k=1}^{\infty} \int_{E_k} \frac{1}{k^2 m_k} \, d\omega = \sum_{k=1}^{\infty} \frac{1}{k^2},$$

we can claim $\mathcal{X}(\omega) \in \tilde{L}^2$. However, since $n_k \geq k$,

$$\int_{-\pi}^{\pi} |H(\omega)X(\omega)|^2 \, d\omega \geq \sum_{k=1}^{\infty} \int_{E_k} \frac{n_k^2}{k^2 m_k} \, d\omega = \sum_{k=1}^{\infty} \frac{n_k^2}{k^2} = \infty,$$

which implies that $H(\omega)X(\omega) \notin \tilde{L}^2$. Therefore, $H(\omega)$ must be in $\tilde{L}^\infty$. The theorem is proved. ∎

EXAMPLE 9.2.5: Let H_0 be the moving average filter and x be the signal in Example 9.2.2. Then

$$H_0(z) = \frac{1}{2}(1 + z), \quad X(z) = 1 + 2z + z^2,$$

and

$$Y(z) = H_0(z)X(z) = \frac{1}{2}(1 + 3z + 3z^2 + z^3),$$

which yields

$$y = \frac{1}{2}(\cdots, 0, 1, 3, 3, 1, \cdots).$$

Let H_1 be the moving difference filter. Then

$$H_1(z) = \frac{1}{2}(1 - z)$$

and

$$Z(z) = H_1(z)X(z) = \frac{1}{2}(1 + z - z^2 - z^3),$$

which yields

$$z = \frac{1}{2}(\cdots, 0, 1, 1, -1, -1, \cdots).$$

2.3 Lowpass Filters and Highpass Filters

Filters are often used to extract required frequency components from signals. For example, high-frequency components of a signal usually contain the noise and the fluctuations, which often have to be removed from the signal. We use lowpass filters and highpass filters to decompose signals into their frequency bands. A lowpass filter attenuates high-frequency components of a signal, while a highpass filter does the opposite job. Since a signal represents a frequency-bounded signal with the bound $|\omega| \leq \pi$, its low-frequency region is centered at the origin and the high-frequency region is near π. For example, we can divide the frequency domain into two regions: $|\omega| \leq \pi/2$ and $\pi/2 < |\omega| \leq \pi$, where the former is the low-frequency region and the latter is the high-frequency region. Then the simplest pair of lowpass and highpass filters is defined as follows.

Definition 9.2.5 The *ideal lowpass filter* $\mathbf{h} = (h(k))$ is defined by

$$\mathbf{H}(\omega) = \sum_{k \in \mathbb{Z}} h(k)e^{ik\omega} = \begin{cases} 1, & |\omega| < \pi/2, \\ 0, & \pi/2 \leq |\omega| \leq \pi, \end{cases} \tag{2.6}$$

and the *ideal highpass filter* $\mathbf{g} = (g(k))$ is defined by

$$\mathbf{G}(\omega) = \sum_{k \in \mathbb{Z}} g(k)e^{ik\omega} = \begin{cases} 0, & |\omega| < \pi/2, \\ 1, & \pi/2 \leq |\omega| \leq \pi. \end{cases}$$

By the Shannon Sampling Theorem (Theorem 9.1.2), the coefficients of the ideal lowpass filter are sampling data of $\frac{1}{2}$ sinc $\left(\frac{\pi t}{2}\right)$, i.e.,

$$h(k) = \frac{1}{2} \text{ sinc } \left(\frac{\pi k}{2}\right) = \begin{cases} \dfrac{1}{2}, & k = 0, \\ \dfrac{(-1)^m}{k\pi}, & k = 2m+1, m \in \mathbb{Z}, \\ 0, & k \text{ is even}, \end{cases}$$

which decays very slowly as $|k| \to \infty$, as does $g(k)$. Hence ideal filters are not practical. By Fourier analysis, the slow decay of the ideal filters is caused by the discontinuity of $H(\omega)$ and $G(\omega)$. To design fast-decay lowpass filters and highpass filters, we replace $H(\omega)$ and $G(\omega)$ by smooth symbols.

Let $\mathbf{H}$ be a filter with a continuous symbol $\mathbf{H}(\omega)$. If $\mathbf{H}$ is a lowpass filter, then it is assumed that $\mathbf{H}(0) \neq 0$ and $\mathbf{H}(\pi) = 0$. It is also clear that if $\mathbf{H}$ is a lowpass filter, then the filter $\mathbf{G}$ with the symbol $G(\omega) = H(\omega + \pi)$ is a high filter, since the graph of $|\mathbf{G}(\omega)|$ is the shift of $|\mathbf{H}(\omega)|$ by π.

EXAMPLE 9.2.6: The moving average filter $\mathbf{h}_0$ is a lowpass filter, while the moving difference filter $\mathbf{h}_1$ is a highpass filter. Assume the signal $\mathbf{x}$ is the sampling data of a differentiable function $f(t)$ with the time step h : $x(n) = f(hn)$. When h is very small, $\frac{2}{h}(\mathbf{h}_1 * \mathbf{x})$ is the approximation of the sampling data of $f'(t)$ and $\sum_{k=0}^{n} h(\mathbf{h}_0 * \mathbf{x})$ is the approximation of $\int_0^{n/h} f(t)\,dt$.

EXAMPLE 9.2.7: Let h_{id} be the ideal lowpass filter and x be the sampling data of a function $f(t)$ with the time step T: $x(n) = f(Tn)$. Assume f is a bounded frequency signal with the highest frequency $\frac{\pi}{T}$. Then $x * h_{id}$ represents the low frequency component of f with the Fourier transform $\hat{f}\chi_{[-\frac{\pi}{2T}, \frac{\pi}{2T}]}$.

EXAMPLE 9.2.8: The lowpass filter $H_n = \frac{1}{n}\sum_{k=0}^{n-1} S^k$ is used to obtain averages of n terms in a signal. Assume x is the record of the daily prices of a stock. Then (H_{50}) x is the data of 50-day averages and (H_{200}) x is the data of 200-day averages for the stock. They indicate the mid-term and long-term tendency of the stock.

The following theorem is an important result in wavelet analysis.

Theorem 9.2.4 Let ϕ generate an MRA $\{V_n\}_n$ and ψ be any wavelet in W_0:

$$\psi(t) = 2 \sum_{k \in \mathbb{Z}} g(k)\phi(2t - k).$$

Then the mask of ϕ is a lowpass filter and the mask of ψ : $\mathbf{g} = (g(k))$ is a highpass filter.

Proof: We leave the proof as an exercise. ∎

2.4 Dual Filters

Let L be a linear transformation from $\mathbb{R}^n$ to $\mathbb{R}^n$ and A be its standard matrix: $Ax = L(x)$, for all $x \in \mathbb{R}^n$. Then, from linear algebra, we know that for each $y \in \mathbb{R}^n$, $\langle L(x), y \rangle = \langle x A^T y \rangle$. Hence, the transpose of A defines another linear transformation called the *dual* of L. For a filter in l^2, we can introduce a similar notion.

Definition 9.2.6 Let **H** be a filter from l^2 to l^2. A filter $\mathbf{H}^T$ is called the *dual filter* of H if

$$\langle Hx, y \rangle = \langle x, H^T y \rangle, \quad \text{for all } x, y \in l^2.$$

Theorem 9.2.5 Assume the matrix form and the sequence form of a filter **H** are T_h and $\mathbf{h} = (h(k))$, respectively. Then the matrix form and the sequence form of its dual filter $\mathbf{H}^T$ are $(T_h)^T$ and $\mathbf{h}^T = (h(-k))$, respectively. Thus,

$$\left(\mathbf{h}^T * \mathbf{x}\right)(n) = \sum_{k \in \mathbb{Z}} h(k - n)x(k) \tag{2.7}$$

and the z-transform of $\mathbf{H}^T$ is $\overline{H(z)}$.

Proof: We leave the proof as an exercise.   ∎

It is also clear that if **H** is a lowpass (or highpass) filter, so is its dual.

2.5 Magnitude and Phase

In Chapter 8, we introduced the phase of a function in $L^2 \cap L^1$. We now define phases for signals and filters.

Definition 9.2.7 Let $x \in l^1$ be a discrete signal and $X(z) = \sum X(n)z^n$ be its symbol. Write

$$X(e^{-i\omega}) = \left| X(e^{-i\omega}) \right| e^{i\phi(\omega)}.$$

Then $\phi(\omega)$ is called the *phase* of **x**. Particularly, for a filter **H**, we write

$$\mathbf{H}(\omega) = |\mathbf{H}(\omega)|e^{-i\phi(\omega)}. \tag{2.8}$$

Then $|\mathbf{H}(\omega)|$ is called the *magnitude* of **H** and $\phi(\omega)$ is called the *phase* of **H**. If $\phi(\omega)$ is linear, then we say **x** has a *linear phase*.

The phase of **H** is not unique; they are congruent with a $2k\pi$ constant. To get the uniqueness of $\phi(\omega)$, we usually require a continuous phase function $\phi(\omega)$. However, in some cases, this requirement cannot be satisfied even by a continuous $\mathbf{H}(\omega)$. For example, let $\mathbf{H}(\omega) = \omega$. Then

$$\mathbf{H}(\omega) = |\omega|\, e^{-i\phi(\omega)},$$

where

$$\phi(\omega) = \begin{cases} \pi, & \omega < 0, \\ 0, & \omega \geq 0, \end{cases}$$

which has a jump π at $x = 0$. It is easy to see that $\phi(\omega)$ always has a jump π at ω_0 as long as $\mathbf{H}(\omega)$ is real in the neighborhood of ω_0 and $\mathbf{H}(\omega)$ changes its sign as ω goes across ω_0. Hence, the phase function $\phi(\omega)$ in (2.8) may have a discontinuity. To avoid it, we introduce the generalized phase function as follows.

Definition 9.2.8 Let $x \in l^1$ be a discrete signal and $x(z) = \sum x(n)z^n$ be its symbol. Write

$$x(e^{-i\omega}) = R(\omega)e^{i\phi(\omega)},$$

where $R(\omega) = \pm|\mathbf{H}(\omega)|$ and the sign $\pm$ is chosen so that $\phi(\omega)$ is continuous. Then $\phi(\omega)$ is called the *generalized phase* of **x**. If the generalized phase is a linear function, then we say that **x** has *generalized linear phase*.

Theorem 9.2.6 If a filter **h** is symmetric or antisymmetric with respect to an integer or half-integer, i.e., $h(k) = h(N - k)$ or $h(k) = -h(N - k)$ for some $N \in \mathbb{Z}$, then it has a generalized linear phase.

Proof: Assume $h(k) = h(N - k)$ for some $N \in \mathbb{Z}$. Then, if $N = 2n$, we have

$$h(n - k) = h(n + k) := c(k)$$

and

$$\mathbf{H}(\omega) = \sum_{k \in \mathbb{Z}} h(k) e^{-ik\omega} = e^{-in\omega} \sum_{k \in \mathbb{Z}} h(k) e^{i(n-k)\omega}$$

$$= e^{-in\omega} \left(h(n) + \sum_{l=-\infty}^{-1} h(n - l) e^{il\omega} + \sum_{l=1}^{\infty} h(n - l) e^{il\omega} \right)$$

$$= e^{-in\omega} \left(h(n) + \sum_{l=1}^{\infty} h(n + l) e^{-il\omega} + \sum_{l=1}^{\infty} h(n - l) e^{il\omega} \right)$$

$$= e^{-in\omega} \left(h(n) + \sum_{l=1}^{\infty} h(n + l) e^{-il\omega} + \sum_{l=1}^{\infty} h(n - l) e^{il\omega} \right)$$

$$= e^{-in\omega} \left(h(n) + 2 \sum_{l=1}^{\infty} c_l \cos l\omega \right).$$

Since the function $h(n) + 2 \sum_{l=1}^{\infty} c_l \cos l\omega$ is real, **h** has generalized linear phase $n\omega$. Similarly, if $N = 2n + 1$, we have

$$h(n + 1 - k) = h(n + k) := d(k)$$

and

$$\mathbf{H}(\omega) = \sum_{k \in \mathbb{Z}} h(k) e^{-ik\omega} = e^{-i(n+\frac{1}{2})\omega} \sum_{k \in \mathbb{Z}} h(k) e^{i(n+\frac{1}{2}-k)\omega}$$

$$= e^{-i(n+\frac{1}{2})\omega} \left(\sum_{l=-\infty}^{-1} h(n - l) e^{i(l+\frac{1}{2})\omega} + \sum_{l=0}^{\infty} h(n - l) e^{i(l+\frac{1}{2})\omega} \right)$$

$$= e^{-i(n+\frac{1}{2})\omega} \left(\sum_{l=1}^{\infty} h(n + l) e^{-i(l-\frac{1}{2})\omega} + \sum_{l=1}^{\infty} h(n + 1 - l) e^{i(l-\frac{1}{2})\omega} \right)$$

$$= e^{-i(n+\frac{1}{2})\omega} \left(2 \sum_{l=1}^{\infty} d_l \cos \left(l - \frac{1}{2} \right) \omega \right),$$

which implies that h has linear phase $(n+\frac{1}{2})\omega$. In the case that h is antisymmetric, the proof is similar. We leave it as an exercise. ∎

2.6 Inverse Filters

A filter is a linear transformation from l^2 to l^2. A natural question is: "Under what conditions is a filter invertible?"

Definition 9.2.9 A filter H from l^2 to l^2 is said to be *invertible* if there exists a filter D from l^2 to l^2 such that $DH = I$. The filter D is called the *inverse filter* of H and is denoted by H^{-1}.

It is clear that if $DH = I$, then $HD = I$. From Theorem 9.2.6, we derive the following.

Theorem 9.2.7 An $l^2 \rightarrow l^2$ filter H is invertible if and only if

$$\text{ess inf }_{\omega \in [-\pi, \pi]} \left| H(\omega) \right| > 0. \tag{2.9}$$

In this case, the symbol of the inverse filter H^{-1} is $\frac{1}{H(\omega)}$.

Proof: If (2.9) holds, then $D(\omega) := \frac{1}{H(\omega)} \in \tilde{L}^\infty$. Let D be the filter, whose symbol is $D(\omega)$. Then it is a filter from l^2 to l^2. Since $D(\omega)H(\omega) = 1$ a.e., D is the inverse of H. The converse is trivial. ∎

By Theorem 9.2.7, we have the following.

Corollary 9.2.1 A filter $h \in l^1$ is invertible if and only if its symbol $H(\omega) \neq 0$ for all $\omega \in \mathbb{R}$.

Proof: We leave the proof as an exercise. ∎

Note that

$$H(\omega) \neq 0 \text{ for all } \omega \in \mathbb{R} \text{ if and only if } H(z) \neq 0 \text{ for all } z \in \Gamma.$$

Hence, in order to determine whether H is invertible, we only need to check whether $H(z) = 0$ for some $z \in \Gamma$. When h is an FIR, $H(z)$ is a Laurent polynomial. Hence, the question is whether $H(z)$ has zeros on Γ.

EXAMPLE 9.2.9: Let $H(z) = \frac{2}{3} + \frac{1}{3}z$ be the z-transform of the filter **H**. The root of $H(z)$ is $z_0 = -2 \notin \Gamma$. Hence **H** is invertible. To find the inverse filter $\mathbf{H}^{-1}$, we expand $1/H(z)$:

$$1/H(z) = \frac{3}{2}\frac{1}{1 + \frac{1}{2}z}$$

$$= \frac{3}{2}\left(1 - \frac{1}{2}z + \frac{1}{4}z^2 - \frac{1}{8}z^3 + \cdots\right)$$

$$= \frac{3}{2}\sum_{k=0}^{\infty}(-1)^k\left(\frac{1}{2}\right)^k z^k.$$

The following result is often used to convert signals.

Theorem 9.2.8 Assume $\phi \in L^2$ is a continuous function which satisfies the interpolation condition. Let $V = \text{span}_{L^2}\{\phi(t - k) \mid k \in \mathbb{Z}\}$ and $\tilde{\phi}$ be the interpolation function in V. Let a function $f \in V$ be represented by

$$f(t) = \sum_{k \in \mathbb{Z}} c_k \phi(t - k)$$

and

$$f(t) = \sum_{k \in \mathbb{Z}} f(k)\tilde{\phi}(t - k).$$

Denote $\mathbf{c} = (c_k)$, $\mathbf{f}_d = (f(k))$, and $\mathbf{h} = (\phi(k))$. Then $\mathbf{h}$ is invertible, $\mathbf{f}_d = \mathbf{h} * \mathbf{c}$, and $\mathbf{c} = \mathbf{h}^{-1} * \mathbf{f}_d$.

Proof: We leave the proof as an exercise. ■

Exercises

1. Prove Lemma 9.2.1: If $\mathbf{h} \in l^1$, then it is a filter from l^2 to l^2, and $\|\mathbf{h} * \mathbf{x}\|_2 \leq \|\mathbf{h}\|_1\|\mathbf{x}\|_2$ for $\mathbf{x} \in l^2$.
2. Prove that if **T** is a linear operator on l^2, then for any $\mathbf{x}_1, \mathbf{x}_2 \in l^2$ and $\lambda, \mu \in \mathbb{R}$, $\mathbf{T}(\lambda\mathbf{x}_1 + \mu\mathbf{x}_2) = \lambda\mathbf{T}\mathbf{x}_1 + \mu\mathbf{T}\mathbf{x}_2$.
3. Prove that if $\mathbf{h} = (h_k)$ is a lowpass filter (highpass filter), then

$$\mathbf{g} = ((-1)^k h_{1-k})$$

defines a highpass filter (lowpass filter).

4. Construct a filter $\mathbf{h}$ from l^2 to l^2, which is not in l^1.

5. Let the filter $\mathbf{h}$ be defined by $h(0) = 1, h(1) = 4, h(2) = 6, h(3) = 4, h(4) = 1$. Find its dual filter.

6. Prove that a filter is a lowpass (highpass) filter if and only if its dual filter is a lowpass (highpass) filter.

7. Prove that the convolution of two lowpass filters is a lowpass filter.

8. Prove Theorem 9.2.4: Let ϕ generate an MRA $\{V_n\}_n$ and ψ be any wavelet in W_0:

$$\psi(t) = 2 \sum g(k)\phi(2t - k).$$

Then the mask of ϕ is a lowpass filter and the mask of $\psi : \mathbf{g} = (g(k))$ is a highpass filter.

9. Prove Theorem 9.2.5: Assume the matrix form and the sequence form of a filter $\mathbf{H}$ are T_h and $\mathbf{h} = (h(k))$, respectively. Then the matrix form and the sequence form of its dual filter $\mathbf{H}^T$ are $(T_h)^T$ and $\mathbf{h}^T = (h(-k))$, respectively, i.e.,

$$\left(\mathbf{h}^T * \mathbf{x}\right)(n) = \sum_{k \in \mathbb{Z}} h(k - n)x(k),$$

and the z-transform of $\mathbf{H}^T$ is $\overline{H(z)}$.

10. Prove that a filter is invertible if and only if its dual is invertible.

11. Prove Corollary 9.2.1: A filter $\mathbf{h} \in l^1$ is invertible if and only if its symbol $\mathbf{H}(\omega) \neq 0$ for all $\omega \in \mathbb{R}$.

12. Find the formula of $g(k)$ for the ideal highpass filter.

13. Prove that a finite filter $\mathbf{s} = (s(k))_{k=0}^N$ has linear phase if and only if $s_k = s_{N-k}$ and the polynomial $S(z) = \sum s(k)z^k$ has zeros only of even multiplicity on the unit circle Γ.

14. Prove that if a filter $\mathbf{h}$ is antisymmetric with respect to an integer or half-integer, i.e., $h(k) = -h(N - k)$ for some $N \in \mathbb{Z}$, then it has a generalized linear phase.

15. Prove that a filter has a linear phase (or generalized linear phase) if and only if its dual filter has a linear phase (or generalized linear phase).

16. Prove that the symbol of the filter $\mathbf{h} * \mathbf{h}^T$ has linear phase.

17. Let $\mathbf{h}$ and $\mathbf{f}$ be filters with linear phases (or generalized linear phases). Prove the following.

    (a) $\mathbf{h} * \mathbf{f}$ has a linear phase (or generalized linear phase).

    (b) If $\mathbf{h}$ is invertible, then $\mathbf{h}^{-1}$ has a linear phase (or generalized linear phase).

18. Write the matrix representation and the z-transform of the filter $\mathbf{S} + \mathbf{S}^{-1}$. Is it invertible?

19. Assume that the impulse response of filter **H** is

$$h(n) = \begin{cases} \alpha^n & n \geq 0, \\ \beta^n & n < 0, \end{cases}$$

where $0 < \alpha < 1$ and $\beta > 1$.

(a) Find the output $y = y(k)$, $k \in \mathbb{Z}$, of **H** for the input $x(n) = \frac{(-1)^n}{n}$.

(b) Is **H** invertible? If it is, find $\mathbf{H}^{-1}$.

20. Prove that neither lowpass filters nor highpass filters are invertible.

21. Let **H** be a finite filter with symbol $H(\omega)$. Assume **H** is not invertible. Prove that T_h has at least a zero eigenvalue, and one of its 0-eigenvectors has the form $\mathbf{v} = (c^k)_{k \in \mathbb{Z}}$, where c is a nonzero complex number.

22. Prove Theorem 9.2.8: Assume $\phi \in L^2$ is a continuous function which satisfies the interpolation condition. Let $V = \mathrm{span}_{L^2}\{\phi(t - k) \mid k \in \mathbb{Z}\}$ and $\tilde{\phi}$ be the interpolation function in V. Let a function $f \in V$ be represented by

$$f(t) = \sum_{k \in \mathbb{Z}} c_k \phi(t - k)$$

and

$$f(t) = \sum_{k \in \mathbb{Z}} f(k) \tilde{\phi}(t - k).$$

Denote $\mathbf{c} = (c_k)$, $\mathbf{f}_d = (f(k))$, and $\mathbf{h} = (\phi(k))$. Then $\mathbf{h}$ is invertible, $\mathbf{f}_d = \mathbf{h} * \mathbf{c}$, and $\mathbf{c} = \mathbf{h}^{-1} * \mathbf{f}_d$.

3 Coding Signals by Wavelet Transform

In this section, we discuss how to use wavelet bases to transform (i.e., code) signals. The purpose of coding is to transform a sampling data $\mathbf{x} \in l^2$ to a new sequence (called the codes of $\mathbf{x}$) in l^2. The codes reveal the hidden information of a signal so that they can be effectively used in signal processing.

3.1 Coding Signals Using Shannon Wavelets

To understand the wavelet coding method, we first study the coding of signals using the Shannon wavelet.

Lemma 9.3.1 The function sinc (t) satisfies the following scaling equation:

$$\text{sinc } (t) = 2 \sum_{k \in \mathbb{Z}} h(k) \text{ sinc } (2t - k), \tag{3.1}$$

where $\mathbf{h} = (h(k))$ is the ideal lowpass filter.

Proof: We leave the proof as an exercise.   ∎

Theorem 9.3.1 The function sinc (t) is an orthonormal MRA generator. Its corresponding orthonormal wavelet can be represented as

$$\psi^s(t) = 2 \sum_{k \in \mathbb{Z}} g(k) \text{ sinc } (2t - k), \tag{3.2}$$

where $\mathbf{g} = (g(k))$ is the ideal highpass filter.

Proof: We leave the proof as an exercise.   ∎

Definition 9.3.1 The function ψ^s in (3.2) is called the *Shannon wavelet* and sinc (t) is called the *Shannon scaling function*.

Let $\{V_n\}$ be the MRA generated by sinc (t) and $\{W_n\}_{n \in \mathbb{Z}}$ be the corresponding wavelet subspaces. Assume the signal x is the sampling data of the function $f(t) \in V_n$, that is $x(k) = f(2^{-n}k)$. By the Shannon Sampling Theorem,

$$f(t) = \sum_{k \in \mathbb{Z}} x(k) \text{ sinc } (2^n t - k).$$

Let

$$f(t) = f_1(t) + g_1(t), \quad f_1(t) \in V_{n-1}, g_1(t) \in W_{n-1}.$$

Then

$$f_1(t) = \sum_{k \in \mathbb{Z}} y(k) \text{ sinc } (2^{n-1}t - k), \quad y(k) = f_1(2^{-(n-1)}k),$$

$$g_1(t) = \sum_{k \in \mathbb{Z}} z(k)\psi^s(2^{n-1}t - k), \quad z(k) = g_1(2^{-(n-1)}k).$$

We can use the FWT algorithm to compute $\{\mathbf{y}, \mathbf{z}\}$ from x and use the FIWT algorithm to recover x from $\{\mathbf{y}, \mathbf{z}\}$. Let $\mathbf{h}$ be the mask of sinc (t) in (3.1) and $\mathbf{g}$

be the mask of ψ^s in (3.2). Using the FWT algorithm, we get

$$y(n) = \sum_{k \in \mathbb{Z}} h(k - 2n)x(k) = \sum_{k \in \mathbb{Z}} h^T(2n - k)x(k)$$
$$z(n) = \sum_{k \in \mathbb{Z}} g(k - 2n)x(k) = \sum_{k \in \mathbb{Z}} g^T(2n - k)x(k)$$
(3.3)

which is also called the coding algorithm. Using the FIWT algorithm, we get

$$x(n) = 2 \left(\sum_{k \in \mathbb{Z}} h(n - 2k)y(k) + \sum_{k \in \mathbb{Z}} g(n - 2k)z(k) \right), \qquad (3.4)$$

which is the decoding algorithm. Formulae (3.3) and (3.4) show that although x may be the sampling data of a particular analog signal with a sampling period T and an underlying function ϕ, the coding and decoding algorithms are independent of T and ϕ. They only involve the filters $\mathbf{h}$ and $\mathbf{g}$. To link these algorithms to convolutions, we introduce the following.

Definition 9.3.2 A *downsampling* (*operator*) $(\downarrow 2)$ on l^2 is defined by

$$(\downarrow 2)x = (\cdots, x(-2), x(0), x(2), \cdots).$$

Correspondingly, an *upsampling* (*operator*) $(\uparrow 2)$ on l^2 is defined by

$$(\uparrow 2)x = (\cdots, 0, x(-2), 0, x(-1), 0, x(0), 0, x(1), 0, x(2), 0, \cdots).$$

By definition, (3.3) now can be written as

$$\mathbf{y} = (\downarrow 2)\mathbf{u} = (\downarrow 2)\left(\mathbf{h}^T * \mathbf{x}\right) = (\downarrow 2)\mathbf{H}^T \mathbf{x}$$
$$\mathbf{z} = (\downarrow 2)\mathbf{v} = (\downarrow 2)\left(\mathbf{g}^T * \mathbf{x}\right) = (\downarrow 2)\mathbf{G}^T \mathbf{x}$$

and (3.4) gives

$$\mathbf{x} = 2(\mathbf{h} * (\uparrow 2)\mathbf{y} + \mathbf{g} * (\uparrow 2)\mathbf{z}) = 2\left(\mathbf{H}(\uparrow 2)\mathbf{y} + \mathbf{G}(\uparrow 2)\mathbf{z}\right),$$

which recovers x from y and z. Combining these two steps together, we have

$$\mathbf{H}(2(\uparrow 2)(\downarrow 2)\mathbf{H}^T) + \mathbf{G}(2(\uparrow 2)(\downarrow 2)\mathbf{G}^T = \mathbf{I}.$$

We now explain the meaning of the coded signals y and z directly. We first establish the following.

Lemma 9.3.2 If a continuous function $v \in L^2$ satisfies supp $\hat{v} \subset [-\pi, -\pi/2] \cup [\pi/2, \pi]$, then

$$v(t) = \sum_{n \in \mathbb{Z}} v(2n) \left(2 \operatorname{sinc}(t - 2n) - \operatorname{sinc}\left(\frac{t}{2} - n\right) \right). \qquad (3.5)$$

Proof: We leave the proof as an exercise.   ∎

Let $\mathbf{u} = \mathbf{h}^T * \mathbf{x}$ and $\mathbf{v} = \mathbf{g}^T * \mathbf{x}$. Since $\mathbf{h}$ is the ideal lowpass filter, $\mathbf{h}^T * \mathbf{x}$ represents a frequency-bounded function $u(t)$ with supp $\hat{u} \subset [-\frac{\pi}{2}, \frac{\pi}{2}]$. Similarly, $\mathbf{v}$ represents a function $v(t)$ with supp $\hat{v} \subset [-\pi, -\pi/2] \cup [\pi/2, \pi]$. By the Shannon Sampling Theorem (Theorem 9.1.2), $u(t)$ can be represented by its sampling data $(u(2n))$. Similarly, by Lemma 9.3.2, $v(t)$ can also be represented by $(v(2n))$. Recall that $y(n) = u(2n)$ and $z(n) = v(2n)$. Hence, $\mathbf{y}$ and $\mathbf{z}$ are the sampling data of $u(t)$ and $v(t)$, where $\mathbf{y}$ is in the low-frequency channel and $\mathbf{z}$ is in the high-frequency channel.

The coded signals reveal the information of $\mathbf{x}$ regarding its frequency bands. If we want to get a blurry version of $\mathbf{x}$ using nearly half of its codes, then $\mathbf{y}$ is much better than $(x(2n))$. Using $\mathbf{y}$ and $\mathbf{z}$ to code $\mathbf{x}$ is called *two-channel coding*.

We now give a summary for the signal flows in these two channels:

(1) In the low-frequency channel:

$$\mathbf{x} \to \mathbf{u}\left(= \mathbf{h}^T * \mathbf{x}\right) \to \tilde{\mathbf{u}}\left(= 2(\uparrow 2)(\downarrow 2)\mathbf{u}\right) \to \mathbf{x}_l(= \mathbf{h} * \tilde{\mathbf{u}}).$$

(2) In the high-frequency channel:

$$\mathbf{x} \to \mathbf{v}\left(= \mathbf{g}^T * \mathbf{x}\right) \to \tilde{\mathbf{v}}\left(= 2(\uparrow 2)(\downarrow 2)\mathbf{v}\right) \to \mathbf{x}_h(= \mathbf{g} * \tilde{\mathbf{v}}).$$

3.2 Alias Cancellation

We now analyze the alias in the foregoing coding/decoding procession. We first analyze the low-frequency channel. The symbol of $\mathbf{u}$ is

$$U(\omega) = \overline{H(\omega)}X(\omega) = \begin{cases} X(\omega), & |\omega| \leq \frac{\pi}{2}, \\ 0, & \frac{\pi}{2} \leq |\omega| \leq \pi. \end{cases}$$

Since $\tilde{\mathbf{u}} = 2(\uparrow 2)(\downarrow 2)\mathbf{u}$, we have

$$\tilde{U}(\omega) = U(\omega) + U(\omega + \pi),$$

where $U(\omega + \pi)$ is in the high-frequency channel. Hence, it is an alias in the low-frequency channel. This alias will be canceled in the next step. In fact, by $\mathbf{x}_l = \mathbf{h} * \tilde{\mathbf{u}}$, we have

$$X_l(\omega) = H(\omega)\tilde{U}(\omega) = H(\omega)U(\omega) = U(\omega).$$

Similarly, the high-frequency channel is as follows. The symbol of $\mathbf{v}$ is

$$V(\omega) = \overline{G(\omega)}X(\omega) = \begin{cases} 0, & |\omega| \leq \frac{\pi}{2}, \\ X(\omega), & \frac{\pi}{2} \leq |\omega| \leq \pi \end{cases}.$$

For $\tilde{\mathbf{v}} = 2(\uparrow 2)(\downarrow 2)\mathbf{v}$, we have

$$\tilde{\mathbf{V}}(\omega) = \mathbf{V}(\omega) + \mathbf{V}(\omega + \pi),$$

where $\mathbf{V}(\omega + \pi)$ is in the low-frequency channel. Hence, it is an alias in the high-frequency channel. This alias will be canceled by $x_h = \mathbf{g} * \tilde{\mathbf{v}}$, since we have

$$\mathbf{X}_h(\omega) = \mathbf{G}(\omega)\,\tilde{\mathbf{V}}(\omega) = \mathbf{G}(\omega)\mathbf{V}(\omega) = \mathbf{V}(\omega).$$

3.3 Coding Signals Using Other Wavelets

The ideal filters $\mathbf{h} = (h(k))$ and $\mathbf{g} = (g(k))$ decay very slow as $|k| \to \infty$. Hence, algorithms (3.3) and (3.4) are not practical. To design fast coding/decoding algorithms, we turn to other wavelets. Let ϕ be an orthonormal scaling function and ψ be its corresponding orthonormal wavelet. Assume $\mathbf{h}$ is the mask of ϕ and $\mathbf{g}$ is the mask of ψ. It is known that $\mathbf{h}$ is a lowpass filter and $\mathbf{g}$ is a highpass filter. Then for a signal x, the FWT algorithm (3.3) decomposes x into y and z, while the FIWT algorithm (3.4) recovers x from y and z. However, the low-frequency channel and high-frequency channel now are different from what we discussed earlier. We still denote the symbol of $\mathbf{h}$ by $\mathbf{H}(\omega)$ and the symbol of $\mathbf{g}$ by $\mathbf{G}(\omega)$. The low-frequency channel in this coding method is enveloped by $\mathbf{H}(\omega)$. That is, a signal is in the low-frequency channel if its symbol can be written as $\mathbf{H}(\omega)\mathbf{S}(\omega)$, where $\mathbf{S}(\omega)$ is a 2π-periodic function. Similarly, the high-frequency channel in this coding method is enveloped by $\mathbf{G}(\omega)$.

Now in the low-frequency channel, the symbol of $\mathbf{u}$ is

$$\mathbf{U}(\omega) = \overline{\mathbf{H}(\omega)}\mathbf{X}(\omega),$$

and for $\tilde{\mathbf{u}} = 2(\uparrow 2)(\downarrow 2)\mathbf{u}$, we have

$$\tilde{\mathbf{U}}(\omega) = \mathbf{U}(\omega) + \mathbf{U}(\omega + \pi) = \overline{\mathbf{H}(\omega)}\mathbf{X}(\omega) + \overline{\mathbf{H}(\omega + \pi)}\mathbf{X}(\omega + \pi).$$

Since $\mathbf{G}(\omega) = e^{-i\omega}\overline{\mathbf{H}(\omega + \pi)}$, we have

$$\mathbf{U}(\omega + \pi) = e^{i\omega}\mathbf{G}(\omega)\mathbf{X}(\omega + \pi),$$

which is in the high-frequency channel and therefore is an alias in the low-frequency channel. This alias will be canceled by $x_l = \mathbf{h} * \tilde{\mathbf{u}}$, for

$$\mathbf{X}_l(\omega) = \mathbf{H}(\omega)\,\tilde{\mathbf{U}}(\omega)$$

is a signal in the low-frequency channel.

Similarly, the high-frequency channel is as follows. The symbol of $\mathbf{v}$ is

$$\mathbf{V}(\omega) = \overline{\mathbf{G}(\omega)}\mathbf{X}(\omega).$$

For $\tilde{\mathbf{v}} = 2(\uparrow 2)(\downarrow 2)\mathbf{v}$, we have

$$\tilde{\mathbf{V}}(\omega) = \mathbf{V}(\omega) + \mathbf{V}(\omega + \pi),$$

where

$$\mathbf{V}(\omega + \pi) = \overline{\mathbf{G}(\omega + \pi)}\mathbf{X}(\omega + \pi) = e^{i\omega}\mathbf{H}(\omega)\mathbf{X}(\omega + \pi)$$

is in the low-frequency channel and therefore is an alias in the high-frequency channel. This alias will be canceled by $\mathbf{x}_h = \mathbf{g} * \tilde{\mathbf{v}}$, since we have

$$\mathbf{X}_h(\omega) = \mathbf{G}(\omega)\,\tilde{\mathbf{V}}(\omega)$$

which is in the high-frequency channel.

We can also use FWT and FIWT algorithms created by biorthogonal scaling functions and wavelets to perform two-channel coding/decoding.

Theorem 9.3.2 Let ϕ and ϕ^* be biorthonormal scaling functions. Let ψ and ψ^* be the corresponding biorthonormal wavelets. Assume the masks of ϕ and ψ are $\mathbf{h}$ and $\mathbf{g}$, and the masks of ϕ^* and ψ^* are $\mathbf{h}^*$ and $\mathbf{g}^*$. Then the algorithms

$$\mathbf{y} = (\downarrow 2)\mathbf{H}^T\mathbf{x}, \quad \mathbf{z} = (\downarrow 2)\mathbf{G}^T\mathbf{x}$$

and

$$\mathbf{x} = 2\left(\mathbf{H}^*(\uparrow 2)\mathbf{y} + \mathbf{G}^*(\uparrow 2)\mathbf{z}\right)$$

perform two-channel coding and decoding, respectively. Then the signal flow in the low-frequency channel is

$$\mathbf{x} \rightarrow \mathbf{u}\left(=\mathbf{h}^T * \mathbf{x}\right) \rightarrow \tilde{\mathbf{u}}\left(= 2(\uparrow 2)(\downarrow 2)\mathbf{u}\right) \rightarrow \mathbf{x}_l(=\mathbf{h}^* * \tilde{\mathbf{u}}),$$

and the signal flow in the high-frequency channel is

$$\mathbf{x} \rightarrow \mathbf{v}\left(=\mathbf{g}^T * \mathbf{x}\right) \rightarrow \tilde{\mathbf{v}}\left(= 2(\uparrow 2)(\downarrow 2)\mathbf{v}\right) \rightarrow \mathbf{x}_h(=\mathbf{g}^* * \tilde{\mathbf{v}}).$$

Proof: We leave the proof as an exercise.   ∎

3.4 Sampling Data Coding

Let ϕ, ϕ^*, ψ, ψ^*, h, g, h^*, and g^* be given as in Theorem 9.3.2. Let

$$U_h = \text{span }_{L^2}\left\{\phi\left(\frac{t}{h} - k\right)\,\middle|\, k \in \mathbb{Z}\right\}, \quad h > 0.$$

Assume a signal x is the sampling data of a function $f \in U_h$: $x(n) = f(hn)$. We now discuss two channel codes of x. Recall that $f \in U_h$ if and only if $f(ht) \in U_1$. Let $x(t) = f(ht)$. Then $x(n)$ is the sampling data of $x(t)$. Let $\tilde{\phi}$ be the Lagrangian interpolation function in U_1. We have

$$x(t) = \sum_{n \in \mathbb{Z}} x(n)\tilde{\phi}(t - n).$$

Let $p = (\phi(k))_{k \in \mathbb{Z}}$. If ϕ is compactly supported, then p is a finite filter. Let $c = (c_n)$ be the sequence in

$$x(t) = \sum_{n \in \mathbb{Z}} c_n \phi(t - n).$$

We have

$$p * c = x, \quad p^{-1} * x = c.$$

Then the two-channel coding for x is the following:

$$c = p^{-1} * x,$$
$$y = (\downarrow 2)H^T c, \quad z = (\downarrow 2)G^T c,$$

and the decoding is

$$c = 2\left(H^*(\uparrow 2)y + G^*(\uparrow 2)z\right),$$
$$x = p * c.$$

The programs are independent of the sampling period h. We point out that c and x become very close as $h \to 0$.

Theorem 9.3.3 Assume $\phi^* \in L^1 \cap L^2$ is an MRA generator (with $\widehat{\phi^*}(0) = 1$). Let $x \in L^2$ be uniformly continuous on $\mathbb{R}$. Then

$$\lim_{h \to 0} \frac{|x(hk) - \langle x, \phi^*(\frac{t}{h} - k)\rangle|}{h} = 0, \tag{3.6}$$

and the limit in (3.6) uniformly holds for $k \in \mathbb{Z}$.

Proof: If $x \in L^2$ is uniformly continuous on $\mathbb{R}$, then $\lim_{|t| \to \infty} x(t) = 0$. It follows that $\max_{t \in \mathbb{R}} |x(t)| < \infty$. Since $\phi^* \in L^1$, for an $\epsilon > 0$, there is an $M > 0$ such that $\int_{|t| > M} |\phi^*(t)| dt < \epsilon$. We have $\int_{-\infty}^{\infty} \frac{1}{h} \phi^*(\frac{t}{h} - k) dt = 1$. Hence,

$$\frac{1}{h} \left| x(hk) - \left\langle x, \phi^* \left(\frac{t}{h} - k \right) \right\rangle \right| = \left| \int_{-\infty}^{\infty} (x(hk) - x(t)) \frac{1}{h} \phi^* \left(\frac{t}{h} - k \right) dt \right|$$

$$\leq \int_{h(k-M)}^{h(k+M)} \left| (x(hk) - x(t)) \frac{1}{h} \phi^* \left(\frac{t}{h} - k \right) \right| dt$$

$$+ \left(\int_{-\infty}^{h(k-M)} + \int_{h(k+M)}^{\infty} \right) \left| (x(hk) - x(t)) \frac{1}{h} \phi^* \left(\frac{t}{h} - k \right) \right| dt.$$

We have

$$\int_{h(k-M)}^{h(k+M)} \left| (x(hk) - x(t)) \frac{1}{h} \phi^* \left(\frac{t}{h} - k \right) \right| dt$$

$$\leq \max_{t \in [h(k-M), h(k+M)]} |x(hk) - x(t)| \int_{-\infty}^{\infty} \frac{1}{h} \left| \phi^* \left(\frac{t}{h} - k \right) \right| dt$$

$$\leq \max_{t \in [h(k-M), h(k+M)]} |x(hk) - x(t)| \, \|\phi^*\|_1,$$

where

$$\lim_{h \to 0} \max_{t \in [h(k-M), h(k+M)]} |x(hk) - x(t)| = 0$$

and the limit uniformly holds for $k \in \mathbb{Z}$. We also have

$$\int_{h(k+M)}^{\infty} \left| (x(hk) - x(t)) \frac{1}{h} \phi^* \left(\frac{t}{h} - k \right) \right| dt$$

$$\leq 2 \max_{t \in \mathbb{R}} |x(t)| \int_{h(k+M)}^{\infty} \frac{1}{h} \left| \phi^* \left(\frac{t}{h} - k \right) \right| dt$$

$$= 2 \max_{t \in \mathbb{R}} |x(t)| \int_{M}^{\infty} |\phi^*(t)| st < 2 \max_{t \in \mathbb{R}} |x(t)| \, \epsilon.$$

Similarly,

$$\int_{-\infty}^{h(k-M)} \left| (x(hk) - x(t)) 2^j \phi^* \left(\frac{t}{h} - k \right) \right| dt < 2 \max_{t \in \mathbb{R}} |x(t)| \, \epsilon.$$

Thus, (3.6) is proved.   ∎

Before we end this section, we briefly explain the advantage of the two-channel coding method. Recall that the analog model of this coding is the wavelet decomposition. Then each code in this coding corresponds to a wavelet atom. As we pointed out in Chapter 7, a wavelet is "local" in both the time domain and the frequency domain. (In signal processing, we say it is a *time–frequency atom*.) Besides, the multilevel structure of wavelet bases provides the "zooming" property of a wavelet. Readers can learn more from [18] and [32]. Hence, this coding method offers well-structured codes for signal processing.

Exercises

1. Prove Lemma 9.3.1: $\operatorname{sinc}(t) = 2\sum_{k\in\mathbb{Z}} h(k) \operatorname{sinc}(2t - k)$, where $\mathbf{h} = (h(k))$ is the ideal lowpass filter.
2. Prove Theorem 9.3.1: (a) The function $\operatorname{sinc}(t)$ is an orthonormal MRA generator. (b) The corresponding orthonormal wavelet of $\operatorname{sinc}(t)$ can be represented as

$$\psi^s(t) = 2\sum_{k\in\mathbb{Z}} g(k) \operatorname{sinc}(2t - k),$$

   where $\mathbf{g} = (g(k))$ is the ideal highpass filter.
3. Prove the Downsampling Shannon Theorem: If a signal x is half banded, i.e.,

$$\mathbf{X}(\omega) = 0, \quad \pi/2 \le |\omega| < \pi,$$

   then $(\downarrow 2)x$ uniquely determines x by the formula

$$x(n) = \sum_{-\infty}^{\infty} x(2k)\frac{\sin((n - 2k)\frac{\pi}{2})}{(n - 2k)\frac{\pi}{2}} = \sum_{-\infty}^{\infty} x(2k) \operatorname{sinc}\left(\frac{1}{2}n - k\right).$$

4. Let x and y be two signals such that

$$\mathbf{Y}(\omega) = \mathbf{X}(\omega)\chi_{[-\pi/2,\pi/2]}(\omega), \omega \in [-\pi, \pi].$$

   Prove $y(n) = \frac{1}{2}\sum_{-\infty}^{\infty} x(k) \operatorname{sinc}\left(\frac{n-k}{2}\right)$.
5. Prove the following. For any $x,y \in l^2$,

$$\langle(\uparrow 2)x, y\rangle = \langle x, (\downarrow 2)y\rangle,$$
$$(\downarrow 2)(\uparrow 2) = I,$$
$$(\uparrow 2)(\downarrow 2)+S^{-1}(\uparrow 2)(\downarrow 2)S = I,$$
$$(\downarrow 2)(x * y) = (\downarrow 2)x * (\downarrow 2)y+(\downarrow 2)Sx * (\downarrow 2)Sy.$$

6. Prove Lemma 9.3.2: If a continuous function $v \in L^2$ satisfies supp $\hat{v} \subset$ $[-\pi, -\pi/2] \cup [\pi/2, \pi]$, then

$$v(t) = \sum v(2n) \left(2 \operatorname{sinc} (t - 2n) - \operatorname{sinc} \left(\frac{t}{2} - n \right) \right).$$

7. Let x be the signal such that $X(\omega) = 0, |\omega| < \pi/2$. Develop the formula to compute $x(n)$ from $(x(2k))_{k \in \mathbb{Z}}$.

8. Let x be the signal such that $X(\omega) = 0, |\omega| < \pi/2$. Can we determine x using $(x(2n + 1))_{n \in \mathbb{Z}}$? If we can, how do we compute $x(n)$ from $(x(2k + 1))_{k \in \mathbb{Z}}$?

9. Let x and y be two signals such that

$$Y(\omega) = X(\omega)\chi_{[-\pi,-/2]\cup[\pi/2,\pi]}(\omega), \omega \in [-\pi, \pi].$$

Prove $y(n) = x(n) - \frac{1}{2} \sum_{-\infty}^{\infty} x(k) \operatorname{sinc} (\frac{n-k}{2})$.

10. Let $f(t)$ be the function with the Fourier transform $\hat{f}(\omega) = |\omega|\chi_{[-\pi,\pi]}(\omega)$. Let x be the signal with $x(n) = f(n)$. Find $(\downarrow 2)x$.

4 Filter Banks

The wavelet coding method leads to filter banks.

Definition 9.4.1 A *filter bank* is a set of filters, linked by downsampling and upsampling. In particular, let L be a lowpass filter and H be a highpass filter. Then the operator on l^2 defined by

$$\begin{bmatrix} (\downarrow 2)\mathbf{L}^T \\ (\downarrow 2)\mathbf{H}^T \end{bmatrix}$$

is called an *analysis filter bank* created by $\{\mathbf{L}, \mathbf{H}\}$, and the operator defined by

$$\begin{bmatrix} 2\mathbf{L}(\uparrow 2) & 2\mathbf{H}(\uparrow 2) \end{bmatrix}$$

is called a *synthesis filter bank* created by $\{\mathbf{L}, \mathbf{H}\}$. In the filter bank, $\mathbf{C} = (\downarrow 2)\mathbf{L}^T$ is called a *lowpass channel* and $\mathbf{D} = (\downarrow 2)\mathbf{H}^T$ is called a *highpass channel*. If an analysis filter bank created by $\{\mathbf{L}, \mathbf{H}\}$ and a synthesis filter bank created by $\{\mathbf{F}, \mathbf{G}\}$ satisfy the condition

$$2\mathbf{F}(\uparrow 2)(\downarrow 2)\mathbf{L}^T + 2\mathbf{G}(\uparrow 2)(\downarrow 2)\mathbf{H}^T = \mathbf{I}, \tag{4.1}$$

then the analysis filter bank and the synthesis filter bank are called *biorthogonal filter banks*. Particularly, if in (4.1) **F=L** and **G=H**, then {**L, H**} is said to create an *orthonormal filter bank*.

For convenience, we often directly call {**L, H**} and {**F, G**} filter banks. By the results in Chapter 7, we have the following theorems.

Theorem 9.4.1 Let ϕ and $\tilde{\phi}$ be biorthonormal MRA generators such that

$$\phi(x) = 2 \sum_k h_k \phi(2x - k),$$

and

$$\tilde{\phi}(x) = 2 \sum_k \tilde{h}_k \tilde{\phi}(2x - k),$$

and let ψ and $\tilde{\psi}$ be corresponding biorthonormal wavelets such that

$$\psi(x) = 2 \sum_k g_k \phi(2x - k),$$

where

$$g_k = (-1)^k \tilde{h}_{1-k}$$

and

$$\tilde{\psi}(x) = 2 \sum_k \tilde{g}_k \tilde{\phi}(2x - k),$$

where

$$\tilde{g}_k = (-1)^k h_{1-k}.$$

Then the filter sets {**h, g**} and {**h̃, g̃**} generate biorthogonal filter banks.

Proof: We leave the proof as an exercise. ■

Similarly, we have the following.

Theorem 9.4.2 Let ϕ be an orthonormal scaling function defined by

$$\phi(x) = 2 \sum_k h_k \phi(2x - k).$$

Let ψ be the corresponding orthonormal wavelet given in

$$\psi(x) = 2 \sum_k g_k \phi(2x - k),$$

where

$$g_k = (-1)^k h_{2l+1-k}, \quad \text{for an } l \in \mathbb{Z}.$$

Then $\{h, g\}$ creates an orthonormal filter bank.

Proof: We leave the proof as an exercise. ■

4.1 Conditions for Biorthogonal Filter Banks

We already know that wavelets lead to filter banks. We now ask the question: Do there exist biorthogonal filter banks other than the masks of biorthonormal scaling functions and wavelets? To answer this question, we derive the conditions for biorthogonal filter banks.

We first analyze the analysis filter bank. Let $\{L, H\}$ be an analysis filter bank, where L is a lowpass filter and H is a highpass filter. Assume x is the signal to be coded. Assume $v_0 = (\downarrow 2)L^T x$ and $v_1 = (\downarrow 2)H^T x$. Then

$$v_0(n) = \sum_k l^T(2n - k)x(k) \tag{4.2}$$

and

$$v_1(n) = \sum_k h^T(2n - k)x(k). \tag{4.3}$$

The z-transforms of (4.2) and (4.3) are

$$V_0(z) = \frac{1}{2}(\overline{L(z^{1/2})}X(z^{1/2}) + \overline{L(-z^{1/2})}X(-z^{1/2})) \tag{4.4}$$

and

$$V_1(z) = \frac{1}{2}(\overline{H(z^{1/2})}X(z^{1/2}) + \overline{H(-z^{1/2})}X(-z^{1/2})). \tag{4.5}$$

We now analyze the synthesis filter bank. Assume $\{F, G\}$ creates the synthesis filter bank, where F is a lowpass filter and G is a highpass filter. Let $y_0 = 2F(\uparrow 2)v_0$ and $y_1 = 2G(\uparrow 2)v_1$. Then after upsampling, the signal $2(\uparrow 2)v_0$ has the z-transform

$$V_0(z^2) = (\overline{L(z)}X(z) + \overline{L(-z)}X(-z)),$$

where the second term causes an alias in the low-frequency channel. Similarly, the signal $2(\uparrow 2)v_1$ has the z-transform

$$V_1(z^2) = (\overline{H(z)}X(z) + \overline{H(-z)}X(-z)),$$

where the second term causes an alias in the high-frequency channel. After filtering by **F** and **G**, respectively, we have

$$Y_0(z) = F(z)\overline{L(z)}X(z) + F(z)\overline{L(-z)}X(-z)) \tag{4.6}$$

and

$$Y_1(z) = G(z)\overline{H(z)}X(z) + G(z)\overline{H(-z)}X(-z)). \tag{4.7}$$

If $\{L, H\}$ and $\{F, G\}$ create a biorthogonal filter bank, then

$$X(z) = Y_0(z) + Y_1(z),$$

which yields the *no-distortion condition*

$$F(z)\overline{L(z)} + G(z)\overline{H(z)} = 1, \tag{4.8}$$

and the *alias cancellation condition*

$$F(z)\overline{L(-z)} + G(z)\overline{H(-z)} = 0. \tag{4.9}$$

We summarize these conditions in the following theorem.

Theorem 9.4.3 Let **L** and **F** be two lowpass filters and **H** and **G** be two highpass filters. Then the analysis filter bank $\{L, H\}$ and the synthesis filter bank $\{F, G\}$ are biorthogonal if and only if they satisfy the alias cancellation condition (4.9) and the no-distortion condition (4.8), i.e., the following holds:

$$\begin{bmatrix} \overline{L(z)} & \overline{H(z)} \\ \overline{L(-z)} & \overline{H(-z)} \end{bmatrix} \begin{bmatrix} F(z) \\ G(z) \end{bmatrix} = \begin{bmatrix} 1 \\ 0 \end{bmatrix}. \tag{4.10}$$

We now use (4.10) to derive the relations among **L**, **F**, **H**, and **G**. From (4.10), we have

$$\begin{bmatrix} F(z) \\ G(z) \end{bmatrix} = \begin{bmatrix} \overline{L(z)} & \overline{H(z)} \\ \overline{L(-z)} & \overline{H(-z)} \end{bmatrix}^{-1} \begin{bmatrix} 1 \\ 0 \end{bmatrix}$$

$$= \frac{1}{\Delta(z)} \begin{bmatrix} \overline{H(-z)} & -\overline{H(z)} \\ -\overline{L(-z)} & \overline{L(z)} \end{bmatrix} \begin{bmatrix} 1 \\ 0 \end{bmatrix}$$

where $\Delta(z) = \overline{L(z)H(-z)} - \overline{L(-z)H(z)}$ is an odd Laurent polynomial.

Since both $F(z)$ and $G(z)$ are Laurent polynomials, $\Delta(z)$ must be equal to cz^{2l+1} for a $c \neq 0$ and an $l \in \mathbb{Z}$. Then

$$F(z) = \frac{\overline{H(-z)}}{cz^{2l+1}}, \quad G(z) = -\frac{\overline{L(-z)}}{cz^{2l+1}}, \tag{4.11}$$

where
$$c = \Delta(1) = \overline{L(1)}H(-1) - \overline{L(-1)}H(1) = L(1)H(-1).$$
We now have the following.

Theorem 9.4.4 The FIR filter banks $\{L, H\}$ and $\{F, G\}$ are biorthogonal if and only if

$$F(z)\overline{L(z)} + F(-z)\overline{L(-z)} = 1, \quad z \in \Gamma, \tag{4.12}$$

and

$$H(z) = -\frac{L(1)H(-1)}{2}z^{2l+1}\overline{F(-z)}, \tag{4.13}$$

$$G(z) = -\frac{2}{L(1)H(-1)}z^{-(2l+1)}\overline{L(-z)}, \quad l \in \mathbb{Z}. \tag{4.14}$$

Proof: If $\{L, H\}$ and $\{F, G\}$ are biorthogonal, by (4.11), we have

$$F(z)\overline{L(z)} + F(-z)\overline{L(-z)} = \frac{\overline{H(-z)}}{\Delta(z)}\overline{L(z)} + \frac{\overline{H(z)}}{\Delta(-z)}\overline{L(-z)}$$

$$= \frac{\overline{L(z)H(-z)}}{\Delta(z)} - \frac{\overline{L(-z)H(z)}}{\Delta(z)}$$

$$= 1$$

and

$$H(z) = \overline{F(-z)}\Delta(-z) = -\frac{L(1)H(-1)}{2}z^{2l+1}\overline{F(-z)},$$

$$\overline{G(z)} = -\frac{L(-z)}{\Delta(z)} = -\frac{1}{L(1)H(-1)}z^{-(2l+1)}L(-z).$$

The proof of "only if" is left as an exercise. ∎

The values $L(1)$ and $H(-1)$ in the theorem can be freely selected. A standard selection is $L(1) = H(-1) = 1$. Then $F(1) = G(-1) = 1$.

By Theorem 9.4.4, we have the following.

Corollary 9.4.1 The pair of filters $\{h, g\}$ creates an orthonormal filter bank if and only if $\{h, g\}$ forms a conjugate mirror filter, i.e., $g = ((-1)^k h_{2N-1-k})$ for an $N \in \mathbb{Z}$ and

$$|H(z)|^2 + |H(-z)|^2 = 1.$$

Proof: We leave the proof as an exercise. ∎

Although some conjugate mirror filters may not be the masks of orthonormal scaling functions and wavelets, most useful conjugate mirror filters are. A similar conclusion can be obtained for the biorthogonal case. Hence, the masks of biorthogonal (including orthonormal) scaling functions and wavelets contain the most useful biorthogonal filter banks. Readers can learn more about filter banks from [28] and [31].

Exercises

1. Let H_0 be the moving average filter and H_1 be the moving difference filter. That is, $H_0 = \frac{1}{2}(I+S^{-1})$ and $H_1 = \frac{1}{2}(I-S^{-1})$. Use Definition 9.4.1 to prove that $\{H_0, H_1\}$ create an orthonormal filter bank.

2. Prove Theorem 9.4.1: Let ϕ and $\tilde{\phi}$ be biorthonormal MRA generators such that

$$\phi(x) = 2\sum_k h_k \phi(2x - k)$$

and

$$\tilde{\phi}(x) = 2\sum_k \tilde{h}_k \tilde{\phi}(2x - k),$$

and let ψ and $\tilde{\psi}$ be corresponding biorthonormal wavelets such that

$$\psi(x) = 2\sum_k g_k \phi(2x - k),$$

where

$$g_k = (-1)^k \tilde{h}_{1-k},$$

and

$$\tilde{\psi}(x) = 2\sum_k \tilde{g}_k \tilde{\phi}(2x - k),$$

where

$$\tilde{g}_k = (-1)^k h_{1-k}.$$

Then the filter sets $\{h, g\}$ and $\{\tilde{h}, \tilde{g}\}$ generate biorthogonal filter banks.

3. Prove Theorem 9.4.2: Let ϕ be an orthonormal scaling function defined by

$$\phi(x) = 2\sum_k h_k \phi(2x - k).$$

Let ψ be the corresponding orthonormal wavelet given in

$$\psi(x) = 2\sum_k g_k \phi(2x - k),$$

where

$$g_k = (-1)^k h_{2l+1-k}, \quad \text{for an } l \in \mathbb{Z}.$$

Then $\{h, g\}$ creates an orthonormal filter bank.

4. Assume $\{h, g\}$ creates an orthonormal filter bank. Let $h_k = S^{2k} h$ and $g_k = S^{2k} g$. Prove that $\{h_k, g_k\}_{k \in \mathbb{Z}}$ forms an orthonormal basis of l^2.

5. Let I be the identity filter and S be the time-delay operator. Prove that the pair $\{I, S\}$ satisfies

$$I(\uparrow 2)(\downarrow 2)I^T + S(\uparrow 2)(\downarrow 2)S^T = I.$$

6. Give an example of an orthonormal filter bank that is not created by the masks of any orthonormal scaling function and wavelet.

7. A signal $x = \sum x(n)\delta_n$ can be written as $x = \sum x(2n)\delta_{2n} + \sum x(2n+1)\delta_{2n+1}$, which yields a decomposition of x: $x = x_e + x_o$, where $x_e = (\uparrow 2)(\downarrow 2)x$ and $x_o = (\uparrow 2)(\downarrow 2)S^T x$. We call this decomposition the *polyphase* of x. The z-transform of the polyphase form of x is

$$X(z) = X_e(z^2) + zX_o(z^2).$$

Write the coding/decoding procession in the polyphase form.

8. Write the conditions of the biorthogonal filter bank in the polyphase form.

Appendix

List of Symbols

$\mathbb{Z}$: the set of integers.

$\mathbb{Z}^+$: the set of all nonnegative integers.

$\mathbb{N}$: the set of natural numbers.

$\mathbb{R}$: the set of real numbers.

$\mathbb{C}$: the set of complex numbers.

$\mathcal{R}_0$: the class (ring) of intervals $(a, d]$.

$\mathcal{I}$: the class of all intervals (open, closed, or half-open and half-closed) of $\mathbb{R}$.

$\mathcal{P}_n$: the vector space of all polynomials of degree n or less.

$\mathcal{P}(A)$ or 2^A: the power set of A.

$AC[a, b]$: the class of absolutely continuous functions on $[a, b]$.

$\mathcal{B}_{\mathbb{R}}$: the class of Borel sets.

$BV[a, b]$: the class of functions of bounded variation on $[a, b]$.

$C[a, b]$: the class of continuous functions on $[a.b]$.

$f_n \to f$ a.e.: f_n converges to f almost everywhere.

$f_n \rightrightarrows f$: f_n converges to f uniformly.

$f_n \xrightarrow{m} f$: f_n converges to f in measure.

$m(E)$: the Lebesgue measure of set E.

$T_a^b(f)$: the total variation of f on $[a, b]$.

$\sigma_r(\mathcal{C})$: the σ-ring generated by the class $\mathcal{C}$.

$\sigma_a(\mathcal{C})$: the σ-algebra generated by the class $\mathcal{C}$.

$\mathcal{M}$: the class of all Lebesgue measurable sets.

$\Gamma := \{z \in \mathbb{C} \mid |z| = 1\}$: the unit circle on the complex plane.

L^2 (or $L^2(\mathbb{R})) := \{f \mid \int_{\mathbb{R}} |f(x)|^2 dx < \infty\}$.

$\|f\|$ (or $\|f\|_{L^2}$): $= \left(\int_{\mathbb{R}} |f(x)|^2 dx \right)^{1/2}$.

$\langle f, g \rangle$ (or $\langle f, g \rangle_{L^2}$): $= \int_{\mathbb{R}} f(x)\overline{g(x)}dx$.

$\|f\|_p := \begin{cases} \left(\int_{\mathbb{R}} |f(x)|^p dx \right)^{1/p}, & \text{if } 1 < p < \infty, \\ \text{ess sup}_{x \in \mathbb{R}} |f(x)|, & \text{if } p = \infty. \end{cases}$

L^p (or $L^p(\mathbb{R})$): $= \{f \mid \|f\|_p < \infty\}$.

$\|f\|_I = \left(\int_0^1 |f(x)|^2 dx \right)^{1/2}$.

$\langle f, g \rangle_I := \int_0^1 f(x)\overline{g(x)}dx$.

$\|f\|_{L^2} = \left(\frac{1}{2\pi} \int_{-\pi}^{\pi} |f(x)|^2 dx \right)^{1/2}$.

$\langle f, g \rangle_{\tilde{L}^2} := \frac{1}{2\pi} \int_{-\pi}^{\pi} f(x)\overline{g(x)}dx$.

$\tilde{L}^2$ (or $\tilde{L}^2_{2\pi}$): $= \{f \mid f(x) = f(x + 2\pi), \|f\|_{\tilde{L}^2} < \infty\}$.

$\tilde{L}^2_I := \{f \mid f(x) = f(x + 1), \|f\|_I < \infty\}$.

$l = \{(c_n)_{n \in \mathbb{Z}} \mid c_n \in \mathbb{C}\}$.

$l^p = \begin{cases} \{(c_n) \in l \mid \left(\sum_{n \in \mathbb{Z}} |c_n|^p \right)^{1/p} < \infty\}, & 1 \leq p < \infty, \\ \sup_{n \in \mathbb{Z}} |c_n| < \infty, & p = \infty. \end{cases}$

span $V := \{\sum_{k \in \mathbb{Z}} c_k v_i(x) \mid (c_k) \in l, v_i \in V\}$, where
$V = \{v_i, i \in \mathbb{Z}\}$, and the series $\sum_{k \in \mathbb{Z}} c_k v_i(x)$ is a.e. convergent.

span $_{L^p} V := (\text{span } V) \cap L^p$.

$\hat{f}(\omega)$: the Fourier transform of the function f defined by

$$\hat{f}(\omega) := \int_{\mathbb{R}} f(x)e^{-ix\omega}dx.$$

$\overset{\vee}{F}(\omega)$: the inverse Fourier transform of F defined by

$$\overset{\vee}{F}(\omega) := \frac{1}{2\pi} \int_{\mathbb{R}} F(\omega)e^{ix\omega}d\omega.$$

$\mathcal{F}f(\omega)$: the normalized Fourier transform defined by

$$\mathcal{F}f(\omega) := \frac{1}{\sqrt{2\pi}} \int_{\mathbb{R}} f(x)e^{-ix\omega}dx.$$

$\mathcal{F}^{-1}F(x)$: the normalized inverse Fourier transform defined by

$$\mathcal{F}^{-1}F(x) = \frac{1}{\sqrt{2\pi}} \int_{\mathbb{R}} F(\omega)e^{ix\omega}d\omega.$$

$f * g(x)$: the convolution of f and g defined by

$$f * g(x) = \int_{\mathbb{R}} f(x - y)g(y)dx.$$

$L_{loc}^p = \{f \mid f \in L^p(a,b) \text{ for any finite interval } (a,b) \subset \mathbb{R}\}.$

$AC_{loc} = \{f \mid f(x) = \int_0^x \phi(t)dt, \phi \in L_{loc}^1\}.$

$AC_{loc}^s = \{f \in C \mid f^{(k)} \in AC_{loc}, 0 \leq k \leq s\}.$

$W_{L^p}^r = \{f \in L^p \mid f = \phi, a.e, \phi \in AC_{loc}^{r-1}, \phi^{(k)} \in L^p, 0 \leq k \leq r\}.$

$W[L^p;(i\omega)^r] = \{f \in L^p \mid (i\omega)^r \hat{f}(\omega) = \hat{g}(\omega), g \in L^p\}.$

Bibliography

A. Textbooks and Standard References

[1] C. Blatter, *Wavelets, A Primer*, A. K. Peters, Natick, MA, 1998.

[2] P. L. Butzer and R.J. Nessel, *Fourier Analysis and Approximation*, Vol. 1, Birkhäuser, 1971.

[3] C.K. Chui, *Multivariate Splines*, SIAM, Philadelphia, 1988.

[4] C.K. Chui, *An Introduction to Wavelets*, Academic Press, San Diego, 1992.

[5] C.K. Chui, editor. *Wavelets: A Tutorial in Theory and Applications*, Academic Press, New York, 1992.

[6] I. Daubechies, *Ten Lectures on Wavelets*, SIAM Publications, 1992.

[7] L. Debnath and P. Minkusiński, *Introduction to Hilbert Spaces with Applications*, 2nd Ed., Academic Press, San Diego, 1998.

[8] C. de Boor, *A Practical Guide to Splines*, Springer-Verlag, 1978.

[9] P.R. Halmos, *Measure Theory*, reprint, Springer-Verlag, New York, 1974.

[10] E. Hernández and G. Weiss, *A First Course on Wavelets*, CRC Press, New York, 1996.

[11] E. Hewitt and K. Stromberg, *Real and Abstract Analysis*, Springer-Verlag, 1969.

[12] B.B. Hubbard, *The World According to Wavelets*, A. K. Peter, Wellesley, MA, 1996.

[13] K. Hrbacek and T. Jeck, *Introduction to Set Theory*, 2nd Ed., Marcel Dekker, New York, 1984.

[14] J.L. Kelley, *General Topology*, reprint, Springer-Verlag, New York, 1975.

[15] I. Kaplansky, *Set Theory and Metric Spaces*, 2nd Ed., Chelsea Publishing, New York, 1977.

[16] J.R. Kirkwood, *An Introduction to Analysis*, 2nd Ed., PWS Publishing Co., Boston, 1995.

[17] S. Lang, *Introduction to Linear Algebra*, 2nd Ed., Springer-Verlag, New York, 1986.

[18] S. Mallat, *A Wavelet Tour of Signal Processing*, Academic Press, San Diego, 1998.

[19] J.N. McDonald and N. Weiss, *A Course in Real Analysis*, Academic Press, San Diego, 1999.

[20] Y. Meyer, *Ondelettes et Opérateurs I: Ondelettes*, Hermann, Paris, 1990.

[21] Y. Meyer, *Wavelets: Algorithms and Applications*, SIAM, 1993. Translated and revised by R.D. Ryan.

[22] D. Pollard, *A User's Guide to Measure Theoretic Probability*, Cambridge University Press, Cambridge, UK, 2002.

[23] H.L Royden, *Real Analysis*, 3rd Ed., Prentice-Hall, Englewood Cliffs, NJ, 1988.

[24] M. Reed and B. Simon, *Methods of Mathematical Physics I: Functional Analysis* (revised and enlarged edition), Academic Press, San Diego, 1980.

[25] W. Rudin, *Principles of Mathematical Analysis*, 3rd Ed., McGraw-Hill, New York, 1976.

[26] W. Rudin, *Real and Complex Analysis*, 3rd Ed., McGraw-Hill, New York, 1987.

[27] I.J Schoenberg, *Cardinal Spline Interpolation*, SIAM, Philadelphia, 1973.

[28] G. Strang and T. Nguyen, *Wavelets and Filter Banks*, Wellesley-Cambridge Press, 1996.

[29] C. Swartz, *An Introduction to Functional Analysis*, Marcel Dekker, New York, 1992.

[30] A.E. Taylor, *General Theory of Functions and Integration*, Blaisdell Publishing Co., Waltham, MA, 1965.

[31] P. P. Vaidyanathan, *Multirate Systems and Filter Banks*, Prentice-Hall, Englewood Cliffs, NJ, 1993.

[32] M.V. Wickerhauser, *Adapted Wavelet Analysis from Theory to Software*, A.K. Peters, 1994.

[33] R.M. Young, *An Introduction to Nonharmonic Fourier Series* (revised edition), Academic Press, San Diego, 2001.

B. Articles

[34] G.A. Battle, A block spin construction of ondelettes, Part I: Lemarié functions, *Comm. Math. Phys.* **110** (1987), 601–615.

[35] G.A. Battle, Wavelets of Federbush-Lemarié type, *J. Math. Phys.* **34** (1993), 1095–2002.

[36] G. Beylkin, R.R. Coinfman, and V. Rokhlin, Fast wavelet transforms and numerical algorithms I, *Comm. Pure Appl. Math.* **44** (1991), 141–183.

[37] C.K. Chui and J.Z. Wang, On compactly supported spline wavelets and a duality principle, *Trans. Amer. Math. Soc.* **330** (1992), 903–916.

[38] C.K. Chui and J.Z. Wang, High order orthonormal scaling functions and wavelets give poor time-frequency location, *Fourier Anal. Appl.* **2** (1996), 415–426.

[39] C.K. Chui and J.Z. Wang, A study of asymptotically optimal time-frequency windows of scaling function and wavelets, *Ann. Num. Math.* **4** (1997), 193–216.

[40] A. Cohen, I. Daubechies, and J.C. Feauveau, Biorthogonal basis of compactly supported wavelets, *Comm. Pure Appl. Math.* **45** (1992), 485–560.

[41] A. Durand, On Mahler's measure of a polynomial, *Proc. Am. Math. Soc.* **83**(1) (1981), 75–67.

[42] R.Q. Jia and J.Z. Wang, Stability and linear independence associated with wavelet decompositions, *Proc. Am. Math. Soc.* **117** (1993), 1115–1124.

[43] P.G. Lemarié, Ondelettes á localisation exponentielles, *J. Math. Pure Appl.* **67** (1988), 227–236.

[44] P.G. Lemarié, Fonctions a support compact dans les analysis multi-résolutions, *Rev. Math. Iberoamericana* **7** (1991), 157–182.

[45] S. Mallat, Multiresolution approximations and wavelet orthonormal bases of $L^2(\mathbb{R})$, *Trans. Am. Math. Soc.* **315** (1989), 69–87.

Index

F_σ set, 53
G_δ set, 53
$L(E)$, 74, 83
L^2, 132
L^p, 146
L^1, 176
L^2, 190
$\tilde{L}^1$, 157
δ-neighborhood, 21
μ^*-measurable set, 48
σ-algebra of sets, 36
σ-finite measure, 72
σ-ring of sets, 35
$\tilde{W}_1^r$, 163
l^2, 117, 141
l^2, 169
z-transform, 308

A

Abelian group, 115
absolute continuity of Lebesgue
    integrals, 76
absolutely continuous function, 183
algebra of sets, 35

alias cancellation, 349
almost everywhere, 57
analog signal, 315
    energy, 316
    finite energy, 316
    periodic, 316
analysis scaling function, 268
analysis scaling wavelet, 268
Archimedean Axiom, 10
autocorrelation, 190, 201
Axiom of Choice, 9
Axiom of Completeness, 10
Axioms of Addition, 10
Axioms of Multiplication, 10
Axioms of Order, 10

B

Baire space, 28
Banach space, 145
basis, 118
Battle-Lemarié wavelets, 237
Bernstein polynomial, 31
Bernstein Theorem, 14
Bessel's inequality, 129

Printed and bound by CPI Group (UK) Ltd, Croydon, CR0 4YY

03/10/2024

01040410-0008